Science of Rice Chemistry and Nutrition

Jinhu Tian • Yukiharu Ogawa • Jaspreet Singh •
Lovedeep Kaur
Editors

Science of Rice Chemistry and Nutrition

Springer

Editors
Jinhu Tian
Food Science & Nutrition
Zhejiang University
Hangzhou, Zhejiang, China

Yukiharu Ogawa
Graduate School of Horticulture
Chiba University
Matsudo, Chiba, Japan

Jaspreet Singh
Riddet Institute
Massey University
Palmerston North, New Zealand

Lovedeep Kaur
Riddet Institute
Massey University
Palmerston North, New Zealand

ISBN 978-981-99-3223-8 ISBN 978-981-99-3224-5 (eBook)
https://doi.org/10.1007/978-981-99-3224-5

This Springer imprint is published by the registered company Springer Nature Singapore Pte Ltd.
The registered company address is: 152 Beach Road, #21-01/04 Gateway East, Singapore 189721, Singapore

Preface

Rice (*Oryza sativa* L.) is the predominant staple food in diets for more than half of the population worldwide and is pivotal for human nutrition, energy supply, and food security. According to International Rice Research Inst. statistics, the world production of paddy rice has reached 481 million tons (http://ricestat.irri.org:8080/wrsv3/entrypoint.htm). Attributing to its high content of digestible starch, rice is generally considered as a high glycemic index (GI) food and positively associated with type II diabetes. However, the GI value of rice is dependent on varietal, compositional, processing, and accompaniment factors; even though there are many studies focused on the GI of rice, a systematic summary concerning rice GI is still needed. As commonly known, the fresh paddy would suffer pre-treatments (e.g., drying, dehulling and milling,, processing) before the circulation on the market as well as multiply cooking (e.g., steaming, stir-frying, boiling, microwaving) and processing (e.g., puffing, extrusion) before consumption. During the process, the by-products (e.g., bran, endosperm, and broken rice) will be generated and the rice structure and chemistry will be changed significantly, and this subsequently alters the rice chemistry and nutrition.

The book *Science of Rice Chemistry and Nutrition* will present the knowledge on the rice chemistry and nutrition, including identification, analysis, and application of bioactive chemical components of rice, the starch digestibility after pre-treatment, domestic cooking or processing. The authors will provide a brief description of the rice bioactive components as well as their role during processing. Particularly, novel applications of rice starch, protein, polyphenols, and modification and next generation of healthy rice products will be summarized and the underline mechanisms of starch digestibility will also be revealed. This work will provide students, researchers, and technologists working in food science and others with most recent information and state-of-the-art technology.

Hangzhou, Zhejiang, China — Jinhu Tian
Matsudo, Chiba, Japan — Yukiharu Ogawa
Palmerston North, New Zealand — Jaspreet Singh
Palmerston North, New Zealand — Lovedeep Kaur

Contents

Chapter 1
Rice Chemistry, Nutrition, and Processing: An-Introduction

Jaspreet Singh, Lovedeep Kaur, Jinhu Tian, and Yukiharu Ogawa

Abstract As one of the three major cereal crops in the world. Rice (*Oryza sativa* L.) is the most widely consumed staple food for more than half of the world's population. In this chapter, an overview of rice production in the world, rice component constitutes, rice processing, and the health benefits of rice are introduced. Particularly, the importance of the rice industry as well as the future directions are highlighted.

Keywords Rice chemistry · Health benefits · Rice processing

Rice (*Oryza sativa* L.) is one of the three major cereal crops in the world. Rice is the most widely consumed staple food for more than half of the world's population. In recent years, more than 500 million tons of milled rice annually is produced globally, of which nearly 450 million tons are being produced in Asia alone (FAO 2018). Population growth in developing countries has increased the demand and importance of rice as food and as a nutritional resource.

The role and importance of chemistry in determining the product quality and nutritional value of rice is well known. Since rice is consumed from ancient times, several varieties are grown around the world, which have usually been accommodated to cooking and consuming styles in various cultures. Paddy is the final product of rice harvesting and threshing, which is used for milling to obtain rice grains. The

---

J. Singh (✉)
Riddet Institute, Massey University, Palmerston North, New Zealand
e-mail: J.X.Singh@massey.ac.nz

L. Kaur
Massey University, Palmerston North, New Zealand
e-mail: L.Kaur@massey.ac.nz

J. Tian
College of Biosystem and Food Science, Zhejiang University, Hangzhou, Zhejiang, China

Y. Ogawa
Graduate School of Horticulture, Chiba University, Matsudo, Japan
e-mail: ogwy@faculty.chiba-u.jp

J. Tian et al. (eds.), *Science of Rice Chemistry and Nutrition*,
https://doi.org/10.1007/978-981-99-3224-5_1

cooking methods of rice depend on the type of rice and the scale of operation. The most common method of rice cooking in many developing countries is boiling rice in water; however, the use of rice cookers and microwave ovens is also getting popular due to an increase in the economic affordability of populations. Because of the variation in cooking and consumption habits, several methods have been used to evaluate the processing and nutritional quality of rice. A critical component in rice is starch, which affects its cooked texture and nutritional properties significantly. The sensory characteristics of rice, such as texture, which is connected with physico-chemical and structural attributes of rice, are critical in relation to the eating quality of cooked rice and its processed products. Therefore, the role of food chemistry in determining rice quality and nutritional properties is important and has been progressively researched by food chemists, nutritionists, and technologists over the years.

There have been a number of developments in the area of cooked grain texture, rice gastro-intestinal digestion (in vitro and in vivo), glycemic index and micronutrients of rice, and the role of rice in human nutrition. The increasing popularity of whole grain consumption has increased the consumption of unmilled or lightly milled, unpolished brown, red, or black rice. They contain unique phyto-chemicals, such as tocopherols, tocotrienols, vitamin B, γ-oryzanols, and phenolic compounds, which are associated with reducing the risks of several chronic diseases. All these new developments signify the increased importance of rice and its products in the human diet.

The book volume on "Rice Chemistry, Nutrition and Technology" presents critical and recent developments in areas such as rice composition and its functional components including rice protein and dietary fiber, commercial processing and quality optimization, boiled and parboiled rice, starch digestibility and nutritional value, new methods of characterization, recent developments in rice starch chemistry, and the use of rice and its byproducts. This book volume will also help in shaping the future research direction in the area of rice chemistry, nutrition, and technology of rice and be useful for the rice processing industry, researchers, and graduate students.

Reference

FAO (2018) Rice market monitor. Food Agric Organ United Nations XXI(1) https://www.fao.org/3/I9243EN/i9243en.pd

Chapter 2
The Structure of Rice Starch and its Application

Wenwen Yu and Cheng Li

Abstract Multiple scales of starch structure are introduced from micro to macro in detail. First, the fine molecular structures of rice starch are illustrated including the wholebranched rice starch molecular structure and debranched rice starch molecular structure. And then the rice starch structure at 2–100 mm scale as well as the growth rings, blocklets, and amorphous and crystalline lamellae are described. After that, the physicochemical properties of rice starch with different amylose contents, such as gelatinization, retrogradation, swelling and pasting properties, and digestion are discussed.

Keywords Starch · Structure · Gelatinization · Retro-gelatinization · Digestion

Abbreviations

AAC	Apparent amylose content
BD	Breakdown
CLD	Chain-length distributions
CPV	Cold paste viscosity
CS	Consistency
DP	Degree of polymerization
DRI	Differential refractive index
DSC	Differential scanning calorimetry
ECL	External chain length
FACE	Fluorophore-assisted carbohydrate electrophoresis
GPC	Gel-permeation chromatography

W. Yu (✉)
Department of Food Science and Engineering, Jinan University, Guangzhou, China
e-mail: yuwenwen@jnu.edu.cn

C. Li
University of Shanghai for Science and Technology, Shanghai, China
e-mail: licheng@usst.edu.cn

J. Tian et al. (eds.), *Science of Rice Chemistry and Nutrition*,
https://doi.org/10.1007/978-981-99-3224-5_2

GT	Gelatinization temperature
HPACE	High-performance anion exchange chromatography
HPV	Hot paste viscosity
MALLS	Multiple-angle laser scattering
NMR	Nuclear magnetic resonance
PT	Pasting temperature
PV	Peak viscosity
Rh	Hydrodynamic radius
RVA	Rapid visco-analyzer
SB	Setback
SEC	Size-exclusion chromatography
TAC	True amylose content
TV	Trough viscosities

2.1 Introduction

Rice (*Oryza sativa, L.*) is the second most widely produced cereal crop in the world, following corn with a global paddy production of $720 million metric tons in 2012 (Patindol et al. 2015). It is the main source of food energy for half the world's population, as more than 90% of the world's rice is grown and consumed in Asia, where 60% of the Earth's population lives (FAOSTAT 2018).

Rice is usually consumed after milling to white rice, the process that leads to a significant loss of nutritional components, especially micronutrients such as minerals and vitamins, as listed in Table 2.1. Of milled rice, the carbohydrates constitute 90% on a dry weight basis, and is thus an excellent source of dietary energy. Rice also contains a protein content of ~16% on a dry weight basis, most of which are located in the outer coatings (bran) of the starchy endosperm and the germ, which is removed during milling (Hoogenkamp et al. 2017). Accordingly, the protein content of milled rice in a germplasm collection reportedly ranged between 4.5% and 15.9%, while the milled rice usually contains a significantly lower content of edible protein between 6.7% and 7.1% (Table 2.1). Except for the carbohydrates and protein, milled rice also contains a small amount of lipids (usually <5% on a dry weight basis), with waxy rice types containing considerably less protein (0.03%) (Juliano 1984). Of rice lipids, the main fatty acids are oleic and linoleic acids, followed by palmitic acid, with a total proportion > 90% of the total fatty acid content of rice lipids (Kitta et al. 2005). In this chapter, only the starch of milled rice will be discussed.

Table 2.1 Comparison of nutrient composition of brown rice, milled, and rice bran (drawn by the author)

	Amount (100 g)		
Property	Brown rice	Milled rice	Rice bran
Moisture (g)	14	14	14
Energy content (kJ)	1480–1610	1460–1560	1670–1990
Energy content(kcal)	355–385	349–373	399–476
Crude protein (g)	7.1–8.3	6.3–7.1	11.3–14.9
Crude fat (g)	1.6–2.8	0.3–0.6	15.0–19.7
Crude fiber (g)	0.6–1.0	0.2–0.5	7.0–11.4
Crude ash (g)	1.0–1.5	0.3–0.8	6.6–9.9
Available carbohydrate (g)	73–87	77–89	34–62
Total dietary fiber (g)	2.9–4.4	0.7–2.7	24–29
Water insoluble fiber (g)	2	0.5	15-27
Sugars (g)	0.8–1.9	0.1–0.5	5.5–6.9
Thiamin (mg)	0.4–0.6	0.07–0.17	1.2–2.5
Riboflavin (mg)	0.04–0.14	0.02–0.06	0.18–0.43
Niacin (mg)	3.5–6.2	1.3–2.5	27–50
Pantothenic acid (mg)	1.4–1.6	0.8–1.3	20–60
Vitamin B6 (mg)	0.5–0.7	0.1–0.4	3.7
Folate (μg)	16–20	4–9	40–140
Vitamin E, alpha-tocopherol (mg)	0.8–2.5	0.1–0.3	3–15
Calcium (mg)	10–50	10–30	30–120
Phosphorus (mg)	0.17–0.43	0.08–0.15	1.1–2.5
Phytic acid (mg)	0.13–0.27	0.02–0.07	0.9–2.2
Iron (mg)	1.4–5.2	0.3–0.8	8.6–43
Zinc (mg)	1.9–2.8	0.8–2.3	4.3–26

2.2 Rice Starch

Cereal crops accumulate starch in the seed endosperm as an energy source. The pathway of starch biosynthesis for different cereals is unique and requires enzyme isoforms not present in other cereals (Li et al. 2017; Zhu et al. 2020). Starch is the most abundant component of rice, comprising 72–82% of the dry weight of brown rice (Frei et al. 2003), and approximately 90% of milled rice (Bao 2019; Fitzgerald et al. 2009a, b). Starch is, therefore, the main factor in determining both the functional and nutritional properties of rice, such as its eating, cooking, and brewing properties (Li et al. 2016; Tao et al. 2019). Starch is a complex branched glucose polymer, with $(1 \rightarrow 4)$-α linear links and $(1 \rightarrow 6)$-α branch points. Starch comprises two types of glucans: amylose (of moderate molecular weight and largely linear with a few long-chain branches) (Yasuhito et al. 1987) and amylopectin (of much higher molecular weight with a vast number of short-chain branches). Amylose usually exists either in the amorphous lamellae, amorphous growth ring, and/or interspersed or co-crystallized with amylopectin molecules (Hoover et al. 2010).

The linear nature confers unique properties on amylose, among which is the ability to form complexes with iodine, resulting in the measurement of starch clear amylose content (AAC) through measuring the content of iodine–amylose complex in the starch sample (Perez and Juliano 1978). However, there are usually significant differences between AAC and the true amylose content (TAC), especially for high-amylose content starches. This is mainly because the long branch chains of amylopectin also overlap with short amylose chains, and thereby developing color with iodine (Bao 2019). Except for the iodine method, size-exclusion chromatography (SEC, also sometimes termed as GPC, gel-permeation chromatography), coupled with DRI (differential refractive index detector) (SEC-DRI) has been successfully used to measure the AAC of starch (Fitzgerald et al. 2009a, b). Nevertheless, for SEC, because of the overlap between short amylose and long amylopectin branches, the AAC obtained by SEC will be sometimes underestimated, especially for high-amylose starches (Huang et al. 2019). Generally, the AAC in waxy rice is <2%, the common rice has an AAC ranging from very low (5–12%), low (12–20%), intermediate (20–25%) to high (25–33%), while the AAC of amylose extender (ae) mutant can be as high as 35% (Li et al. 2020a, b, c; Tao et al. 2018; Tao et al. 2019).

2.3 Rice Starch Structure at Multiple Scales

There are, in general, six complex hierarchical levels of starch structures (Wu et al. 2013) as shown in Fig. 2.1. The first level is the individual starch chain and the second level is the whole fully branched amylose and amylopectin molecules.

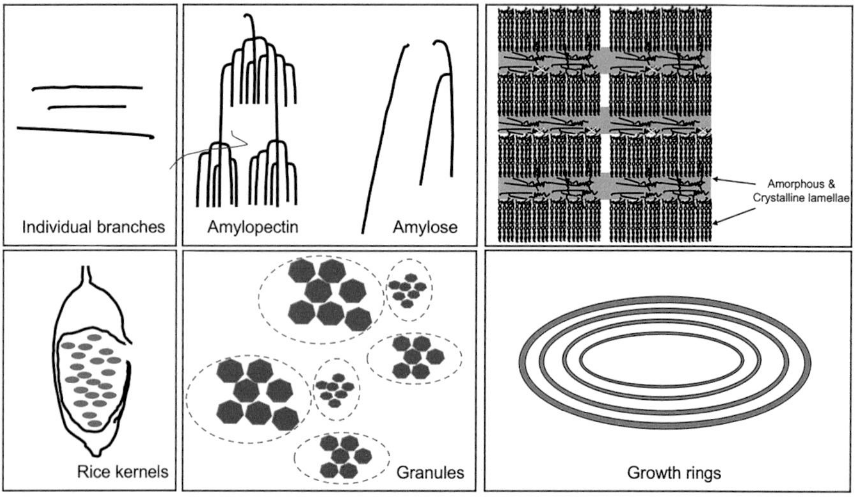

Fig. 2.1 The hierarchical levels of rice starch structures from that of the individual chains to the whole grain (Reprinted with permission from Alex C. Wu et al. (2013))

Higher levels of starch structures involve crystalline and amorphous lamella (3rd), blocklets (4th), granules (5th), and eventually the whole grain (6th). Varied techniques have been correspondingly developed to measure each starch structure level, with detailed information given below. The first two levels of starch structure, including the single starch chain (length), the whole fully branched amylose and amylopectin molecules, are usually termed as starch fine molecular structures or the nanostructure (Dang and Copeland 2003; Patindol et al. 2015). Correspondingly, the micro- and macro-level structures represent higher levels of starch structures.

2.3.1 Rice Starch Fine Molecular Structures

Compared to higher starch structural levels that usually have dimensions ranging from 7 Å (glucose molecule) to ~400 nm (the granule growth ring thickness) (Dang and Copeland 2003; Patindol et al. 2015), the fine molecular structure of rice starch is the most complex. For fully branched molecules, the definitions of starch fine molecular structures mainly comprise the amylose content (both AAC and TAC), the weight-averaged molecular weight ($\bar{M}w$), the (z-averaged) radius of gyration ($\bar{R}g$), degree of branching (DB), and hydrodynamic molecular sizes (R_h). For debranched starches, the first structure level, this usually includes the degree of polymerization (DP) of the individual chains, the number of individual chains with a given DP, the chain-length distributions (CLD). Nuclear magnetic resonance (NMR), SEC coupled with different detectors such as DRI and multiple-angle laser scattering (MALLS), fluorophore-assisted carbohydrate electrophoresis (FACE), high-performance anion-exchange chromatography (HPAEC), and MALLS are the main techniques for characterizing starch fine molecular structures, with each technique has its advantages and drawbacks (Li et al. 2020a, b, c; Yu et al. 2019a, b).

For the last decades, the amylose content has remained a major predictor of quality. For example, studies have shown that compared with high-amylose rice, low-amylose rice cultivars tend to have lower hardness but higher stickiness, contributing to better eating qualities and consumer preferences (Nakamura et al. 2016; Tao et al. 2019). Except for the amylose content, the amylose fine molecular structures are also of significant importance in controlling the functional and nutritional properties of rice products. Scientific studies have compared rice cultivars with similar amylose contents, and those containing smaller amylose molecules and more amylose chains with medium-long chain length (DP ranges between 1000 and 2000), tend to have a harder texture after cooking (Li et al. 2016; Tao et al. 2019), but faster digestibility (Syahariza et al. 2013; Yu et al. 2018). The role of amylopectin in determining the functional (e.g., the eating quality of cooked rice) and nutritional (e.g., the digestibility) of rice products has also been investigated. For example, it is now generally believed that the fine molecular structures of rice amylopectin are responsible for stickiness, whereas amylose seems to be more involved in controlling the hardness of cooked rice (Li et al. 2016; Tao et al. 2019). Both the eating quality and the nutritional value of rice products are important in determining consumer choice for a particular rice variety (Mestres et al. 2019).

2.3.2 Whole Branched Rice Starch Molecular Structure

The characteristics of foods containing starch are understood largely by the mass ratio of amylose/amylopectin, and by the molecular weight of amylose (Wani et al. 2012), both of which vary from genotype to genotype, depending on the genetics and environmental conditions where the plant is grown in. As a consequence, a comparison of the $\bar{M}w$ of whole starch molecules of rice from different studies is thus very difficult as there is a wide range of $\bar{M}w$ reported, for example, ranging from 10^6 to 10^9.

The $\bar{M}w$ of the whole range rice starch was found between (0.52 ~ 1.96) × 10^8 g/mol and between (1.8 × 4.8) × 10^5 g/mol (Syahariza et al. 2013). In some other studies, the $\bar{M}w$ of long, medium/short, and waxy rice starches were 0.786 × 10^8, 1.186 × 10^8, and 1.876 × 10^8 g/mol, respectively (Park et al. 2007). However, because of the difficulty in separating amylose and amylopectin molecules by current techniques, the $\bar{M}w$ of the whole starch instead of only amylopectin or amylose is more generally used. Compared to the $\bar{M}w$ of the whole ranged starches and with those of the fractionated components, the $\bar{M}w$ of starches measured by HPSEC-MALLS are lower than the calculated values corresponding to the amylopectin/amylose. Moreover, evidence has also shown that there is a negative correlation between the $\bar{M}w$ and amylose content, that is, a higher $\bar{M}w$ of rice starch tends to have a lower amylose content, similar to other cereal starches such as barley (Yu et al. 2019a, b).

The $\bar{M}w$ of amylopectin and of amylose is (4.0–5.5) × 10^7 and (3.1 ~ 3.4) × 10^6, respectively, and vary from cultivar to cultivar (Zhong et al. 2006). The $\bar{M}w$ of amylopectin was 1.10 × 10^8 (long grain), 1.81 × 10^8 (short/medium grain), and 2.47 × 10 ^{8} (waxy grain), respectively, and of amylose, was 3.90 × 10^5 (long grain) and 3.73 × 10^5 (short–/medium-grain), respectively (Park et al. 2010). In some other studies, the range of $\bar{M}w$ rice amylopectin is between (0.82 and 2.51) × 10^8 g/mol, and of rice amylose ranges between (2.20 and 8.32) × 10^5 g/mol.

The $\bar{R}g$ of rice starch extracted from different genotypes measured by SEC-MALLS was in the range 119–220 nm (Syahariza et al. 2013), and 177–225 nm. The $\bar{R}g$ of amylose and amylopectin ranged between (184 and 319) nm and (32.1 and 36.5) nm, respectively. Usually, the $\bar{R}g$ of starch molecules from a high M_w fraction tends to become smaller with increased amylose content.

The hydrodynamic radius (denoted as R_h) of whole starch molecules or of amylose/amylopectin individually are also frequently used. The SEC coupled with DRI is currently the most frequently used technique to determine the R_h of starch molecules. However, because of the poor resolution of amylose and amylopectin molecules, especially for high-amylose content starches, a precise measurement of the R_h of either amylose or of amylopectin molecules is difficult, especially when the

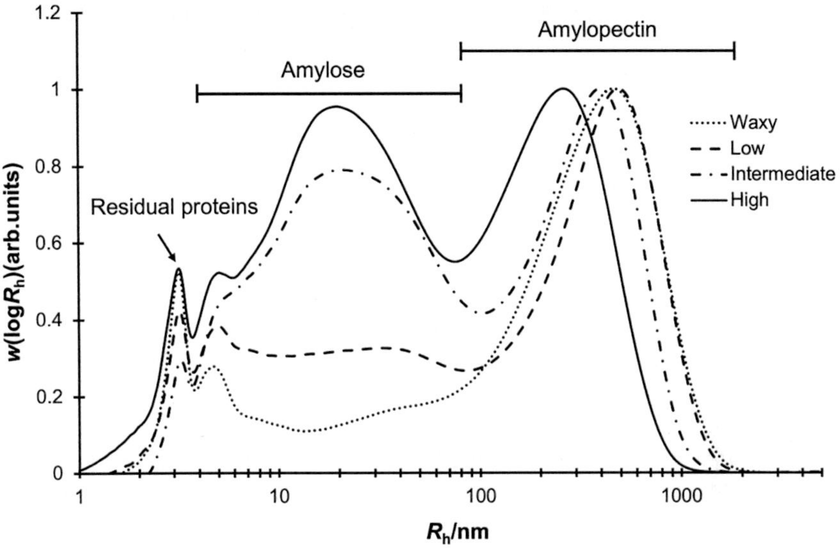

Fig. 2.2 The size-exclusion chromatography (SEC) weight distributions ($w(logR_h)$) of fully branched rice starches with different amylose contents as a function of the hydrodynamic radius (R_h). (Reprinted with permission from Syahariza et al. (2013))

inherent SEC band broadening makes the measurement of amylopectin R_h only semi-quantitative (Tao et al. 2018) (Fig. 2.2). To resolve this, all starch samples must be analyzed at the same time to ensure the machine has the same conditions. Moreover, this issue, to some extent, could be partially solved by calculating the averaged R_h (denoted as $\bar{R_h}$) of starch molecules or by mathematical models (Nada et al. 2017). The $\bar{R_h}$ of rice starch ranges between (58.9 and 79.5) nm as reported by Tao et al. (2018), and between (47.1 and 67.8) nm as reported by Gong et al. (2019).

Because of the overlap between large amylose and small amylopectin molecules (Fig. 2.2), the measurement of the R_h of amylose and amylopectin molecules individually is thus very difficult, especially for high-amylose starches (Li et al. 2019; Tao et al. 2018; Tao et al. 2019). The average molecular size of amylose (denoted as $\overline{R}_{\mathrm{h,Am}}$) was thus more frequently calculated over a certain range, such as the range between 10 and 50 nm (Tao et al. 2018) or between 10 and 100 nm (Wang et al. 2015a, b), which was usually regarded as the range of amylose molecules and the $R_h = 50$ or $R_h = 100$ nm was correspondingly the lowest point for amylose and amylopectin molecules, as shown in Fig. 2.2 (Syahariza et al. 2013). In this respect, the $\overline{R}_{\mathrm{h,Am}}$ of rice starch ranges between (7.9 and 10.5) (Syahariza et al. 2013) or between (16.4 and 18.2) nm (Tao et al. 2018), while in some other studies, the $\overline{R}_{\mathrm{h,Am}}$ ranges between (13.4 and 14.6) nm (Li et al. 2020a, b, c). Unlike amylose, because of the unavoidable shear scission that occurs in SEC (Cave et al. 2009), the amylopectin component of the whole-molecule SEC distribution is therefore not quantitatively reliable. The $\bar{R_h}$ of rice amylopectin (denoted as $\overline{R}_{\mathrm{h,Ap}}$) ranges between

(66.1 and 83.8) nm (Gong et al. 2019) or between (238 and 368) nm (Tao et al. 2019).

2.3.3 Debranched Rice Starch Molecular Structure

After debranching using enzymes (e.g., isoamylase and pullulanase), the molecular weight distribution of the resulting linear individual starch branches could be measured through a number of different techniques. The FACE and HPAEC are the main ones and could directly measure the $N_{de}(X)$ of debranched starch chains, especially of the amylopectin branches with relatively shorter chain length (generally < DP 100), compared to that of the amylose. Figure 2.3 shows a typical FACE elugram of rice starches with different amylose contents (from waxy to high amylose). The FACE elugram is converted to $N_{de}(X)$ directly by integrating the area under each peak and applying a velocity correction. However, these two are confined to relatively low DP <180 for FACE (Wu et al. 2014), whereas HPAEC suffers from a mass bias, which is very laborious to correct (Wong and Jane 1997).

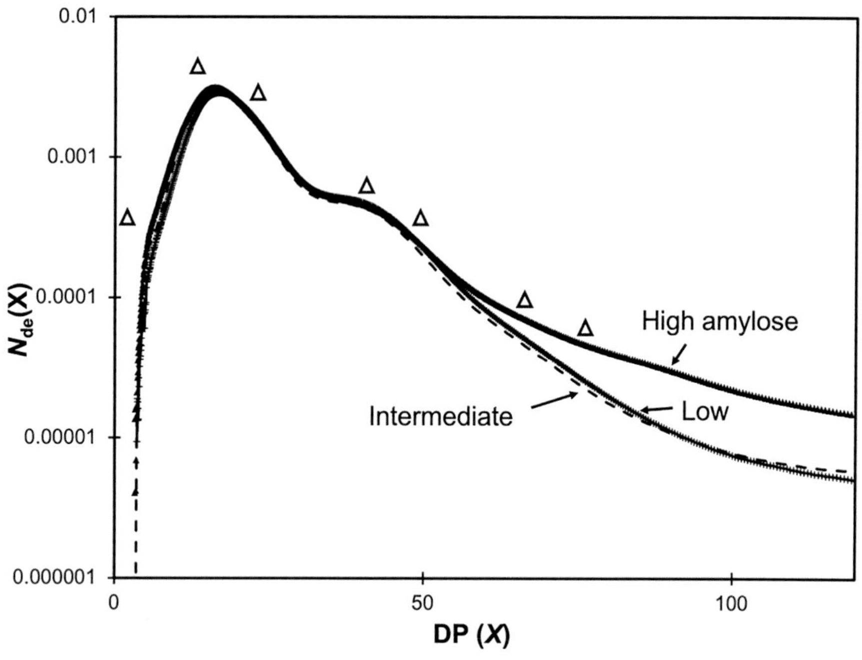

Fig. 2.3 The number chain-length distribution ($N_{de}(X)$) (on a logarithmic axis) as a function of degree of polymerization (DP, X) of debranched rice starches with different amylose contents obtained by FACE, the triangles indicate different features. (Reprinted with permission from Wu et al. (2014))

The SEC coupled with multiple detectors (e.g., DRI and MALLS) has also been successfully developed and used to measure the CLD of linear starch glucans. The MALLS cannot be used directly for insignificant molecules such as very short branches of amylopectin or amylose short chains (O'Shea et al. 1998; Yu et al. 2019a, b). Theoretically, SEC can go to virtually any DP of starch glucan; however, it suffers from the band broadening and thus requires calibration and assumptions (the use of the Mark–Houwink relation) to convert the elution time/volume to the DP for all except very long chains, when the DP can be measured directly by light scattering (Yu et al. 2019a, b). For SEC, the CLD can be expressed equivalently as the corresponding weight distribution of chains, *w*(*logX*). Here, a logarithmic dependent variable (*logX*) is used because it is the natural axis to use for a common method of obtaining this quantity. One has $w(logX) = X^2 \times N_{de}(X)$ (Castro et al. 2005). One has $X = \left(1.7 \times 10^3 \times R_h^{1.72} - 18\right)/162$ (162 is the molecular weight of the anhydroglucose monomer, 18 that of a water molecule, and R_h is in nm).

Figure 2.4 shows a typical example in terms of converting the R_h of linear starch glucans to the CLD. In this respect, rice amylopectin usually has a number average degree of polymerization (denoted as DP_n) of 8200–12,800 (Yasuhito et al. 1987) or

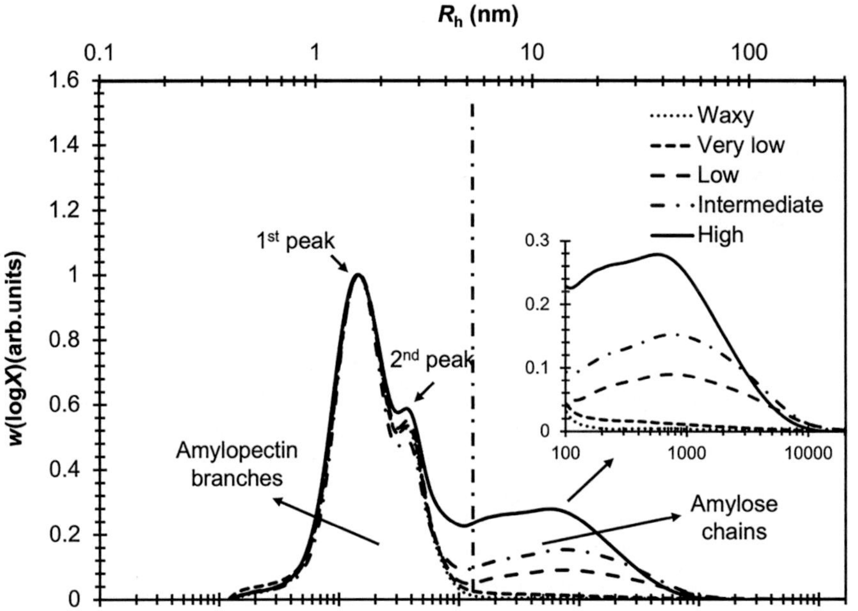

Fig. 2.4 The size-exclusion chromatography (SEC) weight distributions of debranched rice starches with different amylose contents (from waxy to high) as a function of the degree of polymerization (DP, *X*). The amylose and amylopectin chains show a separation at DP = ~ 100 (Reprinted with permission from Yasuhito et al. (1987))

in the range of 9600–15,900, and for large (DP 13400 ~ 26,500), medium (DP 4400 ~ 8400), and for small (DP 1700 ~ 1200) rice species, respectively (Takeda et al. 2003). Unsurprisingly, the average DP of amylopectin (denoted as $\bar{X}, A_p$) varied from study to study. For example, Takeda et al. (2003) reported that of rice starch, the $\bar{X}, A_p$ ranges 19–23, whereas this was found in the range of 13.2–15.1 in the study of Park et al. (2007). The $\bar{X}, A_m$ of rice starch was in the range of 2000–3000, while Takeda et al. (1986) reported that the $\bar{X}, A_m$ of rice starch was in the range of 2750–3320, larger than that reported by Okuda et al. (2005), where the $\bar{X}, A_m$ was only 900–1600, respectively.

2.3.4 Starch Structure at the 2–100 mm Scale

Starch granules from different cereals have different shapes such as round, oval, polyhedral, and particle sizes (2–100 mm) or particle size distributions (unimodal, bimodal, trimodal). Of rice, several polyhedral small (2–7 mm) granules are produced in one amyloplast (Vandeputte and Delcour 2004) and form parts of compound granules. Compared with other cereals, rice starch granules are the smallest with particle size ranges between 3 and 8 mm (Ayoub et al. 2006), usually with different types of shapes (round or polygonal). Larger void spaces could be observed using SEM for round rice starch granules, compared with those of polygonal starch granules (Kang et al. 2015). Study has also shown that rice starch isolated from different accessions of rice from northeast India showed a typical polygonal morphology with a size ranging from 3.4 to 6.4 μm (Chrungoo and Devi 2015).

2.3.5 Growth Rings

In addition to the effects of the amount of water present, the inner structure of starch granules depends strongly on the hydrogen bonds between amylopectin side chains as they are usually considered to play an important role in the formation of these structures (Ayoub et al. 2006). Under the observation using optical microscopy, starch granules consist of alternating semi-crystalline and amorphous growth rings, which grow by apposition from the hilum of the granule (French 1984; Vandeputte and Delcour 2004). The semi-crystalline growth rings have a thickness between 120 and 400 nm, and the amorphous growth ring is at least as thick as the semi-crystalline one (French 1984). Both the number and the size of starch growth rings depend on the botanical origin of starch and the amount of carbohydrates available during starch biosynthesis. The growth rings of rice starch are approximately 400 nm apart (Dang and Copeland 2003).

2.3.6 Blocklets

Starch is biosynthesized as semi-crystalline granules with varying polymorphic types and degrees of crystallinity. The level of order between the growth rings and the lamellae, i.e., blocklets, is, however, not as commonly mentioned in the literature (Buléon et al. 1998; Vandeputte and Delcour 2004). Under the observation of AMF, the 10–300 nm structures on the surface of starch granules are believed to be carbohydrates in nature and correspond to "blockset" structures, and mainly composed of groups of amylopectin side chain clusters that are present at the granule surface (Dang and Copeland 2003). Low-voltage scanning electron microscopy (LVSEM) also suggests that the semi-crystalline and amorphous growth rings could be subdivided into large (20–500 nm in diameter) and small (25 nm in diameter) spherical blocklets, respectively.

One blockset in the semi-crystalline growth ring contains several amorphous and crystalline lamellae (Fig. 2.5b). On average, two end-to-end blocklets constitute a single semicrystalline growth ring. The cross striations within the growth rings of rice starch correspond to the amorphous and crystalline lamellae blockers and have an average size of 100 nm in diameter and are proposed to contain 280 amylopectin side chain clusters (Dang and Copeland 2003).

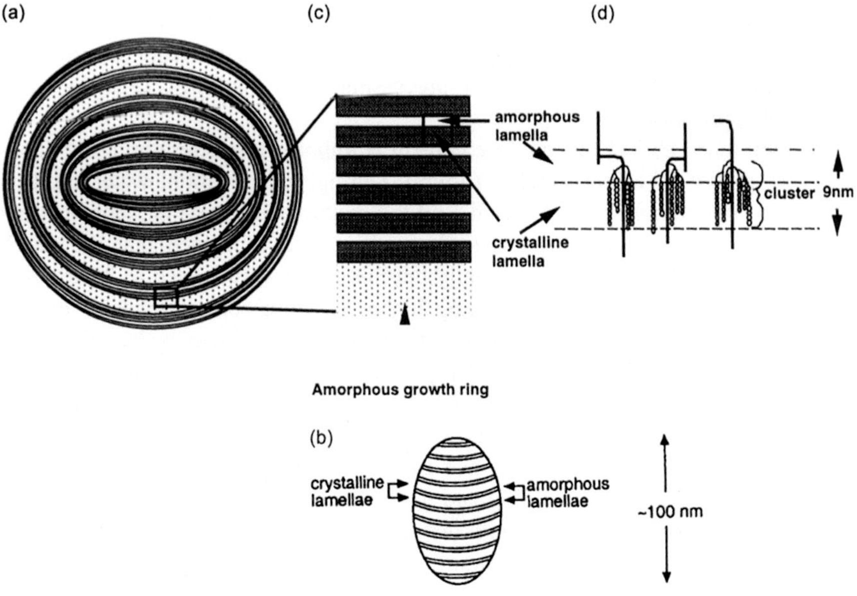

Fig. 2.5 Schematic representation of the starch granule: (**a**) amorphous and semi-crystalline growth rings in a starch granule, (**b**) blocklet, (**c**) amorphous and crystalline lamellae in a stack and part of an amorphous growth ring, (**d**) aligned double helices (from amylopectin side chains) within a crystalline lamella and amylopectin branch points within an amorphous lamella (**a**), (**c**), and (**d**) (Reprinted with permission from Yasuhito et al. (1987))

2.3.7 *Amorphous and Crystalline Lamellae*

One semi-crystalline growth ring (120–400 nm thick) consists of a stack, e.g., about 16 repeats of alternating amorphous (2–5 nm thick) and crystalline (5–6 nm thick) lamellae (Cameron and Donald 1992; French 1984) (Fig. 2.5c). It is commonly believed that despite of botanical resources, the repeat distances of an amorphous and crystalline lamellae of cereal starches are about 9 nm (Jenkins et al. 1993). Amorphous lamellae contain branch points of the amylopectin side chains and possibly some amylose, whereas semi-crystalline lamellae are constituted of amylopectin double helices (Fig. 2.5d). Amorphous growth rings are at least as thick as semicrystalline growth rings (Cameron and Donald 1992), containing amylose and probably less ordered amylopectin.

2.4 Physicochemical Properties of Rice Starch with Different Amylose Contents

Compared with other starches, there have been basic properties associated with rice starch, giving its advantages over other starches (e.g., the hypoallergenicity, digestibility, consumer acceptance, bland flavor, small granular sizes, greater freeze-thaw stability of pastes, and greater acid resistance) (Juliano 1984). The changes that starch undergoes during gelatinization and retrogradation are the major determinants of its functional properties for food processing (e.g., cooking), digestion, and industrial applications.

As mentioned above, the amylose content is important in determining the functional properties of cereal starches (flours). During gelatinization, water sorption occurs and the granule starts to swell, and the double helical and crystalline structures of the granules will be disrupted. Studies have confirmed that the starch gelatinization properties are significantly but not exclusively related to its amylose content. For example, starch containing higher amylose content tends to have a higher onset gelatinization temperature (T_O), whereas a lower value of crystal melting enthalpy (Δh) (Tao et al. 2018). Except for amylose, studies have also shown that the swelling behavior of starch is primarily controlled by the content of amylopectin, while amylose acts as a diluent and/or an inhibitor of swelling, especially in the presence of lipids that can form insoluble complexes with some amylose chains during the swelling and gelatinization process. In this section, the relationships between rice starch with different amylose contents and its functional properties will be discussed as follows.

2.4.1 Gelatinization

Of starch, the gelatinization temperatures including the onset, peak, and conclusion (denoted as T_O, T_P, and T_C, respectively) represent the heat stability of the crystallites, relating to the length of double helices, while the crystal melting enthalpy (Δh) reflects the energy that is needed to disrupt the molecular order, and is usually associated with the amount of molecular order and crystallinity. The gelatinization properties of rice starch determine the time for rice cooking.

Usually, the AAC and gelatinization temperature (GT) combination for each rice accession is not a random feature but may be under genetic control (Bao 2019; Bao et al. 2010). (1) For low-AAC rice accessions (AAC <20%), GT could be divided into high-GT and low-GT classes. While the high-GT rice had a T_P ranging between 77.6 and 79.8 °C, the low-GT rice ranges between 65.8 and 71.0 °C. (2) For high-AAC rice accessions, the GT could also be divided into two groups, including the intermediate GT group with GT ranging between 72.8 and 76.6 °C, and the low-GT group with GT ranging from 63.2 to 67.7 °C. These two low-GT groups with contrasting AAC differed in the range of temperature, with the high-AAC rice having much lower GT than the low-AAC rice (Yang et al. 2014). (3) For waxy rice, their GT could also be divided into high- and low-GT classes as reported (Bao et al. 2010). The high-GT waxy rice had rangedT_P between 76.2 and 79 °C and the low-GT rice ranging between 66.9 and 70.5 °C (Bao 2019; Bao et al. 2010; Li et al. 2020a, b, c).

The role of amylose and amylopectin on starch gelatinization properties has already been well-studied (Fitzgerald et al. 2009a, b; Wang et al. 2015a, b) with a focus on cultivated rice varieties containing lower amylose content. The stable double helices (more crystalline regions) of amylopectin molecules are the main reason in terms of why amylopectin could significantly affect the starch gelatinization temperatures. Although the double helices formed by amylose may also affect the starch gelatinization properties, its contribution is not as big as that of amylopectin. Nevertheless, many other studies have reported that amylose predominantly but not exclusively determines starch gelatinization temperatures. For example, rice cultivars containing higher amylose contents usually tend to have higher T_O, T_P and T_C, but lower Δh (Tikapunya et al. 2017; Wang et al. 2015a, b), which is probably because that for rice starches containing higher amylose content, they tend to have more crystalline and less amorphous regions (Laurence et al. 2011).

The effects of the fine molecular structures of amylose and amylopectin on determining starch gelatinization temperatures have also been well-studied. For example, of rice starches, very short (DP < 12) amylopectin chains have been found related negatively, whereas somewhat longer amylopectin chains with DP ranging between 12 and 24 related positively to T_O, T_P, and T_C (Bao et al. 2009; Noda et al. 2003). Previous study also found that the proportion of amylopectin chains with DP 10 compared to those of DP 24 was negatively correlated with the T_O, while Cuevas et al. (2010) reported that the intermediate GT of rice is characterized by more chains of with DP ranging between 24 and 35. Similar results have also been found by Li and Gong (2020), where T_O, T_P, and the maximum gelatinization rate of rice starches was significantly and negatively correlated with the relative length of amylopectin short chains, whereas T_C was related to the relative length of amylopectin medium chains. It is of the utmost importance that whether the amylopectin chains have a positive or negative influence on the gelatinization temperatures depends on the way they are packed into the lamellar structure within the granules and the exterior amylopectin chains are more involved in determining the gelatinization temperatures as seen for waxy rice starch (Vandeputte and Delcour 2004).

Except for the amylose content, amylose fine molecular structures are also important in determining the gelatinization properties of starches. Li and Gong (2020) reported that although the amylose CLD did not affect the gelatinization temperatures of rice starches, the relative length of amylose medium chains significantly and negatively correlated with the gelatinization temperature range. Tao et al. (2018) reported that the $\overline{R}_{\mathrm{h,Am}}$ of rice starch significantly and negatively correlated with the T_C ($p < 0.01$). They also concluded that of high-amylose rice starch, the gelatinization temperatures will be reduced with increased amylopectin chains with DP < 60, whereas the Δh increases owing to the increased amylopectin content.

2.4.2 *Retrogradation*

When starch is heated in the presence of water and subsequently cooled, the disrupted amylose and amylopectin chains can gradually reassociate into a different ordered structure; the process is termed as retrogradation (Wang et al. 2015a, b). The starch retrogradation process is an ongoing process, initially involving a rapid recrystallization of amylose molecules followed by a slow recrystallization of amylopectin molecules. Starch retrogradation is of great importance in determining not only the functional but also the nutritional properties of rice products. During the initial retrogradation period, double helices are formed between amylose molecules, forming a continuous matrix, while amylopectin short chains recrystallize in a much slower process within the amylose matrix (Gong et al. 2019; Wang et al. 2015b). Because of the different retrogradation rates of amylose and amylopectin, and thus short- and long-term retrogradation occurs depending on how long the gelatinized starch has been left to cool down.

It is believed that the amylose content is an indicator of the rate of starch retrogradation, especially for short-term retrogradation, while amylopectin is more significantly involved within long-term retrogradation. For example, starch retrogradation enthalpy (denoted as ΔH_r) of cooked rice determined by differential scanning calorimetry (DSC) usually significantly correlates with the amylose content ($0.603 \leq r \leq 0.822$, $p < 0.01$) during storage (Yu et al. 2009). Studies have also shown that during the cooling stage of cooked rice, short-term starch retrogradation occurs, which not only affects the textural properties (e.g., hardness and stickiness) (Tao et al. 2019), it also significantly affects the digestibility of cooked rice

(Syahariza et al. 2013). A recent study has also found that the significant correlations between rice amylose content and its digestibility are mainly attributed to the fact that compared to amylopectin, amylose molecules have a much faster retrogradation rate during a short time period, and thus higher amylose content rice usually tends to have a slower digestibility (Li et al. 2020a, b, c). This also explains, to the most extent, that in most retrograde starches, the amylose content is significantly and negatively correlated with its digestibility, but no significant correlation was found between the amylose content and the in vitro digestibility of rice starches at the fully gelatinized stage, where the starch retrogradation has been inhibited (Li et al. 2020a, b, c).

Well defined populations of amylopectin chains either favor (DP 12–22) or inhibit (DP 6–9; DP $<$ 25) amylopectin retrogradation, depending on their ability or lack thereof to form double helices during retrogradation (Lai et al. 2001; Vandeputte and Delcour 2004). Starches with higher proportions of amylopectin chains with DP $<$ 12 most likely less easily regain a partially crystalline polymer system, thereby resulting in lower $T_{O,\ Retro}$, $T_{P,\ Retro}$, $T_{C,\ Retro}$, and Δh_{Retro} or $\Delta h_{AP,\ Retro}$. As at least 10 glucose units are required for malto-oligosaccharides to form double helices, this, therefore, suggests that very short amylopectin chains are unlikely to form longer double helices during the retrogradation process (Michael et al. 1987). This in turn suggests that the minimum DP for amylopectin to retrograde is DP $\geq$ 12 (Vandeputte and Delcour 2004). In contrast, a positive correlation between the relative amounts of very long amylopectin chains with DP $\geq$ 37 was reported (Patindol and Wang 2002), while the external chain length (ECL) of amylopectin branches is also important in controlling the retrogradation rate of rice starch. For example, it was reported that when the average ECL of rice amylopectin was <DP 11.6, the amylopectin retrogradation rate was essentially inhibited (Vandeputte and Delcour 2004).

In addition to amylopectin, smaller amylose molecules, higher amounts of short-medium chains, and shorter amylose medium chains all result in smaller cell diameter and thinner wall thickness of retrograded rice starches suggesting that the amylose molecules with these molecular structures could more rapidly form thermodynamical stable interactions during retrogradation (Gong et al. 2019). Casual relations between amylose fine molecular structures with retrograde amylose networks and with the in vitro digestibility of the retrograde rice starches haves been proposed by the authors (Gong et al. 2019).

2.4.3 Swelling and Pasting Properties

When starch is mixed with enough water and heated up to its gelatinization temperature, pasting viscosities form. During this process, the starch granule swells to several times of its initial size resulting from the loss of the crystalline order and the absorption of water inside the granular structure (Bao 2019). The pasting viscosity during swelling and gelatinization is usually recorded as a Brabender visco-amylograph by a rapid visco-analyzer (RVA). RVA records the viscosity continuously as the temperature is increased, held constant for a time, and then decreased. The RVA measures the pasting viscosity parameters of peak viscosity (PV), hot paste viscosity (HPV), cold paste viscosity (CPV), breakdown (BD), setback (SB), consistency (CS), and pasting temperature (PT) for starches.

Significant correlations among starch structures and the pasting or other physicochemical properties of rice starch have been reported. For example, in the study of Tao et al. (2018), correlations between the PV and trough viscosities (TV) and the amount of shorter amylose chains were significantly positive, whereas with the amylose content correlation was significantly negative. This is because that short amylose chains leach out ahead of the other starch molecules during the gelatinization process (Palav and Seetharaman 2006). After leaching out, those chains in the solution will link with each other and with the rest granules as well, thereby increasing the viscosity, although this linking will be less effective with shorter amylopectin chains. Studies have also shown that starches with longer amylose chains tend to have lower PV and TV, probably resulting from the lower water-holding capacity and granule swelling ability (Matveev et al. 2001). This may also probably be because that longer amylose chains may be involved in more clusters and associated with amylopectin, contributing to the integrity of the starch granules and restricts the starch expansion, reducing the water-holding capacity and granule swelling ability (Tao et al. 2018).

2.4.4 Digestion

Although rice serves as a staple food for about half of the world's people, it is commonly accepted that rice starch products tend to have a high digestibility, and thus rice products are usually considered as not that healthy as other cereal products such as oat and barley. Maintaining a stable blood glucose homeostasis after consuming starch or starchy foods is of great importance for human health. Accordingly, the challenge now facing the rice industry and rice breeder programs is to develop rice varieties with slower starch digestibility to reduce the glycemic response of white rice to a healthy range. The digestibility of rice products such as cooked rice and rice noodles plays an important role in its usage when consumed as human food products.

The relationship between the in vitro digestibility of cooked rice and its starch fine molecular structures has been well-studied previously (Syahariza et al. 2013; Yu et al. 2018). The results showed that both $\overline{R}_{\mathrm{h,Am}}$, and amylose content significantly and negatively correlated with cooked rice digestibility, although further evidence confirmed that the significantly negative relations between amylose content and starch digestibility are mainly driven by the retrodration amylose molecules during the cooling stage (Gong et al. 2019).

The effects of starch structures, from nano to granules level, on starch digestibility varies depending on the digestion conditions and on the physical properties of the starches (flours). At the raw stage, the flour particle size distributions play an essential role in determining the starch digestibility, whereas at the retrograded stage, the amylose content, the fine molecular structures of amylose and amylopectin are all of important in determining the in vitro digestibility of rice starch. Moreover, when at the fully gelatinized stage, the enzyme–substrates interactions are the main factor controlling starch digestibility. More importantly, as for humans, freshly cooked rice products such as rice noodles, rice cakes, and cooked rice all need to be cooled down to room temperature before consumption, and thus factors that could affect the short-term retrogradation rate of starch could all affect the digestibility of rice products as mentioned above.

References

Ayoub A, Ohtani T, Sugiyama S (2006) Atomic force microscopy investigation of disorder process on rice starch granule surface. Starch - Stärke 58(9):475–479

Bao J (2019) 3 - Rice starch. In: Bao J (ed) Rice (fourth edition). AACC International Press, pp 55–108

Bao J, Corke H, Sun M (2010) Genetic diversity in the physicochemical properties of waxy rice (Oryza sativa L) starch. J Sci Food Agric 84(11):1299–1306

Bao J, Xiao P, Hiratsuka M, Sun M, Umemoto T (2009) Granule-bound SSIIa protein content and its relationship with amylopectin structure and gelatinization temperature of rice starch. Starch - Stärke 61(8):431–437

Buléon A, Colonna P, Planchot V, Ball S (1998) Starch granules: structure and biosynthesis. Int J Biol Macromol 23(2):85–112

Cameron RE, Donald AM (1992) A small-angle X-ray scattering study of the annealing and gelatinisation of starch. Polymer 33(12):2628–2635

Castro JV, Ward RM, Gilbert RG, Fitzgerald MA (2005) Measurement of the molecular weight distribution of debranched starch. Biomacromolecules 6(4):2260–2270

Cave RA, Seabrook SA, Gidley MJ, Gilbert RG (2009) Characterization of starch by size-exclusion chromatography: the limitations imposed by shear scission. Biomacromolecules 10(8): 2245–2253

Chrungoo NK, Devi AG (2015) Morphological and rheological properties of starches separated from cultivars of rice (Oryza sativaL.) from North East India. Am J Plant Sci 6(12):2019–2031

Cuevas RP, Daygon VD, Corpuz HM, Nora L, Reinke RF, Waters D, Fitzgerald MA (2010) Melting the secrets of gelatinisation temperature in rice. Funct Plant Biol 37(5):439–447

Dang JMC, Copeland L (2003) Imaging rice grains using atomic force microscopy. J Cereal Sci 37(2):165–170

FAOSTAT (2018) Food and Agricultural Data. http://www.fao.org/faostat/en/

Fitzgerald MA, Bergman CJ, Resurreccion AP, Möller J, Jimenez R, Reinke RF, Martin M, Blanco P, Molina F, Chen M-H, Kuri V, Romero MV, Habibi F, Umemoto T, Jongdee S, Graterol E, Reddy KR, Bassinello PZ, Sivakami R, Rani NS, Das S, Wang YJ, Indrasari SD, Ramli A, Ahmad R, Dipti SS, Xie L, Lang NT, Singh P, Toro DC, Tavasoli F, Mestres C (2009a) Addressing the dilemmas of measuring amylose in rice. Cereal Chem 86(5):492–498

Fitzgerald MA, Mccouch SR, Hall RD (2009b) Not just a grain of rice: the quest for quality. Trends Plant Sci 14(3):133–139

Frei M, Siddhuraju P, Becker K (2003) Studies on the in vitro starch digestibility and the glycemic index of six different indigenous rice cultivars from The Philippines. Food Chem 83(3):395–402

French D (1984) Starch: chemistry and technology. Food Sci Technol:183–247. https://doi.org/10.1016/B978-0-12-746270-7.50013-6

Gong B, Cheng L, Gilbert RG, Li C (2019) Distribution of short to medium amylose chains are major controllers of in vitro digestion of retrograded rice starch. Food Hydrocoll 96:634–643

Hoogenkamp H, Kumagai H, Wanasundara JPD (2017) Chapter 3 - Rice protein and rice protein products. Sustainable Protein Sources:47–65. https://doi.org/10.1016/B978-0-12-802778-3.00003-2

Hoover R, Hughes T, Chung HJ, Liu Q (2010) Composition, molecular structure, properties, and modification of pulse starches: a review. Food Res Int 43(2):399–413

Huang Z, Zeng Z, Gao Y, Liu C, Wu J, Hu X (2019) Crystallization of short-chain amylose: effect of the precipitant. Starch-Starke 71:1900007

Jenkins PJ, Cameron RE, Donald AM (1993) A universal feature in the structure of starch granules from different botanical sources. Starch - Stärke 45(12):417–420

Juliano BO (1984) Chapter XVI—Rice starch: production, properties, and uses. In: Whistler RL, Bemiller JN, Paschall EF (eds) Starch: chemistry and technology (second edition). Academic Press, San Diego, pp 507–528. https://doi.org/10.1016/B978-0-12-746270-7.50022-7

Kang T-Y, Sohn KH, Yoon M-R, Lee J-S, Ko S (2015) Effect of the shape of rice starch granules on flour characteristics and gluten-free bread quality. Int J Food Sci Technol 50(8):1743–1749

Kitta K, Ebihara M, Iizuka T, Yoshikawa R, Isshiki K, Kawamoto S (2005) Variations in lipid content and fatty acid composition of major non-glutinous rice cultivars in Japan. J Food Compos Anal 18(4):269–278

Lai V, Shen MC, Yeh AI, Juliano BO, Lii CY (2001) Molecular and gelatinization properties of rice starches from IR24 and sinandomeng cultivars. Cereal Chem 78(5):596–602

Laurence L, Dongzhi W, Hyun-Jung C, Qiang, Liu (2011) Relationship between the structure, physicochemical properties and in vitro digestibility of rice starches with different amylose contents. Food Hydrocoll 25(5):968–975

Li C, Gong B (2020) Insights into chain-length distributions of amylopectin and amylose molecules on the gelatinization property of rice starches. Int J Biol Macromol 155:721–729

Li C, Gong B, Huang T, Yu W-W (2020a) In vitro digestion rate of fully gelatinized rice starches is driven by molecular size and amylopectin medium-long chains. Carbohydr Polym 254(8):117275

Li C, Luo J-X, Zhang C-Q, Yu W-W (2020b) Causal relations among starch chain-length distributions, short-term retrogradation and cooked rice texture. Food Hydrocoll 108:106064

Li C, Powell PO, Gilbert RG (2017) Recent progress toward understanding the role of starch biosynthetic enzymes in the cereal endosperm. Amylase 1(1):59–74

Li H, Dhital S, Slade AJ, Yu W, Gilbert RG, Gidley MJ (2019) Altering starch branching enzymes in wheat generates high-amylose starch with novel molecular structure and functional properties. Food Hydrocoll 92:51–59

Li H, Prakash S, Nicholson TM, Fitzgerald MA, Gilbert RG (2016) The importance of amylose and amylopectin fine structure for textural properties of cooked rice grains. Food Chem 196:702–711

Li H, Yan S, Yang L, Xu M, Ji J, Mao H, Song Y, Wang J, Sun B (2020c) Starch gelatinization in the surface layer of rice grains is crucial in reducing the stickiness of parboiled rice. Food Chem 341(Pt 2):128202

Matveev YI, van Soest JJG, Nieman C, Wasserman LA, Protserov VA, Ezernitskaja M, Yuryev VP (2001) The relationship between thermodynamic and structural properties of low and high amylose maize starches. Carbohydr Polym 44(2):151–160

Mestres C, Briffaz A, Valentin D (2019) 12 - Rice cooking and sensory quality. In: Bao J (ed) Rice (fourth edition). AACC International Press, pp 385–426

Michael J, Gidley, Paul V, Bulpin (1987) Crystallisation of malto-oligosaccharides as models of the crystalline forms of starch: minimum chain-length requirement for the formation of double helices. Carbohydr Res 161(2):291–300

Nada SS, Zou W, Li C, Gilbert RG (2017) Parameterizing amylose chain-length distributions for biosynthesis-structure-property relations. Anal Bioanal Chem 409:6813–6819

Nakamura S, Cui J, Zhang X, Yang F, Xu X, Sheng H, Ohtsubo KI (2016) Comparison of eating quality and physicochemical properties between Japanese and Chinese rice cultivars. Biosci Biotechnol Biochem 80(12):2437–2449

Noda T, Nishiba Y, Sato T, Suda I (2003) Properties of starches from several low-amylose rice cultivars. Cereal Chem 80(2):193–197

O'Shea M, Samuel MS, Konik CM, Morell MK (1998) Fluorophore-assisted carbohydrate electrophoresis (FACE) of oligosaccharides: efficiency of labelling and high-resolution separation. Carbohydr Res 105(1-2):1–12

Okuda M, Aramaki I, Koseki T, Satoh H, Hashizume K (2005) Structural characteristics, properties, and in vitro digestibility of rice. Cereal Chem 82(4):361–368

Palav T, Seetharaman K (2006) Mechanism of starch gelatinization and polymer leaching during microwave heating. Carbohydr Polym 65(3):364–370

Park IM, Ibáñez A, Shoemaker CF (2010) Rice starch molecular size and its relationship with amylose content. Starch - Stärke 59(2):69–77

Park IM, Ibáñez AM, Shoemaker CF (2007) Rice starch molecular size and its relationship with amylose content. Starch - Stärke 59(2):69–77

Patindol J, Wang YJ (2002) Fine structures of starches from long-grain rice cultivars with different functionality. Cereal Chem 79(3):465–469

Patindol JA, Siebenmorgen TJ, Wang Y-J (2015) Impact of environmental factors on rice starch structure: a review. Starch - Stärke 67(1-2):42–54

Perez M, Juliano BO (1978) Modification of the simplified amylose test for milled rice. Starch - Stärke 30(12):424–426

Syahariza ZA, Sar S, Hasjim J, Tizzotti MJ, Gilbert RG (2013) The importance of amylose and amylopectin fine structures for starch digestibility in cooked rice grains. Food Chem 136(2): 742–749

Takeda Y, Hizukuri S, Juliano BO (1986) Purification and structure of amylose from rice starch. Carbohydr Res 148(2):299–308

Takeda Y, Shibahara S, Hanashiro I (2003) Examination of the structure of amylopectin molecules by fluorescent labeling. Carbohydr Res 338(5):471–475

Tao K, Li C, Yu W, Gilbert RG, Li E (2018) How amylose molecular fine structure of rice starch affects functional properties. Carbohydr Polym 204:24–31

Tao K, Yu W, Prakash S, Gilbert RG (2019) High-amylose rice: starch molecular structural features controlling cooked rice texture and preference. Carbohydr Polym 219:251–260

Tikapunya T, Zou W, Yu W, Powell PO, Fox GP, Furtado A, Henry RJ, Gilbert RG (2017) Molecular structures and properties of starches of Australian wild rice. Carbohydr Polym 172: 213–222

Vandeputte GE, Delcour JA (2004) From sucrose to starch granule to starch physical behaviour: a focus on rice starch. Carbohydr Polym 58(3):245–266

Wang K, Wambugu PW, Zhang B, Wu AC, Henry RJ, Gilbert RG (2015a) The biosynthesis, structure and gelatinization properties of starches from wild and cultivated African rice species (Oryza barthii and Oryza glaberrima). Carbohydr Polym 129:92–100

Wang S, Li C, Copeland L, Niu Q, Wang S (2015b) Starch retrogradation: a comprehensive review. Compr Rev Food Sci Food Saf 14(5):568–585

Wani AA, Singh P, Shah MA, Schweiggert-Weisz U, Gul K, Wani IA (2012) Rice starch diversity: effects on structural, morphological, thermal, and physicochemical properties—a review. Compr Rev Food Sci Food Saf 11(5):417–436

Wong KS, Jane JJ (1997) Quantitative analysis of debranched amylopectin by HPAEC-PAD with a postcolumn enzyme reactor. J Liq Chromatogr Relat Technol 20(2):297–310

Wu AC, Li EP, Gilbert RG (2014) Exploring extraction/dissolution procedures for analysis of starch chain-length distributions. Carbohydr Polym 114:36–42

Wu AC, Witt T, Gilbert RG (2013) Characterization methods for starch-based materials: state of the art and perspectives. Aust J Chem 66(12):1550–1563

Yang F, Chen Y, Tong C, Huang Y, Xu F, Li K, Corke H, Sun M, Bao J (2014) Association mapping of starch physicochemical properties with starch synthesis-related gene markers in nonwaxy rice (Oryza sativa L.). Mol Breed 34(4):1747–1763

Yasuhito T, Susumu H, Bienvenido O, Juliano. (1987) Structures of rice amylopectins with low and high affinities for iodine. Carbohydr Res 168(1):79–88

Yu S, Ma Y, Sun DW (2009) Impact of amylose content on starch retrogradation and texture of cooked milled rice during storage. J Cereal Sci 50(2):139–144

Yu W, Li H, Zou W, Tao K, Zhu J, Gilbert RG (2019a) Using starch molecular fine structure to understand biosynthesis-structure-property relations. Trends Food Sci Technol 86:530–536

Yu W, Tao K, Gidley MJ, Fox GP, Gilbert RG (2019b) Molecular brewing: molecular structural effects involved in barley malting and mashing. Carbohydr Polym 206:583–592

Yu W, Tao K, Gilbert RG (2018) Improved methodology for analyzing relations between starch digestion kinetics and molecular structure. Food Chem 264:284–292

Zhong F, Yokoyama W, Wang Q, Shoemaker CF (2006) Rice starch, amylopectin, and amylose: molecular weight and solubility in dimethyl sulfoxide-based solvents. J Agric Food Chem 54(6):2320–2326

Zhu J, Yu W, Zhang C, Zhu Y, Xu J, Li E, Gilbert RG, Liu Q (2020) New insights into amylose and amylopectin biosynthesis in rice endosperm. Carbohydr Polym 230:115656

Chapter 3
Rice Proteins and their Applications

Sunantha Ketnawa

Abstract A brief introduction to rice protein is described and then the rice protein classification and the rice properties related to protein were discussed. After that, the novel methods of rice protein extractions and the bioactive peptides from rice protein hydrolysis are described. Finally, the applications of protein from rice in infant food formulations, bakery products, snacks, natural antioxidants as well as functional ingredients are discussed, particularly, the future direction of rice protein applications is also illuminated.

Keywords Rice protein · Extractions · Peptides · Application

Abbreviations

ACE	Angiotensin-converting enzyme
BHT	Butylated hydroxyanisole
CDR	Carbohydrate-digested rice
COX-2	Cyclooxygenase-2
DB	Dry basis
DDGS	Distiller's Dried Grains with Solubles
DH	Degree of hydrolysis
FRAP	Reducing power assay
GF	Gluten-free
O/W	Oil-in-water
RBPC	Rice bran protein concentrate
RPH	Rice protein hydrolysate
TNF-α	Tumor necrosis factor-α

S. Ketnawa (✉)
Graduate School of Horticulture, Chiba University, Matsudo, Chiba, Japan

Program of Food Science and Technology, School of Agro-Industry, Mae Fah Luang University, Chiang Rai, Thailand

Functional Food Research Center for Well-being, Chiang Mai University, Chiang Mai, Thailand

J. Tian et al. (eds.), *Science of Rice Chemistry and Nutrition*,
https://doi.org/10.1007/978-981-99-3224-5_3

UAD Ultrasound-assisted α-amylase degradation

3.1 Introduction

Rice is one of the first three major food crops in the world after corn and wheat as well forms the staple diet for about half of the world's population mostly consumed in Asian countries. Furthermore, rice is also a valuable source of nutritional and nutraceutical substances for human health. Knowledge of the nutritional value and health benefits of rice is essential to contribute to these groups and initiate rice consumption in people's daily diets. Generally, whole rice grain comprises carbohydrates 63.6–73.2%, fat 1.5–2.3%, protein 5.8–7.7%, fiber 7.2–10.4%, and ash 2.9–5.2% (Table 3.1) (Juliano and Bechtel 1985). Freshly harvested rice grains consist of carbohydrates 80%, including starch, glucose, sucrose, dextrin, etc. (Juliano 1993). Polished rice grains consist mainly of carbohydrates 77–89% and protein 6.3–7.3% with minor fractions of lipids 0.3–0.5%, ash 0.2–0.5%, fiber, sugars, and free amino acids. Polished rice grains have low-protein content and contain few lipids due to the removal of the bran, but showed higher digestibility of protein than brown rice ones (Saleh et al. 2019).

The composition, nutritive values, and bioactive compounds of rice vary with internal factors viz. cultivar and breed, and environmental factors such as soil fertility, the application of fertilizers, growth duration, and weather (Juliano 1993; Verma and Srivastav 2020). Protein content tends to increase in response to high nitrogen (N) fertilizer application, especially flowering stage (Juliano 1993). The protein content of grain during grain development increases in the wet season due to cloudy weather and short growth duration (Juliano 1993). Besides, environmental stress such as drought, salinity, alkalinity, high or low-temperature diseases, or pests may escalate the protein content. From the literature of Juliano (1993), environmental factors increase protein content and ash but do not affect the fat content of brown rice. Soil organic matter such as total nitrogen, exchangeable calcium, available copper and molybdenum, and total chlorine all tend to increase rice grain protein content as well (Juliano 1993). Rice bran has the highest and hull has the lowest protein content among the milling fractions, as depicted in Table 3.1.

Although whole rice grains are considered non-rich protein seeds which cannot be used as the main source for protein extraction. However, brown rice, germinated rice, and other sidestream products from milling processing such as bran and broken rice kernels can be potential sources due to being relatively rich in protein. They have mostly been under-valued and under-utilized in an excessive number through the years. The industry commonly processes broken rice kernels or rice flour for starch and flour, as well as the residual protein (Kamara et al. 2010). The processing of rice proteins begins getting attention from industry, rice research facilities, and the public because of its exceptionally healthy, hypo-allergenicity, anti-mutagenicity, high digestibility, and human consumption suitability (Verma and Srivastav 2020;

Table 3.1 Proximate analysis of rough rice and its milling fractions at 14% moisture (Juliano 1993)

Rice fraction	Crude Protein (g N× 5.95)	Crude lipids (g)	Crude fibre (g)	Crude ash (g)	Available carbohydrates (g)	Fibre (g)	Energy content	
							kJ	kCal
Rough rice	5.8–7.7	1.5–2.3	7.2–10.4	2.9–5.2	64–73	16.4–19.2	1580	378
Brown rice	7.1–8.3	1.6–2.8	0.6–1.0	1.0–1.5	73–87	2.9–3.9	1520–1610	363–385
Milled rice	6.3–7.1	0.3–0.5	0.2–0.5	0.3–0.8	77–89	0.7–2.3	1460–1560	349–373
Rice bran	11.3–14.9	15.0–19.7	7.0–11.4	6.6–9.9	34–62	24–29	670–1990	399–476
Rice hull	2.0–2.8	0.3–0.08	34.5–45.9	13.2–21.0	22–34	66–74	110–1390	265–332

Table 3.2 Rice protein classifications and characteristics (Fabian and Ju 2011)

Protein classification	Composition (%)	Solubility	Size (kDa)	Nutrition
Albumin	34	Water soluble (possess lower amount of disulfide crosslinking)	10–20(Glycoprotein as one of major constituents)	• Readily digestible and absorbable. • Exhibits antioxidant activity.
Globulin	15–26	Salt soluble offer co-extracted with albumin due to the presence of salts in the plant materials	25 and 16(Polypeptides stabilized by interchain disulfide bonds)	Moderately high amounts of cysteine and methionine
Glutelin	11–27	Alkali soluble extensive aggregation disulfide bonding and glycosylation contribution to a difficult extraction and solubilization	45–150	Some glutelin types are rich in lysine
Prolamin	4	Alcohol soluble 60–70% ethanol needed for solubilization	12–17	High levels of glutamic acid or glutamine, alanine, glycine and arginine

Rodríguez-Restrepo et al. 2020; Kawakatsu and Takaiwa 2019; Pantoa et al. 2020a, b). Rice protein is highly digestible (93%) due to its higher lysine concentration (~4%), with a high biological value (74%) and a protein efficiency ratio (2.02%–2.04%). Thus, rice proteins have the potential for many applications such as the usage as functional food ingredients in infants, aging formulas, and sports nutrition or options for alternative plant-based food (Verma and Srivastav 2020; Amagliani et al. 2017a, b). Additionally, an increasing number of animal-related disease cases have driven attention to the consumption of rice protein.

According to the classification of protein by solubility by Osborne (1924), rice proteins are categorized into the four main protein fractions including glutelin (alkali/acid-soluble), globulin (salt-soluble), albumin (water-soluble), and prolamin (alcohol-soluble) (Osborne 1924) (Table 3.2). Glutelin (about 80%) and globulin (about 12%) are the major rice protein components (Kawakatsu and Takaiwa 2019; Amagliani et al. 2017a, b). Hence, water and alkaline solutions are the solvents of choice for the processing of rice proteins. However, depending on the cultivars or sources of rice, the effect of solvent, various processing designs, and methods to extract protein have been developed by many researchers through the decade. Besides, novel methods, for example, solely or in a combination of enzymatical assistance (Braspaiboon et al. 2020), microwave (Phongthai et al. 2016a, b, c; Bedin et al. 2019), ultrasonication (Phongthai et al. 2016a, b, c; Singh et al. 2019; Yang et al. 2018; Ly et al. 2018), subcritical water (Sereewatthanawut et al. 2008), supercritical carbon dioxide and dry fractionation (Silventoinen et al. 2019), ultrafiltration (Doan and Lai 2021), and liquid partitioning (Wang et al. 2020a, b) and mechanical disruption through bead milling (Larsson 2020) and freeze-milling (Wang et al. 2015). In addition to the extraction of rice protein, there are some extra components known as bioactive compounds, which are typically presented in different parts of rice (viz. bran, dreg, germ, and endosperm) and have performed different biological activities when produced as protein hydrolysates. These compounds are very beneficial for human health such as in vitro α-amylase (Uraipong and Zhao 2016), β-glucosidase (Uraipong and Zhao 2016), DPP-IV (Pooja et al. 2017), and ACE inhibition activities (Uraipong and Zhao 2016; Suwannapan et al. 2020), antioxidant properties (Zaky et al. 2019; Rani et al. 2018; Pantoa et al. 2020a, b), neuroprotective properties (Pannangrong et al. 2018), lipid-lowering action (Yang et al. 2012), anti-inflammatory (Wang et al. 2019a, b; Feng et al. 2021; Wen et al. 2016), and anti-cancer (Kannan et al. 2009).

3.2 Rice Protein Sciences and Classifications

Carbohydrates (over 70%) are mainly present in rice grains followed by protein (usually found as 5–12%) (Juliano 1993). Compared to other major cereal proteins, rice protein itself is acknowledged as a hypoallergenic, highly digestible, and high nutritional protein (Silventoinen et al. 2019). The edible components of rice grains contain a total protein of about 5–16% (Table 3.1), which are located in the aleurone and sub-aleurone layers (bran) and the starchy germ and endosperm (Hoogenkamp et al. 2017). According to Juliano (1993), the rice varieties that contain protein contents higher than 10% and 7% have been considered as high and low-protein rice, respectively. Rice flour protein content mostly depends on cultivars, maturity stage, nitrogen fertilization, and water availability during rice grain growth (Bryant et al. 2013). Pantoa and Baricevic-Jones (Pantoa et al. 2020a, b) reported protein contents of rice grains in various maturity development stages (flowery, milky, dough, and mature stages) ranging from 7.18 ± 1.10 to 11.48 ± 0.90% (dry basis, DB) in white rice (Pathum Thani 1 variety) and 8.41 ± 1.96 to 9.19 ± 0.04% (DB) in colored rice (Riceberry variety), and 8.9 ± 0.05% (DB) in rice bran. The highest protein content was found at the early development stage (flowery-to-milky) as 9.84 ± 0.04% in the flowery stage of rice berry and 11.48 ± 0.90% in the milky stage of Pathum Thani 1. Higher protein content in young rice than in mature rice was partly related to an increase in metabolic enzymes during grain filling (Olivares Díaz et al. 2019). Interestingly, young rice protein is also easy to digest and showed hypoallergenic potential enzyme digestion proved by protein pattern by SDS-PAGE. Rice allergens in the mature stages of white rice (13–14 and 52 kDa) and colored rice (13–14 and 37–39 kDa) were considered indigestible, vice versa rapidly decreased and completely disappeared after digestion in young rice. Additionally, the digestibility of the dough-to-mature stage was much slower, especially for colored rice (Pantoa et al. 2020a, b). Colored rice flours (rice berry flour, brown jasmine rice flour, and red jasmine rice flour) provide higher nutrients than white rice flour because they contain higher values of ash, crude fat, crude protein, and crude fiber (Kraithong et al. 2018). Also, protein is highly concentrated in the outer bran layer of rice grain, which makes the protein concentration of brown rice higher than that of milled rice (Saleh et al. 2019). Several studies have determined the protein and amino acid contents of brown rice and reported protein concentrations ranging from 5.4% to 11.9% among brown rice accessions (Bryant et al. 2013; Kim et al. 2013; Meesook and Yoonhwa 2002). External factors such as the cultivation environment viz. soil type and fertility, ambient air temperature, solar radiation, day length, relative humidity, and panicle characteristics affect protein content influencing the rice grain filling and nutritional content (Olivares Díaz et al. 2019; Kinoshita et al. 2017).

3.2.1 Classification of Rice Proteins

The oldest protein classification system in foods first identified by Thomas B. Osborne (1859–1929) is still in use nowadays and based on the solubility of proteins in various solutions (Osborne 1924; Nadathur et al. 2017). The Osborne classification system has evolved over the years and has been identified into 5 protein groups (Nadathur et al. 2017), which are (1) albumins, are readily soluble in water and dilute salt solutions. Albumins do not have a specific amino acid composition but are abundant in acidic amino acids. Albumins are widely found in body fluids and plant storage organs such as seeds; (2) globulins, are poorly soluble in water, but readily soluble in dilute salt solutions, usually rich in glutamates and aspartates. Globulins are abundant in plant and animal cells; (3) histones, are soluble in acidic salt solutions and rich in basic amino acids with a small amount of S-containing amino acid. Histones make complexes with nucleic acids and are components of the nuclei of eukaryotic cells; (4) prolamins, are insoluble in water and absolute ethanol, but soluble in ethanol–water mixtures, and commonly have high arginine, proline, glutamate, and asparagine levels; and (5) scleroproteins, are insoluble in water and salt solutions and include fibrous proteins of animal connective tissues rich in glycyl, prolyl, and alanyl residues (Osborne 1924).

Rice protein is approximately composed of glutelin (60–80%), albumin (4–22%), globulin (5–13%), and prolamin (1–5%) (Hoogenkamp et al. 2017; Kim et al. 2013). The ratio of these compositions was also influenced by cultivars, plantation environment, milling degree, protein fractionation methods, and protein extraction solvents and methods (Wang et al. 2014). Korean rice cultivars (1 non-waxy and 8 waxy rice grains) were produced applying a nitrogen fertilizer level of 7 kg/10 acre (Kim et al. 2013). They found that the crude protein contents of waxy cultivars were higher than that of the non-waxy ones. Crude protein contents among waxy cultivars, ranged from 8.6% to 9.7% in brown rice and from 8.1% to 8.5% in milled rice. The protein fraction of waxy brown rice was found as glutelin: albumin: globulin: prolamin of waxy brown rice was 7.31:1.86: 0.50:0.05, whereas those of milled rice were 8.40: 0.50: 0.30: 0.1. From the study of Meesook and Yoonhwa (2002), rice proteins were extracted from brown and milled rice of five Korean varieties and protein patterns under study by SDS-PAGE. The ratios of glutelin: albumin: globulin: prolamin in the brown rice were 68.6–72.8:10.8–14.1:12.4–16.4: 3.6–5.3 and in milled rice were 75.7–79.8:4.4–5.6:10.6–12.0:3.9–5.4 that glutelin, albumin, globulin, and prolamin were observed at MWs ranging from 14.4 to 57.4 kDa (7 bands), 14.9 to 96.8 kDa (7 bands), 14.4 to 56.9 kDa (4 bands), 14.4 kDa (1 band), respectively (Meesook and Yoonhwa 2002).

Glutelin or oryzenin is rich in hydrophobic amino acids, such as Phe, Val, Tyr, Pro, and Leu. Glutelin is soluble in acid below pH 3 and base above pH 10 and has limited solubility in water due to its molecules being covalently linked from two subunits by an extensive disulfide bond and hydrophobic interactions resulting in the highest molecular weight (MW) molecules, that range from 64 to 500 kDa (Cao et al. 2009). Glutelin came from a 57-kDa polypeptide precursor that cleaves into a large acidic subunit (30–39 kDa) and a small basic subunit (2–25 kDa) (Paraman et al. 2006). Due to the structure, glutelin may partly account for the difficulty in protein extraction. Chanput et al. (2009) and Xia et al. (2012) found that rice glutelin is composed of high MW protein ranging from 10 to 60 kDa.

Albumin is normally known as a water-soluble protein. Albumin in rice grains ranges about 6.24–9.73% and about 30.9–42.7% in rice bran (Cao et al. 2009; Chanput et al. 2009). The extraction of albumin with water always results in globulin contamination since minerals present in the rice grain that dissolve in a water solvent. Rice albumins were separated into 3–5 sub-fractions using gel filtration chromatography proving MWs range from 18 to 50 kDa when analyzed by size-exclusion HPLC (Wang et al. 2014; Hamada 2000). Furthermore, Cao et al. (2009) reported that albumins have major polypeptides with MWs of 18 20 kDa, while Chanput and Theerakulkait (2009) observed two main bands of albumin between 30 and 50 kDa using SDS-PAGE. From the study by Wang et al. (2014), albumin appeared in five bands with MWs of 32, 31, 22, 17, and 14 kDa.

Globulin, salt-soluble and sulfur-rich proteins, approximately found 12.5–24.9% of the storage proteins in rice grains (Juliano and Bechtel 1985; Cao et al. 2009; Chanput et al. 2009) contain the different polypeptide chains that were stabilized by disulfide linkages (Hamada 2000). The globulin of the rice endosperm could be separated by gel filtration chromatography into four sub-fractions; MWs ranged from 16 to 130 kDa (Kawakatsu and Takaiwa 2019). Xia et al. (2012) reported that MWs of globulin in broken rice was 26 kDa. Furthermore, the globulin of the Thai rice variety also had the MWs at about 15, 30, and 50 kDa (Chanput et al. 2009). From the study by Wang et al. (2014), globulin in defatted rice bran appeared in five major bands with MWs of 63, 53, 49, 36, and 22 kDa.

Prolamin is an alcohol-soluble protein in rice grains. It is usually extracted after the removal of albumin and globulin (Juliano and Bechtel 1985). It may be extracted with 70% ethanol or 50% propanol. Prolamin presents as 20–25% of the rice endosperm protein (Hoogenkamp et al. 2017). In rice bran, prolamin presents only about 3.24–11.6% of the total protein (Cao et al. 2009; Chanput et al. 2009). Prolamin is usually not focused on the solvent or conditions in the protein extraction because of its very low content and ability to be co-extracted when alkaline extraction is applied. The prolamin fraction contains proteins in the MWs range of

10–17 kDa (Cao et al. 2009). Prolamin 10–13 kDa was found in the dough and mature stages of Phatumthani rice (Pantoa et al. 2020a, b). Prolamin (13–14 kDa), a potential allergen in rice protein, remained after commercial protease digestion, especially in the dough-to-mature stage (Pantoa et al. 2020a, b). From the study by Wang et al. (2014), prolamin in rice bran appeared in only 2 bands with MWs of 13 and 22 kDa, while 3 major bands of the prolamin fraction at MWs of 10, 15, and 25 kDa were documented by Chanput et al. (2009). Meanwhile, a similar result by Cao et al. (2009) also confirmed that prolamin consisted of 3 polypeptide subunits with clear MWs of 10, 13, and 16 kDa.

3.2.2 Protein-Related Rice Properties

Apart from the nutritional aspect, protein is one of the composition which a vital role in rice nutritional quality, processing, palatability of cooking rice (Kraithong and Rawdkuen 2019), and rice flour properties (Parada and Santos 2016). According to Parada and Santos (2016) . Hydrophobic and hydrophilic groups, as well as protein side chains, can associate with amylose, known as amylose–protein complex. When the temperature is changed, heat-resistant capacity, hardness, and stickiness of rice flour gel are increased while pasting temperature is decreased influenced by the protein content of rice flour. High amounts of protein in rice flour also supported peak viscosity due to the higher water-holding ability of its hydrophilic groups (Kraithong and Rawdkuen 2019; Hsu et al. 2015).

However, when protein is extracted from rice grain makes the properties change. Rice grains contain a high MW protein known as glutelin (about 80%), which is composed of subunits bound by disulfide linkages and which is soluble only in dilute acid or alkali (Paraman et al. 2006). The application of rice protein in the food industry is limited by certain techno-functional properties such as the poor solubility induced by its compact conformation. The solubilization of rice protein isolates has been regarded as one of the critical and challenging aspects affecting commercial applications. The solubility-based classification is most relevant to food processing and commercial manufacturing of food proteins. To improve techno-functional properties of proteins viz. water and oil solubility, foaming ability, emulsion ability, various approaches to modify the protein structure such as physical treatment, chemical modification, and enzymatic treatment, genetic modification, green technology viz. germination, and fermentation have recently been practiced (Wang et al. 2015; Cao et al. 2009; Paraman et al. 2006; Li et al. 2019; Zhang et al. 2012; Li et al. 2016; Cao et al. 2010).

Li et al. (2016) reported the change in rice protein including the exposure of hydrophobic groups, redistribution of the secondary structure, and changes in the microstructures by pretreating rice protein with ultrasound and ultrasound-assisted alkali pretreatment. The result showed that the latter pretreatment was better than the first one only. The accessibility of the attacking enzyme is substantially improved by the pretreatment of ultrasound-assisted alkaline, thus, decreasing or disrupting the structure of protein more effectively (Li et al. 2016). In another study, simultaneous treatment with freezing and milling (freeze–milling) combined with alkali pretreatment was investigated by Wang et al. (2015) to solve the solubility limitations of rice protein. Freeze–milling combined with alkali pretreatment can remarkably increase the maximum amounts of soluble rice protein isolate up to 42 times due to unfolding with ice crystals forming inside protein bodies by freeze–milling. Because of that, the hydrophobic surface was disrupted and possessed and hydrophilic inner groups were exposed, contributing to disaggregation and strong water–protein interactions (Wang et al. 2015). In another study by Li et al. (2019), electron beam irradiation induced the amino acid oxidation, as well and destroyed the secondary structure and microscopic morphology of rice protein hydrolysates. Interestingly, the amino acid content of rice protein increased with enzymatic hydrolysis. Although irradiation reduced the total amino acids and total essential amino acids of the hydrolysates, the antioxidant activity was significantly improved, possibly due to changes in hydrophobic regions and secondary structures (Li et al. 2019). Silventoinen, Rommi (Silventoinen et al. 2019) successfully improved the protein solubility and colloidal stability of rice bran protein using supercritical carbon dioxide. Besides, the protein flour obtained by dry fractionation from the same research group was also enriched with insoluble dietary fiber whereas starch content was decreased (Wen et al. 2016).

Generally, brown rice protein is difficult to dissolve due to extensive disulfide bonding and aggregation. Green and economical approaches such as germination can increase soluble protein yield in brown rice (Cao et al. 2010). Gaining protein about 20% higher than that from non-germinated brown rice was observed. The researcher found that phytase enzyme activity increased within the first day of soaking that makes a soluble protein increase in the germinating grain, and then phytate was decomposed to relax protein aggregation. From the study, the highest soluble protein yields were obtained from germinated brown rice at an extraction condition of 41.5 °C for 2 h using a solid to solvent ratio of 1:10 at pH 8.5. Germination greatly increased the solubility of the protein, particularly at acidic ($pH < 4$) and alkaline ($pH > 7$) conditions (Cao et al. 2010). The lowest solubility of brown rice protein and germinated brown rice was at pH of 5.0 and 4.0, respectively (Cao et al. 2010). In the study according to Parkman, Hettiarachchy (Paraman et al. 2006), revealed that enzymatic hydrolysis of endosperm protein from rice improved its solubility and emulsification properties.

3.3 Rice Protein Extraction and Utilizations

3.3.1 The Novel Methods of Rice Protein Extraction

Apart from the conventional alkaline extraction, rice protein is also produced by enzyme or physical treatment from rice flour followed by precipitation. Then centrifugation or filtration is often used to recover rice protein after precipitation

(Amagliani et al. 2017a, b). Phongthai, D'Amico (Phongthai et al. 2016a, b, c) raised the point about allowing proteins to precipitate after pH adjustment under cold conditions instead of centrifugation. This could be possible to reduce the cost of operation and power consumption when practicability for an industrial scale. However, the product yield may be less than one with centrifugation (Phongthai et al. 2016a, b, c). With conventional alkaline extraction, the protein yield is generally low because of the degradation at high pH conditions (Braspaiboon et al. 2020). High pH conditions could also lead to undesirable results, including molecular cross-linking and rearrangements, resulting in a decrease in nutritive value and formation of toxic compounds such as lysinoalanine, which is a carcinogen (De Groot and Slump 1969). Furthermore, the remaining alkali needs to be washed thoroughly from the product, leading to the generation of a large volume of undesirable wastewater (Braspaiboon et al. 2020). The whole process consuming a long time makes the process are not economically efficient and sustainable.

Recently, there were several attempts to harvest rice protein from sidestream products such as bran (Bedin et al. 2019; Phongthai et al. 2016a, b, c; Ly et al. 2018; Sereewatthanawut et al. 2008; Larsson 2020; Zhang et al. 2019; Herrero et al. 2006), dreg (Yang et al. 2018; Zhang et al. 2021), endosperm, water from rice starch production (Doan and Lai 2021), distillery byproducts (Singh et al. 2019), and electrolyte drink byproducts (Braspaiboon et al. 2020) with the target to fully use sidestream products as well as yield the maximum of protein recovery. Numerous studies have been published on rice sidestream product protein extraction using solely conventional or novel methods or a combination of certain novel techniques used for rice protein extraction, as summarized in Table 3.3.

According to Braspaiboon et al. (2020), conventional solvent extraction compared to enzymatic extraction was studied to extract protein from carbohydrate-digested rice (CDR), the residue derived from waste of electrolyte drinks production. The extraction of protein from CDR was investigated by comparing alkaline (different concentrations of NaOH) and enzymatic (Alcalase® or Flavourzyme®) extraction. From the study, the maximum protein recovery of 30.04%) was obtained from Alcalase® hydrolysis the less protein recovery was gained from the extraction with 0.1 M NaOH (55 °C) (27.43%) or hydrolysis with Flavourzyme®. Moreover, protein solubility by extraction with alkaline was much lower (23.46%) than that achieved via the enzymatic method (100%). Braspaiboon et al. (2020) summarized that Alcalase® extraction facilitated high protein recovery and the highest amount of total soluble protein. Thus, Alcalase® extraction should be applied for the commercial production of protein powder or drinks from CDR (Braspaiboon et al. 2020). Besides, the application of co-solvent in the techniques like three-phase partitioning (TPP) was applied to extract protein from rice bran using $(NH_4)_2SO_4$ and *t*-butanol to precipitate protein and lipids, respectively (Wang et al. 2020a, b). The optimal conditions of $(NH_4)_2SO_4$ concentration 28% (w/v), slurry to *t*-butanol ratio 1:1.1 (v/v), pH 5.10, extraction temperature 40 °C and time 1 h provided the highest extraction yields of rice bran protein (6.81%) (Wang et al. 2020a, b). Another source of protein from byproducts, Distiller's Dried Grains with Solubles (DDGS) and Wet Solids (WS) from distillery byproducts are rich in fermented proteins (Singh et al.

Table 3.3 Novel methods for protein extraction: recovery and efficiency

Methods/ Raw materials	Conditions	Protein recovery/ yield (%)	Protein content	References
Enzyme assisted extraction				
Commercial Rice bran	• Alcalase (0.013 μM/mg/min) at 1% of protein content of rice bran at optimal conditions of pH (7.5) and temperature (60 °C) for a period of 30 min. • Alcalase treated RB was treated with the endoglucanase from *Aspergillus aculeatus* (1.4 unit/mg/min) for 1 h for hydrolyzing cellulose present in rice bran at 1% level at 55 °C and pH 5.	100%	N/A	Vallabha et al. (2015)
Rice bran	0.4 M NaCl, 80% ethanol, and 0.01 M NaOH as extraction solvents to fractionate albumin, globulin, prolamin, and glutelin	13.8% purity (45–47%)	N/A	Wang et al. (2014)
Novel technology				
Ultrasound				
Rice bran (*Oryza sativa* OM488)	• First series: The sonication power was changed: 0 (control), 5, 10, 15, 20 and 25 W/g of material dry mass. The sonication time was 3 min. After the ultrasonic treatment, the time of the additional extraction was 30 min. • Second series: The sonication time was varied: 0, 1, 2, 3, 4 and 5 min. • The selected ultrasonic power was 15 W/g. • After the ultrasonic treatment, the time of the additional extraction was fixed at 30 min. • Third series: The sonication power and time were set at 15 W/g and 2 min, respectively. After the ultrasonic treatment, the time of the additional extraction was changed: 0, 5, 10, 20, 30, 40, 50 and 60 min.	25.6%, maximum of 64.5% when the additional extraction time was 20 min	N/A	Ly et al. (2018)

(continued)

Table 3.3 (continued)

Methods/ Raw materials	Conditions	Protein recovery/ yield (%)	Protein content	References
Organic rice bran	Optimal condition: 76% sonication amplitude, 18 min extraction and 0.99 g/10 mL solid–liquid ratio	4.73.		Phongthai et al. (2016a, 2016b, 2016c)
Microwave				
Organic rice bran	1000 W of microwave power, 90 s of extraction time, and a solid to liquid ratio of 0.89 g rice Bran/10 mL of distilled water	4.37		Phongthai et al. (2016a, 2016b, 2016c)
Combination				
Rice bran	Mechanical disruption through Bead milling in combination with solubilization using pH shift.	30.2	33%	Larsson (2020)
Rice bran	Ultrasound combined with 0.9% sulfuric-acid soaking at 90 °C.(0.9 wt % H_2SO_4 soaking at 90 °C and 300 W ultrasonication for 30 min)	72.8%		Zhang et al. (2019)
Milled rice bran	Supercritical carbon dioxide–extracted	38.92% 48.11%		Silventoinen et al. (2019)
Rice bran (*Oryza sativa* L)	Alkaline extraction 60 min and MAE DRB:Water 0.5:10 MAE were 2 min	Conventional: 12.85–15.68 MAE 75.32–79.98	6.19%	Bedin et al. (2019)

2019). Solvent extraction, ultrasonication, and protease-mediated extraction have been carried out for extracting fermented protein. Aqueous-ethanol extraction resulted in the highest yield of 47.7 and 59.8 (%, w/w) for DDGS and WS, respectively. Glutelin and prolamin are the predominant fractions with a recovery of 81.75% and 75.02%, respectively. The maximum recovery of proteins has been achieved in the order of aqueous-ethanol > alkaline-ethanol > protease-mediated > ultrasonication > pH shifting techniques (Singh et al. 2019). In another study, Sereewatthanawut et al. (2008) also reported a higher amount of protein and amino acids from a novel method such as subcritical water than those obtained by conventional alkali extraction. Applying subcritical water at a temperature of 200 °C for 30 minutes to hydrolyze defiled rice bran provided the highest yield of protein and amino acids were 219 ± 26 and 8.0 ± 1.6 mg/g of dry bran (Sereewatthanawut et al. 2008). The yields generally increased with increased temperature and hydrolysis time. They also summarized that thermal degradation of the product was observed when hydrolysis was carried out at higher temperatures for extended periods (Sereewatthanawut et al. 2008). Silventoinen et al. (2019) studied a novel approach using non-heated supercritical carbon dioxide to enrich protein from rice bran. Protein content was enriched from original 18.5% up to 25.7% in the fine powder. Besides, the resulting fine rice bran protein powder was also enriched with insoluble dietary fiber whereas starch content was decreased with improved protein solubility and colloidal stability (Silventoinen et al. 2019).

The extraction of rice protein from rice dreg powder using ultrasound-assisted α-amylase degradation (UAD) treatment was studied by Yang et al. (2018). Compared to α-amylase degradation treatment and conventional extraction, UAD treatment significantly increased protein extraction rate ($p < 0.05$). The highest protein extraction rate of 89.58% and protein purity of 92.99% were obtained under 20/35 kHz of sequential frequency ultrasound treatment (Yang et al. 2018). They found that the sequential frequency treatment gave a higher protein extraction rate than that of simultaneous or single frequency ones (Yang et al. 2018). In another study of protein extracted from dreg, Zhang et al. (2021) reported that rice dreg protein isolate treated by micro-fluidization (0, 40, 80, 120, and 160 MPa) yielded rice dreg protein isolate with better surface hydrophobicity and solubility. Additionally, the DPPH radical scavenging activity of rice dreg protein was efficiently improved by the optimal micro-fluidization of 40 MPa (Zhang et al. 2021). For bran sources, defatted rice bran was pretreated by ultrasound combined with 0.9% sulfuric acid soaking at 90 °C according to the study by Zhang et al. (2019). With the pretreatment, the yield of protein was increased from 53.1% to 72.8%. The higher yield of protein resulting from the chemical structure was significantly induced and the thermal stability was reduced from the destruction of the matrix structure in the pretreated rice bran protein isolate. The destruction of the matrix structure improves the accessibility of the solvent to protein (Zhang et al. 2019). Other novel methods such as ultrasonication (UAE) and microwave (MAE) using response surface methodology were applied to extract protein from organic rice bran according to the study of Phongthai et al. (2016b), respectively. The protein yield of 4.73 ± 0.03% and MWs 11–75 kDa was derived from the optimal UAE condition of 76% sonication amplitude, 18-min extraction, and 0.99 g/10 mL solid-liquid ratio (Phongthai et al. 2016a, b, c). The positive result of UAE was also reported by Li et al. (2016) and Ly et al. (2018) that the initial extraction rate, extraction constant, and protein extraction yield in UAE were significantly higher than those of the conventional extraction. Furthermore, both methods yield proteins with a similar protein profile, and techno-functional properties such as water and oil absorption capacities, emulsifying capacity, and emulsion stability. Phongthai et al. (2016b) also studied the extraction of rice bran protein from MAE and reported the protein recovery was higher by 1.54-fold than that of alkaline extraction under the optimal condition of 1000 W of microwave power, 90 seconds of extraction time, and a solid to liquid ratio of 0.89 g rice bran/ 10 mL of distilled water. The digestibility of MAE rice bran protein was proved to be better than commercial soy protein concentrates (Phongthai et al. 2016a, b, c). Similar trend from Bedin et al. (2019) that MAE provides production yield of 22.04%, protein content of 6.19% higher and 30 times faster compared to the values obtained from conventional alkaline extraction, but the characterization of the extracts did not affected by microwaves (Bedin et al. 2019).

Interestingly, mechanical techniques such as ultrafiltration and alkali combined with bead milling were also applied to harvest rice protein. The recovery of rice protein from starch production wastewater by ultrafiltration was investigated by Doan and Lai (2021). The highest recovery yield and purity of the protein were obtained by using ultrafiltration (50 kDa) with a permeate flux of 8 bar (Doan and Lai 2021). In the study of Larsson (2020), alkaline conditions along with mechanical disruption (bead milling) were the most efficient method to solubilize and extract rice bran protein.

3.3.2 The Bioactive Peptide from Rice Protein Hydrolysis

Enzymatic treatment has been used to modify both techno-functional such as solubility, foaming, and emulsifying properties and biofunctional properties such as antioxidant, anti-inflammatory, anti-diabetes, and anti-hypertensive, anti-cancer of protein concentrates/isolates/hydrolysates from rice sources for decades. Bioactive peptides/amino acids free through enzymatic hydrolysis of rice protein expose different bio-functional characteristics closely associated with the sources of native proteins, extraction methods, purification methods, degree of hydrolysis (DH), and hydrolysis conditions, the types of proteases used for the hydrolysis, molecular weights (MWs), length of peptides, peptide polar groups and amino acid compositions/sequences, and hydrophobicity (Wen et al. 2016; Thamnarathip et al. 2016). Proteases, especially alkali proteases (such as trypsin, Alcalase®, Flavourzyme®, and Neutrase®) are used to extract high-value rice protein hydrolysates, with bioactivities largely depending on hydrolysis conditions and type of enzyme (Zaky et al. 2020; Liu et al. 2019).

The details of the biofunctions of rice protein and protein hydrolysates are summarized in Table 3.4. Antioxidants are substances that can break or inhibit this chain reaction by donating their hydrogen atoms to free radicals. The antioxidant activity potential of rice protein hydrolysates depends on several aforementioned factors and the distinction of amino acids of peptides, especially as basic amino acids (His, Lys) or acidic amino acid residues (Glu, Asp). Besides, aromatic amino acids including Tyr, Met, Phe, and His were reported as more potent proton donors than other amino acids (Zaky et al. 2020). Additionally, the antioxidant activity of hydrolysates from rice bran proteins could also be influenced by the phenolic hydroxyl groups of tyrosine, that when they are bonded to the aromatic ring provide atoms that end up the radical chain reaction by scavenging free radicals (Thamnarathip et al. 2016). Moreover, the hydrophobic amino acids such as Val, Ala, Pro, and Lue could increase the presence of the peptides at the water–lipid interface, which improves scavenging activity with hydrophobic or free radicals generated in the lipid phase depending on the specificity of proteolytic enzymes to cleavage and release (Liu et al. 2019; Zhang et al. 2014). Additionally, the antioxidant activity of protein hydrolysates was found to have a synergistic effect with phenolic compounds (Pantoa et al. 2020a, b). Aromatic rings with hydroxyl groups

Table 3.4 Bioactive peptides from rice protein: production, characterization and bioproperties

Source	Products	Production	Characteristics	Bio-properties	References
Young rice protein	Protein hydrolysate	Alcalase Flavourzyme Neutrase Optimal condition: Alcalase, 55 °C pH 8, 4 h	Ultrafiltration: 10 kDa	• **Angiotensin I-converting enzyme (ACE) inhibition**: Milky stage of PT1 (96.30 ± 5.90%) and flowery stage of RB (78.50 ± 5.40%). • **Antioxidant**: • 1, 1-Diphenyl-2-picrylhydrazyl (DPPH) • Iron chelating activity. • **Anti-proliferation:** HepG2 cells.	Pantoa et al. (2020a, b)
Rice	Protein hydrolysate	• Alcalase enzyme to substrate ratio was 1: 50 (w/w) pJH 8.0,2 h. • Electron beam irradiated: (5 kGy, 10 kGy, 25 kGy, and 50 kGy) at 5.0 MeV energy level. 0.86 s, 1.72 s, 4.3 s, and 8.6 s, room temperature (25 °C).	EBI induced the amino acid oxidation, destroyed the secondary structure and microscopic morphology	• **Antioxidant**: • DPPH and 2,2′-azino-bis (3-ethylbenzothiazoline-6-sulfonic acid)(ABTS) radical scavenging activity of 96.81% and 92.04%.	Li et al. (2019)
Pigmented rice	Hydrolysate peptide type A-2	N/A	Orally feed to male adult Wistar rats: Various doses ranging from 10, 20, and 40 mg/kg BW, 9 weeks	• **Neuroprotective.** • Cognitive deficits by. • Decreasing the discriminative index, • Reflecting the increase of learning, short-term and long-term recognition.	Pannangrong et al. (2018)

(continued)

Table 3.4 (continued)

Source	Products	Production	Characteristics	Bio-properties	References
Rice protein	Purified peptides	Trypsin digestion Purification: Gel permeation chromatography and IER: Ion-exchange resin	16 amino acid peptides (QRDFLLAGNKRNPQAY)	**Immodulatory** • Impeded the nuclear translocation of p65, down-regulated the transcription of TNF-a, IL-6, and IL-1b, and reduced secretion of TNF-a. • Response in LPS-stimulated RAW264.7 macrophages, and probed their underlying molecular mechanisms of action. **Anti-inflammation** • Suppression activity by inhibiting the release of nitric oxide (NO) and tumour necrosis factor-a (TNF-a). • Attenuated iNOS and repressed the nuclear transcription factor (NF-kB) signaling pathway by impeding the nuclear translocation of p65.	Wen et al. (2016)
Se-enriched brown rice	Purified peptides	Trypsin concentration of 0.2% (w/w, enzyme/substrate) 55 °C, 1 h	Ultrafiltration: A molecular weight cut-off: MWCO: 3.5 and 1.0 kDa (ALLLQAVQSQYEEK)	**Anti-inflammation**: • LPS-induced RAW264.7 cell model. • Inhibiting the production of nitric oxide (NO), prostaglandin E_2 (PGE_2) and pro-inflammation cytokines including interleukin-6 (IL-6),	Feng et al. (2021)

				interleukin-1β (IL-1β), tumor necrosis factor-α (TNF-α). • Suppress mRNA and protein expression of inducible nitric oxide synthase (iNOS), cyclooxygenase-2 (COX-2).	
Commercial defatted rice bran	Protein hydrolysate protein content: 30.3% a protein yield: 56.6%	Alcalase, Flavourzyme and Neutrase Optimal conditions: Alcalase 2 h	Ultrafiltration: 6 kDa	Total phenolic compound: 22.1 mg gallic acid equilavents/g **Antioxidant**: • ABTS: 7.57 mg gallic acid equilavents/g. • FRAP 152 mM Fe^{2+}/g.	Thamnarathip et al. (2016)
Defatted organic rice bran	Protein: Protein yield: 4.73% Protein hydrolysate	Optimal condition: 76% sonication amplitude, 18 min extraction and 0.99 g/10 mL solid–liquid ratio Subtilisin A (SPH) Actinase E (APH) Neutrase 0.8 L (NPH)	<11 to 75 kDa	**Antioxidant**: • DPPH radical scavenging activity. • Reducing power, • Metal chelating. **ACE-inhibitory**	Phongthai et al. (2016a, 2016b, 2016c)
Commercialstabilized defatted ricebran	Protein hydrolysate	Response surface method: 3.5 AU (Anson units of enzyme) Alcalase (3.5 AU, 4970 μL), pH 8.0, 50 °C, 1 h	Ultrafiltration: <5 kDa	**Anti-cancer & anti-tumor:** Colon cancer cells cytotoxicity activities on human colon (HCT-116) and breast (HTB-26) cancer cell lines	Kannan et al. (2009)
Rice bran	Protein hydrolysate	Alcalase	Ultrafiltation: MW < 3 kDa • (Val-Leu-Ala-Ala-Gly-Cys-Pro-Leu) • (Ala-Met-Ser-Fhe-Ala-Glu-Met)	• **Antioxidant.** • **Reactive oxygen species (ROS) inhibition.** • **ACE inhibition.**	Ngamsuk et al. (2020)

(continued)

Table 3.4 (continued)

Source	Products	Production	Characteristics	Bio-properties	References
Raw organic rice bran	Protein hydrolysate	**Optimal condition:** 1000 W of microwave power, 90 s of extraction time, and a solid to liquid ratio of 0.89 g rice bran/10 mL of distilled water	MWs: <11 kDa and 100 kDa	• **Antioxidant:** DPPH radical scavenger. • Ferric reducing agent metal ion chelator	Phongthai et al. (2016a, 2016b, 2016c)
Thai jasmine rice bran: Khao Dawk Mali 105	Peptides	Protease G6 3% of protein in RBP), pH 8.0 a 55 °C, 4 h	Mw: <50 kDa (GSGYF)	• **ACE-inhabitation.** IC_{50} value of 2.11 μg/ mL as captopril (1.15 μg/mL).	Suwannapan et al. (2020)
Rice residue protein	Purified peptides	Alcalase (pH 8.5 at 55 °C), Flavourzyme (pH 6.0 at 50 °C), Protamex (pH 7.0 at 50 °C), pepsin (pH 3.0 at 37 °C), Papain (pH 6.0 at 55 °C), Trypsin (pH 8.0 at 37 °C), amount of protease, 2.5%; concentration of substrate (RRP), 5%, 4 h	• Membrane ultrafiltration followed by size exclusion chromatography and reversed-phase fast protein chromatography. **Four peptides:** • RPNYTDA (835.9 Da). • TSQLLSDQ (891.0 Da). • TRTGDPFF (940.0 Da). • NFHPQ (641.7 Da).	Antioxidant: DPPH and ABTS radical scavenging, FRAP-Fe^{3+} reducing assay,	Yan et al. (2015)
Rice bran	Protein hydrolysate Protein content: 62.90%	Alcalase, trypsin, Protamex, and Flavourzyme, (2, 4, and 6 h), Optimum conditions: Flavourzyme 2 h	Degree of hydrolysis: Protamex > Trypsin > Alcalase > Flavourzyme	• TPC: 893.75 ± 0.57 mg GAE/100 g. • **Antioxidant**:DPPH: 83.68 ± 0.97%, ABTS: 35.11 ± 0.66%, and a metal chelating activity: 53.90 ± 0.13%.	Zaky et al. (2019)

Cold-pressed defatted Thai Hom Mali rice bran	Protein powder yield: 8.8% Protein hydrolysate	Protease G6 Optimum condition:N/A	Orally treated with 8-week-old male db/m and db/db mice: (100 or 500 mg/kg/day) for 8 weeks	• Attenuates the progression of diabetic nephropathy. • Restored renal function by suppressing the expression of proangiogenic and profibrotic proteins. • Inhibiting pro-inflammatory mediators, and restoring the antioxidant and cytoprotective system.	Boonloh et al. (2018)
Defatted Riceberry bran-different maturity stages	Purified peptides	Alcalase® protein hydrolysate (sonicated at amplitude 20%, 30 min and 40%, 60 min	• MW: < 3 kDa. • Ultra-filtration and ion-exchange chromatography & LC-MS/MS. **Two peptides** • (Val-Leu-AlaAla-Gly-Cys-Pro-Leu) • (Ala-MetSer-Phe-Ala-Glu-Met)	• **Antioxidant.** • **ROS inhibition.** • **ACE inhibition.**	Uraipong and Zhao (2016)

and carboxylic acid in phenolic compounds have a natural interaction that increases their affinity to conjugate with other components (such as lipids, carbohydrates, and proteins) as well as provide hydrogen atoms, which terminated the radical chain reaction by scavenging free radicals. These interactions reacted by hydrogen bonding, covalent bonding, hydrophobic interactions, and ionic bonding. Some free amino acids such as Tyr, Met, His, Lys, Trp, and Pro were also reported to have antioxidant activity but act as pro-oxidants at high concentrations. Rice bran protein hydrolysate produced by Alcalase® (MW < 3 kDa) according to Ngamsuk et al. (2020) reported exhibiting antioxidant and reactive oxygen species (ROS) activities (sequenced as Val-Leu-Ala-Ala-Gly-Cys-Pro-Leu). High DH and antioxidant activities were obtained after hydrolysis at the flowery-to-milky stage using Alcalase® at 55 °C, pH 8 for 4 h (Pantoa et al. 2020a, b). Yan et al. (2015) reported 4 peptides and their amino acid sequences with high antioxidant properties were (1) RPNYTDA (835.9 Da), (2) TSQLLSDQ (891.0 Da), (3) TRTGDPFF (940.0 Da), and (4) NFHPQ (641.7 Da), respectively. Furthermore, these four peptides were chemically synthesized and confirmed their antioxidant activities by DPPH and ABTS radical scavenging and Fe^{3+} reducing assay, respectively. In other in vitro antioxidant assays, Zhao et al. (2012) found that rice protein can inhibit lipid oxidation, reducing the formation of lipid hydroperoxides and thiobarbituric acid reactive substances (TBARS). Furthermore, selenium-enriched rice protein hydrolysates exhibit the potential antioxidant activity evaluated by cellular antioxidant activity test according to the research of Fang et al. (2017).

The reducing power assay (FRAP) is often used to evaluate the ability of an antioxidant to be able to reduce the Fe^{3+}/ferric cyanide complex to the ferrous form. The reducing power of the rice dreg protein hydrolysates obtained by five different proteases showed different reducing power activity, depending on its DH (Zhao et al. 2012). The same report from Phongthai et al. (2016a, b, c) mentioned that the difference in reducing power might be caused by the length of the resultant peptides and different amino acid sequences from different hydrolysis conditions. Furthermore, the presence of cysteine in protein hydrolysates contributes to strong reducing properties due to the sulfhydryl group in its structure (Udenigwe and Aluko 2012).

Transition metal ions viz. Fe^{2+} and Cu^{2+} accelerate generations of reactive oxygen species (ROS), which break down volatile compounds in the food storage phase as well as accelerating auto-oxidation causing cell damage in living organisms (Zhang et al. 2014). Chelating agents from peptides or plant extracts could act as secondary antioxidants to stabilize the oxidized form of the metal ions by reducing the redox potential (Phongthai et al. 2016a, b, c). Charged and polar amino acids, the acidic and basic amino acids are preferable at the binding sites for chelating of metal ions preventing the oxidation (Wu et al. 2020). As well, any high-affinity metal-binding groups became completely exposed or recently formed, hence potentially metal chelating (Phongthai et al. 2017). The metal chelating ability could also be also potentially activated by amino acids containing the carboxylic groups such as glutamic and aspartic acids and the imidazole groups such as histidine (Sirikul et al. 2009). Also, the increase in metal chelating activity occurred due to the rising in DH. Phongthai et al. (2016a, b, c) demonstrated that organic rice bran protein hydrolysates showed metal ion chelating activity.

The most potent angiotensin I-converting enzyme (ACE) inhibitors contain hydrophobic amino acid (Trp, Phe, Tyr, or Pro) residues at each of their C-terminus and branched aliphatic amino acids at the N-terminus (Verma and Srivastav 2020). Young rice protein hydrolysates with small peptides of less than five amino acids were reported to exhibit a higher ACE inhibitory activity (Uraipong and Zhao 2016). The differences in ACE inhibitory activities of the rice protein or its hydrolysates might be due to the different MWs, peptide levels, and amino acid sequences generated by enzymatic hydrolysis. There were some studies have reported on the ability of ACE inhibition of rice protein hydrolysates (Uraipong and Zhao 2016; Pantoa et al. 2020a, b; Ngamsuk et al. 2020; Udenigwe 2016). Protein hydrolysates produced from albumin and glutelin of rice bran hydrolyzed with Alcalase and/or Protamax® inhibit ACE, α-amylase, and β-glucosidase activities from the study of Uraipong and Zhao (2016). Rice bran protein hydrolysate is produced by Alcalase® hydrolysis according to Ngamsuk et al. (2020). Active peptide sequenced as Ala-Met-Ser-Phe-Ala-Glu-Me, MW < 3 kDa was identified and provided ACE inhibitory activity (Ngamsuk et al. 2020). The highest ACE inhibitory activity was obtained from protein hydrolysates in the milky stage of the Pathum Thani rice barn and the dough stage of rice berry rice bran (Pantoa et al. 2020a, b).

Besides the antioxidative and ACE inhibitory properties, rice protein can act as other bioactive substances in living systems. From the study according to Udenigwe (2016), oryzacystatins and cysteine protease inhibitors were considered for the efficient production of bioactive peptides from rice bran protein. From the study, in silico evidence showed that aspartate protease (pepsin at pH > 2) and metalloproteinase (thermolysin) have a strong capacity in cleaving the QXVXGX motif of oryzacystatins. The cleaved oryzacystatins were inactivated and released bioactive sequences from the protease inhibitors as well as possess multi-bioactivities, which can be applied in the management of oxidative stress, type 2-Diabetes Mellitus, hypertension, and other aberrant cellular processes. Moreover, di- and tri-peptides with several bioactive potencies were also identified in cleaved oryzacystatin peptide pools (Udenigwe 2016). Usually, the amino acids in the bioactive peptides range from 3–16 amino acids depending upon their activities amino acids like Val, Leu, Pro, His, Tyr, glutamic acid, or aspartic acid have been predominantly shown anti-inflammatory activities (Kannan et al. 2009). Protein hydrolysates exhibited anti-inflammatory based on, first, the indicators of inhibiting the expression of pro-inflammatory mediators such as nitric oxide (NO), tumor necrosis factor-α (TNF-α), and inducible nitric oxide synthase (iNOS). Second, the ability to decrease phagocytosis. Third, the peptides process downregulation of the nuclear factor kappa light chain enhancer of the activated B cells (NF-kB) pathway by impeding the lipopolysaccharide (LPS) induced accumulation of p65 in the nucleus (Feng et al. 2021; Wen et al. 2016; Fang et al. 2017). The rice protein hydrolysate fraction purified by gel permeation followed by anionic ionic chromatography was found to be the best inflammation suppression activity in

LPS-stimulated RAW264.7 macrophages (Wen et al. 2016). The anti-inflammatory effects of rice protein are attributable to the suppression of the activation of NF-κB. The functional groups of the amide I band (1600–1700 cm^{-1}) can be provided from the FTIR spectra and have been widely used to calculate the content of the secondary structure because of its sensitivity to protein conformation changes (Li et al. 2019). The peptide fractions 1.0–3.5 kDa from Selenium-rich rice protein exhibited the most effective anti-inflammatory properties by inhibiting the production of NO, prostaglandin E_2 (PGE_2), and pro-inflammation cytokines including interleukin-6 (IL-6), interleukin-1β (IL-1β), TNF-α in LPS-induced RAW264.7 cell model as well as the mRNA and protein expression of iNOS and cyclooxygenase-2 (COX-2) (Feng et al. 2021).

Anti-cancer peptides usually contain hydrophobic amino acids such as Pro, Leu, Gly, and Ala with Lys, Arg, Ser, Glu, Thr, and Tyr (Chalamaiah et al. 2018). Hydrophobic low MW peptides were found to inhibit human breast cancer and mammary epithelial cells (Chalamaiah et al. 2018). Rice bran protein hydrolysates using Alcalase® MW < 10 kDa inhibited the growth of Caco-2 cells by 80%, while MW < 5 kDa inhibited the growth of HepG2 cells by 50% (Kannan et al. 2009). Other bio-properties about neuroprotective effects were reported that protein hydrolysate types A-2 from pigmented rice showed a neuroprotective effect against β-amyloid (Aβ) injected rats.

The aforementioned studies provide a significant consideration and a direction that can lead to the efficient release of bioactive peptides from rice proteins for functional food applications.

3.4 Applications of Rice Proteins from Rice Grains, Including Rice by-Products

The characterization of rice proteins was studied and clarified for centuries according to the classification of seed storage proteins based on solubility by Osborne (1924). The known physicochemical properties and functionality of rice protein provide us with basic information that contributes to the application. Considering their outstanding rice protein physicochemical properties, rice protein has some unique characteristics, for example, a well-balanced amino acid profile, hypo-allergenicity, high digestibility, and no mutagenicity (Pantoa et al. 2020a, b; Amagliani et al. 2017a, b). Therefore, rice protein concentrates, isolates, or hydrolysates can serve as valuable ingredients in many food and non-food applications. Additionally, protein hydrolysate products from rice proteins have been widely investigated and proven to display health benefits. The valorization of rice or side stream of the rice milling, which is otherwise used for animal feed or fuel could lead to a more sustainable use of natural resources and protein, and help reduce process-induced losses in the production process of white rice.

The basic application of rice proteins in food is an inexpensive source of protein for protein-energy malnutrition in undeveloped and developing countries because traditional protein sources are usually expensive and not always available. Alternatively, there is an increasing demand for alternative inexpensive high-quality protein sources of proteins for plant-based food consumer groups or people who concern about protein from plant-based sources. Therefore, interest in rice proteins surges because they are relatively affordable, widely available, and have health benefits. Rice proteins can be utilized by the food industry to increase the nutritional value of food products at a low cost. Protein products developed from rice bran, broken rice, germinated brown rice, and rice syrup co-product are currently available in the market and mostly in hypoallergenic formulas, sports nutrition, and beverage-type products. According to Fabian and Ju (2011), rice bran protein hydrolysates have been reported to be used in nutritional supplements, functional ingredients, and flavor ingredients (Fabian and Ju 2011). Apart from cosmetic and personal care products, rice proteins are found to be used in soups, sauces, gravies, meat products, soft drinks and juices, confectionery, and other savory applications (Fabian and Ju 2011). The application of proteins produced from rice and rice side streams is summarized below.

3.4.1 Infant Food Formulations

The highlight of unique nutritional and hypoallergenic properties of rice proteins make them suitable ingredients for certain specific groups viz. infant and celiac patients (Koo and Lasekan 2007). Rice bran protein has been used as hypoallergenic milk replacer in infant formulas for a long time ago (Koo and Lasekan 2007). Rice proteins were also used in infant foods. They could also replace allergenic cereal foods and add variety to the restricted diets of children with food allergies (Amagliani et al. 2017a, b; Dupont et al. 2020).

3.4.2 Bakery Products

Jiamyangyuen et al. (2005) and Sadawarte et al. (2007) explored the potential of rice bran protein concentrate in the preparation of wheat bread. Adding rice bran protein concentrate (RBPC) at 1–5% (w/w) levels increased the protein content up to 12.10%–12.3%, 16.5%, and 21.1% corresponding levels of protein. Sensory quality viz. color, tastein. Sensory quality viz. color, taste, flavor, texture, and overall acceptability of bread was similar to wheat bread except for the use of the level of substitution was beyond 5%. Besides bread, rice bran protein concentrate (RBPC) prepared by the wet alkaline extraction method was also incorporated in biscuits (Yadav et al. 2011). Biscuits were added with 5%, 10%, and 15% (w/w) RBPC. The protein content of the biscuits increased from 7.3% (control) up to 15.4% (with 15%

(w/w) RBPC). The biscuits with 5% (w/w) RBPC were evaluated to be similar in color, taste, flavor, texture, and overall acceptability as the control (Yadav et al. 2011). Phongthai et al. (2016a) developed a gluten-free (GF) bread by using RBPC as a replacement for egg albumin to improve the quality of gluten-free bread. They found the advantage of using RBPC to improve both the sensory and physical properties of the bread. The 2% (w/w) enrichment of the RBPC concentrate in gluten bread also reduced staling rate (Phongthai et al. 2016a, b, c).

3.4.3 Snacks

Rice protein hydrolysate was successfully used as egg replacement flaky egg rolls, besides, maintaining their functional and physicochemical properties (Chan et al. 2020). The protein content of egg replacement flaky egg rolls increased from 19.69 to 22.18 (%,w/w), while carbohydrate and sugar contents decreased to 64.12 and 12.26 (%, w/w), respectively. Overall amino acid contents significantly increased. Acceptability for sensory evaluation was acceptable; however, the color of the protein-rich ingredients accounted for Maillard reaction progression and decreased brightness (L^*) and an increase in redness (a^*) (Chan et al. 2020). The enhancement of the protein components of extruded snacks made from rice flour by a mixture of rice, sorghum, and defatted soy flour was studied. The extruded snack produced higher protein content and lysine content of 1.19 ± 0.09 (%,w/w) (Omwamba and Mahungu 2014).

3.4.4 Natural Antioxidants

The potential of rice protein hydrolysate (RPH) as natural antioxidants was exclusively reviewed by Rani et al. (2018). Supawong et al. (2019) reported that RBH-treated fried fish cakes had higher concentrations of total phenolics (63.9 mg GAE/100 g sample) and showed the highest DPPH• and ABTS• + radical scavenging activity. Besides fish cakes, fish patties made from Nile Tilapia adding bioactive rice bran protein and textural additives were investigated by Wongthahan and Thawornchinsombut (2015). Rice bran protein hydrolysate and red jasmine rice bran extract (2%) could be used instead of butylated hydroxyanisole (BHT) (0.02%) to prevent lipid oxidation in the patties (Dupont et al. 2020). Gomes and Kurozawa (2021) studied the lipid oxidation stability and physicochemical properties of linseed oil microparticles obtained through spray-drying an encapsulating material rice protein hydrolysate (Gomes and Kurozawa 2021). Rice protein hydrolysates were used as an emulsifier for the stability of oil-in-water emulsions (Pan et al. 2019). Besides, the antioxidant activities of rice bran protein hydrolysates in soybean oil and oil-in-water (O/W) emulsions were examined by Cheetangdee and Benjakul

(2015). Bonifacino et al. (2020) reveal the stable oil-in-water (O/W) emulsions, including both soluble and insoluble fractions, in acidic and neutral conditions.

3.4.5 Functional Ingredients

Rice bran protein has been proved to be the potential to be used as a natural anti-browning agent in the potato industry. Kubglomsong and Theerakulkait (2014) reported that rice bran protein, which heated to 40 and 60 °C had browning inhibition in potato puree extent similar to unheated Rice bran protein extract ($P > 0.05$) based on the polyphenol oxidase inhibition exhibited a mixed-type inhibition for potato polyphenol oxidase (Omwamba and Mahungu 2014). Rice bran protein extract inhibited browning in potato puree higher than 5, 10, and 20 mM ascorbic acid and 5 and 10 mM citric acid ($P \leq 0.05$) (Kubglomsong and Theerakulkait 2014). Rice flour have been used as ingredient in gels, puddings, and ice creams (Pracham and Thaiudom 2016). Rice protein hydrolysates may also be used as nutritional supplements, functional ingredients, and flavor enhancers in foods, coffee whiteness, confectionery, and in the fortification of soft drinks and juices (Fabian and Ju 2011). A rice drink with a low-protein rice drink fortified with caseinomacropeptide concentrate low-protein and fat contents was developed by Karimidastjerd and Kilic-Akyilmaz (2021). Glutinous rice (*Oryza sativa* L.) protein extract with potent α-amylase inhibitory activity was extracted by Sarnthima et al. (2020) and suggested the development further to be a food supplement for blood sugar and weight controls. Besides that, enzymatically hydrolyzed rice bran protein concentrate was used as a flavoring agent (Arsa and Theerakulkait 2018). Bagnasco et al. (2013) reported that hydrolysate of rice middling yields medium-to-small size polypeptides contributing to the umami taste (Bagnasco et al. 2013).

3.4.6 Other Products

Rice bran protein was employed as an inexpensive, alternative substrate for *L*-lactic acid production according to Alexandri et al. (2019). The utilization of defatted rice bran as the sole carbon and nutrient source to grow and produce high-purity *L*-lactic acid. The *Bacillus coagulans* A107 isolate presented the most promising results in terms of final lactic acid concentration (75.9 g/L), yield (0.90 g/g), and productivity (2.7 g/L/h) (Alexandri et al. 2019). Edible composite films based on rice protein hydrolysates and chitosan were developed (Wang et al. 2020a, b). Rice bran protein aggregates were prepared by heating Rice bran protein solution at 90 °C for 4 h at pH 2, 7, or 11 and used for preparing packaging films. The formation of rice protein aggregates could improve the physical, mechanical, and thermal properties of Rice bran protein-based film, especially at pH 11 (Wang et al. 2019a, b). Pires et al. (2020) used rice protein solution (8%, w/v) coating with different plasticizer types (combined with glycerol, propylene glycol, or sorbitol) to preserve the quality of eggs storage at 20 °C for 6 weeks. The best egg protection results in terms of egg quality are obtained in eggs coated with rice protein and sorbitol (Pires et al. 2020).

3.5 Summary and Future Direction

Rice grains may not be the main source of protein extraction. However, early-stage rice (young rice-flowery to the milky stage) can be considered a protein source. Apart from that side stream, namely rice bran, dreg can be economic source of high-quality plant-based proteins that can exhibit excellent techno-functional properties and interesting biofunctions modified through chemical, enzymatic or physical methods, or combinations. Rice protein and bran have been used as ingredients in many food products such as meat products, snacks, breakfast cereals, biscuits, bread, pasta, noodles, beverages, gluten-free products, and hypoallergenic protein products, and non-food namely edible films. Moreover, hydrolysates from rice proteins can apply in nutraceutical products because of the evidence of bioactive potential properties. The full valorization of the proteins from rice supports the alternative proteins from plants concerning animal welfare, and health awareness, and reduces food loss as well as climate impact from food production. Fully effective processing methods, particularly on the extraction of rice bran proteins with high recovery and efficiency on economical production cost, remain to be developed. Besides, a big attempt still must solubilize rice protein under edible conditions. When chemical or biochemical modifications are applied to rice proteins, it may lose commercial value likely to generate undesirable substances, such as bitter taste, certain essential amino acid disappearance, or toxic compounds. More effort is needed to cope with these pitfalls. More work needs to be done especially, in vivo biological activity tests in animal models or human subjects. Furthermore, more works related to those and product development using rice proteins need to be done and meet the challenge of fully using the unique product from rice proteins.

References

Alexandri M et al (2019) Evaluation of various Bacillus coagulans isolates for the production of high purity L-lactic acid using defatted rice bran hydrolysates. Int J Food Sci Technol 54(4): 1321–1329

Amagliani L et al (2017a) The composition, extraction, functionality and applications of rice proteins: a review. Trends Food Sci Technol 64:1–12

Amagliani L et al (2017b) Composition and protein profile analysis of rice protein ingredients. J Food Compos Anal 59:18–26

Arsa S, Theerakulkait C (2018) Preparation, aroma characteristics and volatile compounds of flavorings from enzymatic hydrolyzed rice bran protein concentrate. J Sci Food Agric 98(12): 4479–4487

Bagnasco L et al (2013) Use of food-grade proteases to recover umami protein–peptide mixtures from rice middlings. Food Res Int 50(1):420–427

Bedin S et al (2019) Implication of microwaves on the extraction process of rice bran protein. Braz J Chem Eng 36(4):1653–1665

Bonifacino C et al (2020) Emulsifying properties of defatted rice bran concentrates enriched in fiber and proteins. J Sci Food Agric 100(3):1336–1343

Boonloh K, Lee ES, Kim HM, Kwon MH, Kim YM, Pannangpetch P, Kongyingyoes B, Kukongviriyapan U, Thawornchinsombut S, Lee EY, Kukongviriyapan V, Chung CH (2018) Rice bran protein hydrolysates attenuate diabetic nephropathy in diabetic animal model. Eur J Nutr 57(2):761–772. https://doi.org/10.1007/s00394-016-1366-y

Braspaiboon S et al (2020) Comparison of the effectiveness of alkaline and enzymatic extraction and the solubility of proteins extracted from carbohydrate-digested rice. Heliyon 6(11):e05403

Bryant RJ et al (2013) Genetic variation and association mapping of protein concentration in brown rice using a diverse rice germplasm collection. Cereal Chem 90(5):445–452

Cao X, Wen H, Li C, Gu Z (2009) Differences in functional properties and biochemical characteristics of congenetic rice proteins. J Cereal Sci 50(2):184–189. https://doi.org/10.1016/j.jcs.2009.04.009

Cao X et al (2010) Extraction technique and characteristics of soluble protein in germinated brown rice. Int J Food Prop 13(4):810–820

Chalamaiah M, Yu W, Wu J (2018) Immunomodulatory and anticancer protein hydrolysates (peptides) from food proteins: a review. Food Chem 245:205–222

Chan Y-J et al (2020) Effect of rice protein hydrolysates as an egg replacement on the physicochemical properties of flaky egg rolls. Foods 9:2

Chanput W, Theerakulkait C, Nakai S (2009) Antioxidative properties of partially purified barley hordein, rice bran protein fractions and their hydrolysates. J Cereal Sci 49(3):422–428. https://doi.org/10.1016/j.jcs.2009.02.001

Cheetangdee N, Benjakul S (2015) Antioxidant activities of rice bran protein hydrolysates in bulk oil and oil-in-water emulsion. J Sci Food Agric 95(7):1461–1468

De Groot AP, Slump P (1969) Effects of severe alkali treatment of proteins on amino acid composition and nutritive value. J Nutr 98(1):45–56

Doan NTT, Lai QD (2021) Ultrafiltration for recovery of rice protein: fouling analysis and technical assessment. Innovative Food Sci Emerg Technol 70:102692

Dupont C et al (2020) Hydrolyzed rice protein-based formulas, a vegetal alternative in cow's milk allergy. Nutrients 12(9):2654

Fabian C, Ju Y-H (2011) A review on rice bran protein: its properties and extraction methods. Crit Rev Food Sci Nutr 51(9):816–827

Fang Y et al (2017) The correlation between in vitro antioxidant activity and immunomodulatory activity of enzymatic hydrolysates from selenium-enriched rice protein. J Food Sci 82(2):517–522

Feng M et al (2021) Anti-inflammatory effects of three selenium-enriched brown rice protein hydrolysates in LPS-induced RAW264.7 macrophages via NF-κB/MAPKs signaling pathways. J Funct Foods 76:104320

Gomes MHG, Kurozawa LE (2021) Influence of rice protein hydrolysate on lipid oxidation stability and physico-chemical properties of linseed oil microparticles obtained through spray-drying. LWT 139:110510

Hamada JS (2000) Characterization and functional properties of rice bran proteins modified by commercial exoproteases and endoproteases. J Food Sci 65(2):305–310

Herrero M, Cifuentes A, Ibañez E (2006) Sub- and supercritical fluid extraction of functional ingredients from different natural sources: plants, food-by-products, algae and microalgae: a review. Food Chem 98(1):136–148

Hoogenkamp H, Kumagai H, Wanasundara JPD (2017) Chapter 3—Rice protein and rice protein products. In: Nadathur SR, Wanasundara JPD, Scanlin L (eds) Sustainable protein sources. Academic Press, San Diego, pp 47–65

Hsu RJC et al (2015) Effects of added water and retrogradation on starch digestibility of cooked rice flours with different amylose content. J Cereal Sci 61:1–7
Jiamyangyuen S, Srijesdaruk V, Harper WJ (2005) Extraction of rice bran protein concentrate and its application in bread. Extraction 27(1):56
Juliano B. Rice in human nutrition. 1993: Int. *Rice Res*. Inst
Juliano B and Bechtel DB (1985) The grain and its gross composition. Rice: *Chemistry and Technology*: p. 17–57
Kamara JS et al (2010) Variation in free amino acid profile among some rice (*Oryza sativa* L.) cultivars. Breed Sci 60(1):46–54
Kannan A, Hettiarachchy N, Narayan S (2009) Colon and breast anti-cancer effects of peptide hydrolysates derived from rice bran. The Open Bioactive Compounds Journal 2:1
Karimidastjerd A, Kilic-Akyilmaz M (2021) Formulation of a low-protein rice drink fortified with caseinomacropeptide concentrate. Food Bioprod Process 125:161–169
Kawakatsu T, Takaiwa F (2019) 4 - Rice proteins and essential amino acids. In: Bao J (ed) Rice, 4th edn. AACC International Press, pp 109–130
Kim JW et al (2013) Protein content and composition of waxy rice grains. Pak J Bot 45(1):151–156
Kinoshita N et al (2017) Identification of quantitative trait loci for rice grain quality and yield-related traits in two closely related *Oryza sativa* L. subsp. *japonica* cultivars grown near the northernmost limit for rice paddy cultivation. Breed Sci 67(3):191–206
Koo WW, Lasekan JB (2007) Rice protein-based infant formula: current status and future development. Minerva Pediatr 59(1):35–41
Kraithong S, Lee S, Rawdkuen S (2018) Physicochemical and functional properties of Thai organic rice flour. J Cereal Sci 79:259–266
Kraithong S, Rawdkuen S (2019) The effects of chemical composition of Rice flour on Physicochemical and functional properties. Prawarun Agric J 16(1):49–70
Kubglomsong S, Theerakulkait C (2014) Effect of rice bran protein extract on enzymatic browning inhibition in potato puree. Int J Food Sci Technol 49(2):551–557
Larsson E (2020) Extraction of Rice Bran Protein Screening of enzymatic digestion, solubilization using pH shift and mechanical disruption through bead milling
Li S, Yang X, Zhang Y, Ma H, Liang Q, Qu W, He R, Zhou C, Mahunu GK (2016) Effects of ultrasound and ultrasound assisted alkaline pretreatments on the enzymolysis and structural characteristics of rice protein. Ultrason Sonochem 31:20–28. https://doi.org/10.1016/j.ultsonch.2015.11.019
Li T et al (2019) Effect of enzymolysis-assisted electron beam irradiation on structural characteristics and antioxidant activity of rice protein. J Cereal Sci 89:102789
Liu YQ et al (2019) Functional peptides derived from rice bran proteins. Crit Rev Food Sci Nutr 59(2):349–356
Ly HL, Tran TMC, Tran TTT, Ton NMN, Le VVM (2018) Application of ultrasound to protein extraction from defatted rice bran. Int Food Res J 25(2):695–701
Meesook K, Yoonhwa J (2002) Extraction and electrophoretic characterization of rice proteins. [Extraction and Electrophoretic Characterization of Rice Proteins]. Prevent Nutr Food Sci 7 (4):437–441. Retrieved from http://www.dbpia.co.kr/journal/articleDetail?nodeId=NODE00637164
Nadathur SR, Wanasundara JPD, Scanlin L (2017) Chapter 1 - proteins in the diet: challenges in feeding the global population. In: Nadathur SR, Wanasundara JPD, Scanlin L (eds) Sustainable protein sources. Academic Press, San Diego, pp 1–19
Ngamsuk S et al (2020) Ultrasonication of milky stage rice milk with bioactive peptides from rice bran: its bioactivities and absorption. Food Bioprocess Technol 13(3):462–474
Olivares Díaz E, Kawamura S, Koseki S (2019) Effect of thickness and maturity on protein content of japonica brown rice collected during postharvest processing. Biosyst Eng 183:160–169
Omwamba M, Mahungu SM (2014) Development of a protein-rich ready-to-eat extruded snack from a composite blend of rice, sorghum and soybean flour. Food Nutr Sci 2014
Osborne TB (1924) The vegetable proteins. Nature 114(2875):822–822

Pan X et al (2019) Effect of enzyme types on the stability of oil-in-water emulsions formed with rice protein hydrolysates. J Sci Food Agric 99(15):6731–6740
Pannangrong W, Welbat JU, Sripanidkulchai B (2018) Effect of protein hydrolysate type A-2 from pigmented rice on memory deficits in A-injected rats. Srinagarind Med J 33(6):566–571
Pantoa T et al (2020a) Young rice protein as a new source of low allergenic plant-base protein. J Cereal Sci 93:102970
Pantoa T et al (2020b) Characterization and bioactivities of young rice protein hydrolysates. J Cereal Sci 95:103049
Parada J, Santos JL (2016) Interactions between starch, lipids, and proteins in foods: microstructure control for Glycemic response modulation. Crit Rev Food Sci Nutr 56(14):2362–2369
Paraman I et al (2006) Physicochemical properties of Rice endosperm proteins extracted by chemical and enzymatic methods. Cereal Chem 83(6):663–667
Phongthai S, Homthawornchoo W, Rawdkuen S (2017) Preparation, properties and application of rice bran protein: a review. Int Food Res J 24(1):25
Phongthai S, D'Amico S, Schoenlechner R, Rawdkuen S (2016a) Comparative study of rice bran protein concentrate and egg albumin on gluten-free bread properties. J Cereal Sci 72:38–45. https://doi.org/10.1016/j.jcs.2016.09.015
Phongthai S, Lim S-T, Rawdkuen S (2016b) Ultrasonic-assisted extraction of rice bran protein using response surface methodology. J Food Biochem 41(2):e12314. https://doi.org/10.1111/jfbc.12314
Phongthai S, Lim S-T, Rawdkuen S (2016c) Optimization of microwave-assisted extraction of rice bran protein and its hydrolysates properties. J Cereal Sci 70:146–154
Pires PGDS et al (2020) Plasticizer types affect quality and shelf life of eggs coated with rice protein. J Food Sci Technol 57(3):971–979
Pooja K et al (2017) Physico-chemical, sensory and toxicity characteristics of dipeptidyl peptidase-IV inhibitory peptides from rice bran-derived globulin using computational approaches. Int J Pept Res Ther 23(4):519–529
Pracham S, Thaiudom S (2016) The effect of protein content in jasmine rice flour on textural and rheological properties of jasmine rice pudding. Int Food Res J 23:4
Rani S, Pooja K, Pal GK (2018) Exploration of rice protein hydrolysates and peptides with special reference to antioxidant potential: computational derived approaches for bio-activity determination. Trends Food Sci Technol 80:61–70
Rodríguez-Restrepo YA et al (2020) Valorization of rice by-products: protein-phenolic based fractions with bioactive potential. J Cereal Sci 95:103039
Sadawarte SK et al (2007) Enrichment of bread with rice bran protein concentrate. J Food Sci Technol 44:195–197
Saleh ASM et al (2019) Brown rice versus white rice: nutritional quality, potential health benefits, development of food products, and preservation technologies. Compr Rev Food Sci Food Saf 18(4):1070–1096
Sarnthima R, Khammuang S, Joompang A (2020) Glutinous rice (Oryza sativa L.) protein extract with potent α-amylase inhibitory activity. J Food Sci Technol 57(8):3157–3163
Sereewatthanawut I et al (2008) Extraction of protein and amino acids from deoiled rin and amino acids from deoiled rice bran by subcritical water hydrolysis. Bioresour Technol 99(3):555–561
Silventoinen P et al (2019) Biochemical and techno-functional properties of protein- and fibre-rich hybrid ingredients produced by dry fractionation from rice bran. Food Bioprocess Technol 12(9):1487–1499
Singh J, Karmakar S, Banerjee R (2019) Extraction and nutritional properties of protein derived from rice (Oryza sativa) based distillery byproducts:a potential substrate for food formulation. Biocatal Agric Biotechnol 22:101364
Sirikul A, Moongngarm A, Khaengkhan P (2009) Comparison of proximate composition, bioactive compounds and antioxidant activity of rice bran and defatted rice bran from organic rice and conventional rice. Asian J Food Agro-Industry 2(4):731–743

Supawong S, Park JW, Thawornchinsombut S (2019) Effect of rice bran hydrolysates on physicochemical and antioxidative characteristics of fried fish cakes during repeated freeze-thaw cycles. Food Biosci 32:100471
Suwannapan O et al (2020) Angiotensin-I-converting enzyme (ACE)-inhibitory peptides from Thai jasmine rice bran protein hydrolysates. Int J Food Sci Technol 55(6):2441–2450
Thamnarathip P et al (2016) Identification of peptide molecular weight from rice bran protein hydrolysate with high antioxidant activity. J Cereal Sci 69:329–335
Udenigwe CC (2016) Towards rice bran protein utilization: in silico insight on the role of oryzacystatins in biologically-active peptide production. Food Chem 191:135–138
Udenigwe CC, Aluko RE (2012) Food protein-derived bioactive peptides: production, processing, and potential health benefits. J Food Sci 77:1
Uraipong C, Zhao J (2016) Rice bran protein hydrolysates exhibit strong in vitro α-amylase, β-glucosidase and ACE-inhibition activities. J Sci Food Agric 96(4):1101–1110
Vallabha VS, Indira TN, Jyothi Lakshmi A, Radha C, Tiku PK (2015) Enzymatic process of rice bran: a stabilized functional food with nutraceuticals and nutrients. J Food Sci Technol 52 (12):8252–8259. https://doi.org/10.1007/s13197-015-1926-9
Verma DK, Srivastav PP (2020) Bioactive compounds of rice (Oryza sativa L.): review on paradigm and its potential benefit in human health. Trends Food Sci Technol 97:355–365
Wang C et al (2014) Comparison of two methods for the extraction of fractionated rice bran protein. J Chem 2014:546345
Wang H et al (2020a) Three phase partitioning for simultaneous extraction of oil, protein and polysaccharide from rice bran. Innovative Food Sci Emerg Technol 65:102447
Wang L et al (2020b) Effect of ultrasonic power on properties of edible composite films based on rice protein hydrolysates and chitosan. Ultrason Sonochem 65:105049
Wang N et al (2019a) Effect of protein aggregates on properties and structure of rice bran protein-based film at different pH. J Food Sci Technol 56(11):5116–5127
Wang T, Zhang H, Wang L, Wang R, Chen Z (2015) Mechanistic insights into solubilization of rice protein isolates by freeze–milling combined with alkali pretreatment. Food Chem 178:82–88. https://doi.org/10.1016/j.foodchem.2015.01.057
Wang Z et al (2019b) Rice protein exerts anti-inflammatory effect in growing and adult rats via suppressing NF-κB pathway. Int J Mol Sci 20(24):6164
Wen L et al (2016) Rice protein hydrolysates (RPHs) inhibit the LPS-stimulated inflammatory response and phagocytosis in RAW264.7 macrophages by regulating the NF-κB signaling pathway. RSC Adv 6(75):71295–71304
Wongthahan P, Thawornchinsombut S (2015) Quality improvement of reduced-salt, phosphate-free fish patties from processed by-products of Nile Tilapia using textural additives and bioactive rice bran compounds. J Texture Stud 46(4):240–253
Wu W et al (2020) Food protein-derived iron-chelating peptides: the binding mode and promotive effects of iron bioavailability. Food Res Int 131:108976
Xia N et al (2012) Characterization and in vitro digestibility of rice protein prepared by enzyme-assisted microfluidization: comparison to alkaline extraction. J Cereal Sci 56(2):482–489
Yadav RB, Yadav BS, Chaudhary D (2011) Extraction, characterization and utilization of rice bran protein concentrate for biscuit making. Br Food J 113(9):1173–1182
Yan QJ et al (2015) Isolation, identification and synthesis of four novel antioxidant peptides from rice residue protein hydrolyzed by multiple proteases. Food Chem 179:290–295
Yang L et al (2012) Rice protein improves adiposity, body weight and reduces lipids level in rats through modification of triglyceride metabolism. Lipids Health Dis 11:24
Yang X et al (2018) Effects of ultrasound-assisted α-amylase degradation treatment with multiple modes on the extraction of rice protein. Ultrason Sonochem 40:890–899
Zaky AA et al (2019) Preparation and assessment of bioactive extracts having antioxidant activity from rice bran protein hydrolysates. J Food Meas Charact 13(4):2542–2548
Zaky AA et al (2020) An overview on antioxidant peptides from rice bran proteins: extraction, identification, and applications. Crit Rev Food Sci Nutr:1–13

Zhang HJ et al (2012) Preparation and functional properties of rice bran proteins from heat-stabilized defatted rice bran. Food Res Int 47(2):359–363

Zhang HJ et al (2014) Antioxidant activities of the fractionated protein hydrolysates from heat stable defatted rice bran. Int J Food Sci Technol 49(5):1330–1336

Zhang L et al (2021) Characteristics of rice dreg protein isolate treated by high-pressure microfluidization with and without proteolysis. Food Chem 358:129861

Zhang Y, Wang B, Zhang W, Xu W, Hu Z (2019) Effects and mechanism of dilute acid soaking with ultrasound pretreatment on rice bran protein extraction. J Cereal Sci 87:318–324. https://doi.org/10.1016/j.jcs.2019.04.018

Zhao Q et al (2012) Enzymatic hydrolysis of rice dreg protein: effects of enzyme type on the functional properties and antioxidant activities of recovered proteins. Food Chem 134(3):1360–1367

Chapter 4
Oil from Rice Bran and its Processing Technology

Jinhu Tian and Changqing Wei

Abstract A general introduction to rice bran is described and the factors that affect the rancidity of rice bran (including enzymes, hydrolysis, oxidation, and micro rancidity) are discussed and the stabilization technologies of rice bran (i.e., extrusion, chemical, hot steam, microwave, infrared, and enzymatic) are introduced. After that, the fatty acid composition, bioactive compounds, and the processing technologies of rice bran oil are described and the refining technology of rice bran oil (degumming, deacidification, dewaxing, etc.) as well as the technologies to remaining the bioactive compounds in rice bran oil are introduced. Finally, the applications of rice bran oil in food processing, medical treatment, and washing cosmetics are described in detail.

Keywords Rice bran oil · Rancidity · Stabilization · Refining

Abbreviations

HDL-C	High-density lipoprotein cholesterol
LDL-C	Low-density lipoprotein cholesterol
MUFA	Monounsaturated fatty acids
PUFA	Polyunsaturated fatty acids
SFA	Saturated fatty acids

J. Tian (✉)
College of Biosystem Engineering & Food Nutrition, Zhejiang University, Hangzhou, China

C. Wei
College of Food, Shihezi University, Shihezi, China

J. Tian et al. (eds.), *Science of Rice Chemistry and Nutrition*,
https://doi.org/10.1007/978-981-99-3224-5_4

4.1 General Introduction of Rice Bran

As the most important grain in the world, the output of rice reached up to 700 million tons per year, among which China's annual output reached up to 200 million tons, ranking first in the world (FAOSTAT 2019). With the increasing strategic status of rice in the national economy and people's livelihoods, more and more attention have been paid to the development of rice processing technology. In some developed countries, the processing technologies of rice are relatively mature, and many enterprises have shifted their focus to the deep processing and comprehensive usage of rice (Garofalo et al. 2021). As the largest rice production and consumption country, the loss of rice during processing is astonishing in China, and the added value of rice consumed as rations is quite low and limited. In addition to the main product of rice processing, the by-products, such as broken rice, rice bran, rice embryo, rice husk, etc. rice, can also be husk, etc., and refined into high-value products rice, but these technologies are still limited in China, and some other developing countries (Moraes et al. 2014).

Rice bran is one of the main by-products during rice processing, which includes the pericarp, seed coat, and aleurone layer, as well as endosperm starch and embryo (Juliano and Tuaño 2019). Fresh rice bran usually has the flavor of rice with a strong ability for water absorption and shows light yellow, gray-yellow, and yellow-white colors. Normally, the composition of rice bran could be affected by the raw material and the processing methods of rice. Generally, rice bran contains about 15% protein, 10% moisture, 3–8% sugar, 16–22% fat with a calorie of 125.1 kJ/g (Spaggiari et al. 2021). It has been proved that rice bran has certain curative effects, such as defecation, reduced cholesterol content, and reduced urinary stones (Kahlon 2009). The protein from rice bran is almost superior to all other plant proteins, and its amino acid composition pattern is quietly close to the recommended pattern of FAO/WHO. The biological potency is also very high, which is only slightly inferior to the protein from milk and eggs. Compared with other proteins, rice bran protein is also less likely to induce allergies, so it is quite suitable for nutrition infants and patients (Phongthai and Rawdkuen 2020). The fatty acids contained in rice bran oil are mostly unsaturated fatty acids necessary for the human body. These fatty acids have been reported to be beneficial in reducing the accumulation of unsaturated cholesterol on the vascular wall and then relieving hyperlipidemia and atherosclerosis effectively (Duansak et al. 2020). In addition to proteins and lipids, there are abundant vitamin E, bioactive lipid enzymes, squalene, sitosterol, sterol, and other bioactive compounds in rice bran. Studies have shown that long-term consumption of lipids from rice bran is conducive to canners, high blood pressure, and other chronic metabolic diseases (Elsaid et al. 2018). As the largest rice producer in the world, the production of rice bran in China also ranks first in the world, but normally the rice bran was used as fodder or directly discarded, resulting in a great waste. Thus, a full description and the usage of rice bran are quite urgently needed.

4.2 Rancidity of Rice Bran

During rice processing, more than 80% of rice bran are scattered, and only 20% are under centralized control. This brings limitations to the full usage of rice bran. As commonly known, fresh rice bran is quite unstable, and the acid value increases sharply within the first 6 h, which will make its application value decreases (Saunders 1985). The main factor affecting the quality of rice bran is the activity of lipase. In intact seeds, lipase is located in the seed cortex, while the oil is located in the aleurone layer, subaleurone layer, and embryo, they are separated in different areas and have no contact between them. However, after the rice is milled, lipase is transferred into the rice bran, and oil hydrolysis occurs quickly. The acid value of fresh rice bran oil rises significantly in the first 6 hours and will bring an undesirable factor for further rice oil processing (Wu et al. 2020).

4.2.1 Enzymes that Induce the Rancidity of Rice Bran

Lipolytic Enzyme

Lipopolytic enzyme is a general term for substances that catalyze the hydrolysis and synthesis of esters, mainly including esterase (EC 3.1.1.1) and lipase (EC 3.1.1.3). Although the sequences of those two enzymes are quite different, most of them contain a conserved catalytic triplet [Ser-histidine (His) aspartic acid (Asp)/glutamic acid (Glu)] with an active site of Serine (Ser), and have a special sequence of -Gx1Sx2G- (where x is any amino acid) (Brenner 1988). Microbial lipases, esterases, and related proteins can be divided into 38 superfamilies and 112 homologous families according to the structures and residues that form the oxygen-containing anion holes (amino acids that stabilize the active sites of the reaction intermediates) and catalytic triads. Esterase catalyzes the hydrolysis of ester bonds of short-chain fatty acids with less than 10 carbons, while lipase catalyzes the hydrolysis of triacylglycerides into glycerol and fatty acids, of which with more than 10 carbons.

The esterase hydrolyzes the water-soluble substrate, while the lipase hydrolyzes the oil–water interface substrate (Park et al. 2007). In addition to lipolytic enzymes, there are other three types of phospholipases in rice bran, namely phospholipase A (A1 and A2), phospholipase C, and phospholipase D. The phase-to-activity ratio of these four types of enzymes was 100:24:35:39, and phospholipase acted on the monomolecular layer of phospholipids and was the key enzyme for the germination and mobilization of seed-stored lipids (Dolui et al. 2020). Lipase-II is the most active lipolytic enzyme, which is a basic protein and catalyzes the triad of Ser175-Asp231-His292.

LOX

LOX is a key enzyme in the metabolism of unsaturated fatty acids, which can transform arachidonic acid, linoleic acid, and other polyunsaturated fatty acids into bioactive metabolites, thus affecting cell structure, metabolism, and signal transduction (Umate 2011). LOX in rice mainly exists in embryos and coat of the seed, and its content is quite low. Linoleic acid, α-linolenic acid, and arachidonic acid in rice are the substrates of the LOX-catalyzed oxidation reaction to produce conjugated diene hydrogen peroxide. This LIPolysis catalyzed by LOX is called lipid hydrogen peroxide. Rice genomes could encode 14 types of LOX, of which five have been cloned and identified, including 3 LOX-isozymes (LOX-1, LOX-2, and LOX-3) and HI-LOX. LOX-1, LOX-2, and LOX-3 have different functions in the lifetime of rice seeds, while others are involved in biological stress responses. Among them, LOX-1 and LOX-2 are key genes influencing seed longevity, and LOX-3 is a key gene influencing rice insect resistance. Among the three kinds of LOX, LOX-3 accounts for 80% of the total LOX activity. The main product of oxidized-unsaturated fatty acids of LOX-1 and LOX-2 is 13-hydroperoxide (HPOD). The main component of the volatile products produced during the modification of rice bran is 1-hexanal, which is the degradation product of 13-HPOD, but the volatile 13-HPOD and its lysate 1-hexanal are toxic to pests. 9-HPOD, the main product of oxidized-unsaturated fatty acids of LOX-3, can be converted from 13-HPOD of LOX-1 and LOX-2 to a substance that is not conducive to the lysolytic enzyme of 13-HP OD and may also compete with LOX-1 and LOX-2 for fatty acids. Therefore, rice bran with low LOX-1 and LOX-2 content can improve storage tolerance, and rice bran with low LOX-3 content can improve pest resistance.

The oil content in brown rice is relatively stable because the lipolysis enzyme in the whole grain is mainly located in the cross cells of the seed coat, while most of the lipolysis enzyme in the whole grain is located in the cross cells of the seed coat.

The sebum is stored in the aleurone layer and the germ, and after grinding operation, the stable state of the mutual isolation is destroyed. The lipid in rice is mainly concentrated in the seed embryo, followed by the aleurone layer of the seed coat, so the lipid is mainly distributed in the rice bran, which is composed of seed embryo and seed coat. Under the action of lipase, esterase, and LOX with strong activity in rice bran (Gul et al. 2015), these lipids undergo hydrolysis, oxidation, and other reactions to form free radicals and volatile carbonyl compounds, resulting in the rapid increase in rice bran acid price and unesterified fat acids give food a bitter and musty taste, while oxidizing compounds produce a rancid odor. The mechanism of rice bran rancidity is shown in Fig. 4.1. Additionally, the rancidity rate of rice bran is also related to microbial action, rice varieties, and contents of total phenolic substances, total flavonoids, and total anthocyanins.

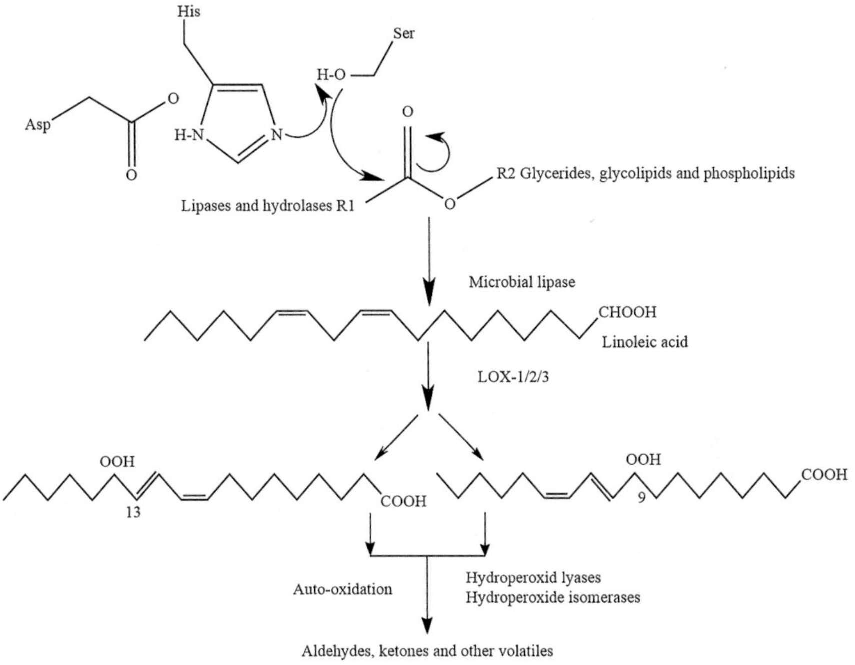

Fig. 4.1 Rancidity mechanism of rice bran

4.2.2 The Hydrolysis Rancidity of Rice Bran

After rice milling, the α-helix cap of the enzymes was opened and the active sites were exposed and contacted with the substrate under the action of hydrophobic and hydrophilic residues. The binding site of lipase-II with the highest activity of lipolytic enzyme in rice bran showed a shallow bowl structure, and the substrate could enter and bind to the active center of Ser. The acylation reaction occurs under the stability of hydrophobic interaction between the formed hydrogen bond and lipase (Nardini and Dijkstara 1999). With the participation of the His residue from the active center, Ser is activated, and the proton hydrogen on the hydroxyl group is transferred to the imidazole ring of the His residue. The hydroxyl residue of Ser then increases its nucleophilicity and attacks the carbonyl group of the substrate to form a tetrahedral intermediate complex, which carries a negative charge on the oxygen of the carbonyl group. The oxygen anion hole forms a stable charge distribution and the state energy of the tetrahedral intermediate compound is reduced, and then deacylation steps occur. The nucleophile attacks the carbon atom of the ester bond to break. His imidazole ring again transfers proton hydrogen to the anion oxygen of Ser, thus releasing a large number of volatile fatty acids and glycerol (Jaeger et al. 1999).

4.2.3 Oxidation Rancidity of Rice Bran

The volatile fatty acids, linoleic acid, linolenic acid, and other unsaturated fatty acids produced by the rancidity hydrolysis of rice bran were further catalyzed by LOX as substrates for oxidative rancidity. LOX catalyst polyunsaturated fatty acids that contain 1, 4 isoprene structure, and generate conjugate double-bond lipid peroxide hydrogen compounds to as lipid hydroperoxide lyase and lipid hydroperoxide Isomerase degradation of volatile hexanal and amyl alcohol, etc., resulting in a rancid flavor (Shen et al. 2002). Unsaturated fatty acids are the major reactants from the oxidation reaction, and the oxide could be induced by light, heat, oxygen, water, metal, and other environmental factors. The lipid oxygen free radicals in rice bran attack new lipid molecules and then chain reactions occur, and it will be terminated until the hydrogen source is not available. The degree of unsaturation determines the oxidation stability of fatty acids. The linolenic acid with three double bonds requires the least activation energy to remove hydrogen atoms and is more easily oxidized. The oxygenation rate-rate ratio of linolenic acid, linoleic acid, oleic acid, and stearic acid is reported to be 2500:1200:100:1 (Bollinedi et al. 2020).

4.2.4 Microbial Rancidity of Rice Bran

Microorganism is also an important cause of the deterioration of rice bran. As commonly known, rice bran is rich in protein, lipids, and other nutrients, and is easily spoiled by humidity, heat, oxygen, mold, etc. The main factors of rice bran mildew, including high moisture content (or high-water activity), heat generation, and improper storage, etc. Moreover, rice bran has a high content of fat, and the dominant molds in rice, such as *Aspergillus flavus*, *Aspergillus Niger*, and *Aspergillus fumigatus*, rapidly proliferate and produce a large number of other enzymes, which decompose lipids with endogenous enzymes of rice bran and accelerate the rancidity (Wang 2006). It has been reported that if the rice bran was stored at 50 °C and relative humidity of 70% for 6 days, it would increase the temperature and mildew, and subsequently result in a significantly deepened color of rice bran oil (Zheng et al. 2015). In addition to causing mildew deterioration, these microorganisms can produce carcinogenic, teratogenic, and mutagenic effect of mycotoxins, which has quite strong tolerance of toxins and are difficult to remove during rice processing.

4.3 The Stabilization Technology of Rice Bran

As mentioned above, enzymes, microorganisms, and other harmful factors contained in rice bran are the root causes of deterioration. In the process of grain milling, enzymes, and lipids contact each other adequately, leading to rice bran rancidity, and deterioration. The free fatty acid value of fresh rice bran can be increased by 7%~8% in the first day and then by increased by 4%~5% per day. Therefore, the key technology to prevent rice bran rancidity is inhibiting the activity of lipozyme. In China, the rice processing enterprises are scattered and small in scale, and the unstable fresh rice bran brings great trouble to the storage and transportation process, large parts of rice bran can only be used as feed in locally. The stabilization technology is the key to fully unitization of rice bran. In present part, the technologies of rice bran stabilization, including dry heating, wet heating, ohmic heating, microwave heating, chemical addition, and extrusion method, etc. the following part will discuss the stabilization technology of rice bran in detail.

4.3.1 Extrusion Technology

Extrusion technology refers to the technology of extruding, heating, puffing, and shaping rice bran under the condition of high temperature and pressure of a spiral extruder, so that peroxidase and lipophylase in rice bran can be inhibited effectively, reducing microbial activity (Zhou 2014). The water-holding capacity of insoluble dietary fiber in rice bran and the content of soluble dietary fiber in rice bran could be significantly improved by extruding technology. Meanwhile, it can also make the water in the rice bran tissue evaporate quickly, producing many tiny holes. Qiao et al. treated rice bran by extruding technology and achieved the optimal process parameters that the lipase inhibition rate was 76% when the water content was 19%, the screw speed was 180 r/min, and the cylinder temperature was 130 °C (Qiao et al. 2012).

4.3.2 Chemical Methods

The chemical method is to change the condition of rice bran with chemical reagents so that the pH is not suitable for the activity of enzymes. Usually, an acid or alkali method is used to treat rice bran, which reduces the activity of lipase in rice bran, thus inhibiting the activity of enzymes and preventing rice bran from rancidity. Compared with other methods, chemical methods have simple equipment and faster processing speed. Prabhakar and Venkatesh (1986) showed that the pH of rice bran after chemical treatment decreased from 6.9 to 4.0. After storing at room temperature for 30 days, the free fatty acid content of untreated rice bran oil was 47.5%.

However, the free fatty acid content of crude oil treated with the chemical method was only 9.3% after 51 days of storage. Wang et al. showed that after treating rice bran with acid, the fatty acid value of rice bran with pH of 3, 4, and 5 increased to 20.2 mg KOH/g, 24.1 mg KOH/g, and 42.9 mg KOH/g, respectively, when the storage time was 35 d. It was concluded that the lower the pH was, the higher the fatty acid value was and the slower rancidity of rice bran.

4.3.3 Hot Steam Treatment Method

Treating with high-temperature steam is a traditional method to stabilize rice bran. The specific operation is to adjust the moisture content to 17%, then increase the temperature to 105 °C and heat with steaming for 2 h, and finally dry the rice bran. Hot steam treatment can slow down the hydrolysis of oil by inhibiting the activity of the enzyme. Cui et al. (2012) studied the effect of treatment of hot steam on rice bran and found that when the rice bran was treated by steaming at 100 °C for 30 s, the enzyme residual activity was 64.0%. While the treatment time increased to 180 s, the enzyme residual activity was only 5%.

4.3.4 Microwave Treatments

Compared to other treatments, the microwave treatment of rice bran shows the advantages of uniform heating, shorter time, and less energy requirements. There are also studies that have reported the microwave stabilization of glutinous millet rice bran and reported that the optimal process conditions for microwaving were 27% moisture content, 90-s treatment, and 600-W microwave power. Feng et al. (2004) investigated the influence of microwave radiation on the lipase activity in rice bran and reported that microwaving had a good effect on reducing the lipase activity in rice bran, and the activity decreased with the extension of microwave time and the enhanced microwave power.

4.3.5 Infrared Heating

Infrared heating is a method that transfers the heat energy into infrared energy, and then radiates to the interior of the object. For the rice bran, it is quite easy to reach the ideal conditions that are required for enzyme inactivation and dehydration by infrared heating. Compared with other methods, infrared heating has the characteristics of fast heating and high safety performance. Irakli et al. (2018) treated the rice bran with a heating temperature of 140 °C and time of 3.5 min, and the rapid rancidity of rice bran was effectively inhibited, while the composition of oryzanol and fatty acid in rice bran was not significantly affected.

4.3.6 Enzymatic Method

Enzymatic treatment is a method that rice bran reacts with protease (with the ability to resist lipase) under certain conditions, then inactivate the enzyme and stabilizes rice bran. The enzymatic method was invented and patented by Rebus Corporation in the United States. Compared with other methods, the enzymatic method has the characteristics of safety, simplicity, and high quality of products, but the cost was much higher than other methods.

4.3.7 Other Methods

In addition to the abovementioned methods, rice bran can be stabilized by low-temperature storage, ohmic heating, and a combination of stabilization methods. For example, Sreenarayanan and Chattopadhyay (1985) used ohmic heating to stabilize rice bran and found that the increase in free fatty acids in rice bran was less than 2% even in storage for 6 weeks.

4.3.8 Industrialization of Main Production Technology

Up to now, there are many methods to stabilize rice bran effectively. However, rice bran is generally treated by the extrusion method in most enterprises for its advantages of safety, no pollution, simple operation, and low cost. During extrusion, steam is injected into the machine to improve the temperature and humidity of the rice bran, and then the pressure drops instantly to make the rice bran become granular material and then dried at 80 °C, so that the moisture content is about 7%. The researchers also reported that under the optimal treatment conditions of feeding humidity of 20%, extrusion temperature of 130 °C, and twin-screw rotation speed of 200 RPM, rice bran could store at room temperature for 8 weeks, and the acid value did not show significant increase.

4.4 Processing of Rice Bran Oil

There are about 15–20% crude oil in rice bran; normally, the rice bran oil is extracted by pressing or solvent extraction methods, such as supercritical carbon dioxide extraction and enzymatic extraction, and then refined by decolorization, deodorization, or other processes. About 85% of rice bran oil consisted of saturated fatty acids and unsaturated fatty acids. Among them, palmitic acid accounts for 13–18%, oleic acid accounts for 40–50%, and linoleic acid accounts for 26–35%. Additionally, there are also some protein, fiber, and other ingredients.

4.4.1 Fatty Acid Composition of Rice Bran Oil

The fatty acid composition of rice bran oil varies according to the place of origin, climate, and process technology. The detailed fatty acid constitutions are as shown in Table 4.1. And a comparation with oil from other plants, oleic acid, linoleic acid, and palmitic acid were the main components of rice bran oil. The ratio of saturated fatty acids (SFA) to monounsaturated fatty acids (MUFA) to polyunsaturated fatty acids (PUFA) was about 1:2.1:1.8, which is close to the 1:1.5:1 recommended by the World Health Organization. Similar to olive oil, corn oil, camellia oil, and other plant oils, rice bran oil contains nearly 80% total unsaturated fatty acids, and unsaturated fatty acids have a preventive effect on atherosclerosis and coronary heart disease, and other chronic diseases. The ratio of linoleic acid to oleic acid is about 1:1.2, which is a typical oleic acid–linoleic acid-type plant oil. Both fatty acids have the function of lowering low-density lipoprotein cholesterol (LDL-C), increasing high-density lipoprotein cholesterol (HDL-C) and lowering serum cholesterol, thus playing a beneficial role in regulating blood lipids.

4.4.2 Bioactive Compounds from Rice Bran Oil

Researchers have reported that rice bran oil was rich in γ-glutamicin, phytosterol (mainly stigmasterol, sitosterol, rapes sterol), vitamin E, squalene, and other bioactive components (Spaggiari et al. 2021). The structure of squalene and the structural framework of glutamyl, phytosterol, and tocotrienol are shown in Fig. 4.2. Glutamate is a unique active ingredient that only existed in rice bran oil. Tetrienol was detected only in rice bran oil and corn oil, and the content of Tetrienol was much higher in rice bran oil. Rice bran oil has the highest total phytosterol content and

Table 4.1 The main fatty acids composition of rice bran oil (drawn by the author)

Fat acids	Content range (%)
Myristic acid (C14:0)	0.18–2.13
Palmitic acid (C16:0)	10.78–19.78
Palmitoleic acid (C16:1)	0.07–2.13
Stearic acid(C18:0)	0.58–2.17
Oleic acid(C18:1)	37.95–46.78
Linoleic acid (C18:2)	31.70–38.52
Linolenic acid(C18:3)	1.24–2.13
Arachidic acid (C20:0)	0.48–0.92
Arachidonic acid (C20:1)	0.44–0.71

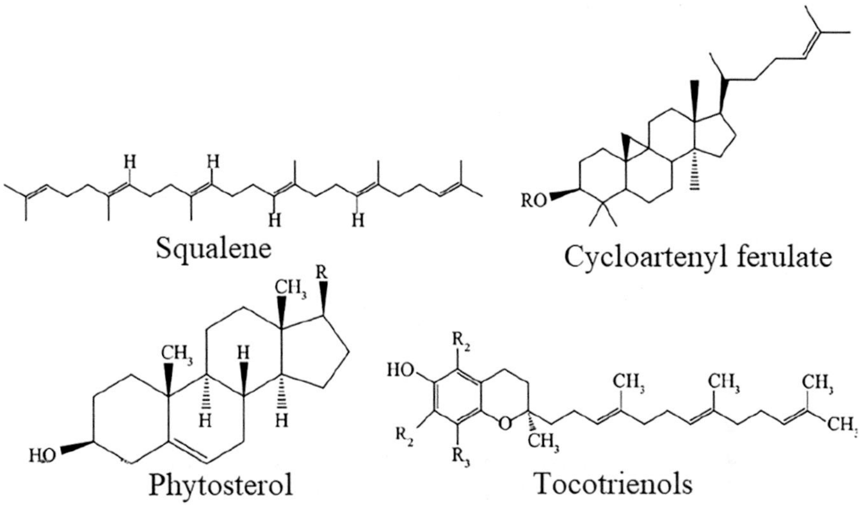

Fig. 4.2 The structural and structure skeleton of biologically active substances in rice bran oil (drawn by the author)

squalene content, which is more abundant than other commercially available plant oils (only less than olive oil). These active ingredients cannot only improve the antioxidant capacity of rice bran oil but also have a high nutritional value and various benefits to the body. For example, glutamate is more prominent in reducing blood glucose, improving gastrointestinal function, inhibiting cholesterol synthesis in the body, and reducing serum cholesterol content. Multiple unsaturated bonds of squalene enable the rice bran oil to have a strong antioxidant activity, improve the body's tolerance to hypoxia, and prevent the occurrence of breast cancer, pancreatic cancer, and other cancers.

4.4.3 Processing Technology of Rice Bran oil

In addition to the health benefits of rice bran oil, the rice bran oil contains many lipid-soluble impurities and metal ion chelates, which made the oil show a dark color and easy to oxidation. Thus, how to remove impurities in rice bran oil and retain nutrients such as oryzin and phytosterol to the maximum extent are a big challenge in the processing of rice bran oil (Garofalo et al. 2021). Recently, super rice oil degumming and enzymatic degumming technologies have been applied in practice. These technologies increased rice bran oil yield, improved rice bran oil quality, and the bioactive components in the refined rice oil were also effectively maintained. In this part, the traditional chemical refining process and modern physical refining process of rice oil were reviewed to provide references for the refining of rice bran oil.

Mechanical Squeezing

Mechanical squeezing is a relatively primitive and traditional method to produce the rice bran oil. Normally, the mechanical squeezing could be divided into a power screw press and a hydraulic press (Noor and Tajul 2011). Its technological processes are as follows: rice bran → screening → steaming → pressing cake → pressing → filtering. The pressing method has low requirements for equipment and technologies and has the advantages of convenient operation and high safety. However, compared with other methods, the yield of rice bran oil by mechanical squeezing is low and the residual oil rate of dry cake is high, which accounts for about 3.0%~4.0%.

Supercritical Fluid Extraction

As a new separation technology, supercritical fluid extraction is central to the extraction of rice bran oil. Supercritical extraction is a technology that separates a component from a mixture by using a supercritical fluid as an extraction agent based on the similar compatibility principle of solvents. CO_2 is the most commonly used supercritical fluid because of its low cost, non-toxic, harmless, and pollution-free advantages (Garcia et al. 1996). The yield of rice bran oil is much higher when using supercritical fluid extraction. The wax, phosphorus, and free fatty acid can be removed, and the loss of glutamate in rice bran oil can also be reduced, and finally the high-quality rice bran oil can be obtained.

Organic Solvent Leaching Method

For the solvent extraction methods, rice bran was mixed with a solvent in a certain proportion, and the effect factors are including temperature, extraction time, extraction solvent, solid-liquid ratio, pH, and so on. The common extraction solvent used in studies and industries is n-hexane, but n-hexane is an air pollutant, so researchers try finding other substances instead of n-hexane. Li et al. (2012) determined the optimal extraction conditions using isoacetone as a solvent and reported that pH 6.0, temperature 60 °C, and solid-liquid ratio 1:5 (g: mL) were the optimal extraction conditions. Proctor and Browan (1996) used hexane and isopropanol as solvents to extract oil from rice bran at room temperature. And reported that the yield of rice bran oil from isopropanol was similar to that from hexane, but the stability of oil was much higher than that from hexane.

Enzyme-Catalyzed Leaching

Treating rice bran with enzymes is a new oil processing technology. In this technology, the cell tissue of rice bran is hydrolyzed with protease and then oil is released to achieve the purpose of oil extraction (Hanmoungjai et al. 2001). In a previous study, Chen (2014) used alkaline protease to extract oil from rice bran and determined the optimal process parameters. In addition to treating rice bran with a single protease, Aprara et al. (2001) believed that the synergistic effect of amylase, cellulase, and protease on extracting rice bran oil could improve the rice bran oil yield.

Rapid Equilibrium Leaching

Rapid equilibrium leaching at room temperature refers to the method whereby rice bran oil is obtained by mixing organic solvent rice bran at room temperature and adopting an appropriate solid-liquid ratio, so that the extraction process can quickly reach the equilibrium state in a very short time (Xie 2013). The factors affecting the oil extraction are temperature, time, extraction solvent, water content of rice bran, grinding the degree of rice bran, ratio of solid to liquid, and pH.

Aqueous Solution Leaching

The aqueous solution leaching method refers to the process whereby an aqueous solution extracts fat from an oil-bearing substance (Ding et al. 2009). The rice bran was separated by centrifugation to obtain an emulsified oil layer. After the oil layer was emulsified, the crude oil was obtained. Normally, rice bran oil with high quality can be extracted with this method.

Other Methods

In addition to the abovementioned methods, there are other methods such as three-phase separation, membrane separation, and other methods to extract oil from rice bran. For example, some researchers have studied the extraction of rice bran oil by enzymatic hydrolysis with a three-phase separation method and obtained the optimal extraction process parameters (Ghosh 2007).

4.5 Refining Technology of Rice Bran oil

As mentioned above, there are many phospholipids, free fatty acids, pigments, waxes, and other substances in the crude rice bran oil, and the rice bran oil could not be used for consumption directly in this state. The crude rice bran oil must go through a series of refining processes before it can meet the requirements of the standard of edible oil; those refining processes include degumming, decolorizing, and deodorizing (Haraldsson 1983), these processing are briefly described as follows.

4.5.1 Degumming

The crude rice bran oil contains certain amounts of phospholipids and other gels, some of which will induce precipitation and cause a great loss of neutral oil. The aiming of the degumming process of rice bran oil is to removing these substances and reduces the loss. Up to now, there are many kinds of degumming methods, such as enzymatic method, hydration method, acid method, and adsorption method. For example, Wang et al. (2004) used the new phospholipase A1 to conduct the degumming experiment on rice bran oil and obtained rice bran oil containing about 25 mg/kg of gum. Additionally, there are also mixed oil-film separation, oil-phase-film separation, and combined degumming methods.

4.5.2 Deacidification

Deacidification is the process of removing free fatty acids from crude rice bran oil. The deacidification methods of rice bran oil include alkali refining, distillation, mixing, re-esterification, and extraction. The alkali refining method has comprehensive effects such as deacidification and decolorization, but the refining loss is much higher. Distillation has a good deacidification effect, but it is easy to loss the nutrients. The mixed refining method can improve the refining rate and obtain rice bran oil with high quality, and could also reduce the loss of solvent. The re-esterification method not only removed the free fatty acids effectively but also increased the content of glycerol in rice bran oil significantly. Additionally, the extraction method also has a good effect on deacidification. Liu (2005) obtained rice bran oil with an acid value reduced from 31.02 to 4.52 mg/g at the ratio of oil to solvent (95% ethanol) of 1:1.8. After alkali refining, the oleic acid value of rice bran oil was reduced to 0.23 mg/g.

4.5.3 Decolorization

The decolorization process cannot only improve the color of rice bran oil but also remove trace metal ions, peroxides, and some odorous substances (He 2010). There are also many ways to decolorize the rice bran oil, such as adsorption, enzyme, and membrane separation. The adsorption method generally uses active clay, silica gel, and other substances with strong adsorption to treat the crude rice bran oil. While membrane separation is using the membrane technology to treat crude rice bran oil during the refining process.

4.5.4 Deodorization

The aiming of the deodorization process is to remove the odorous substances that affect the odor of grease. The general conditions of deodorization are setting the pressure at 0.9~1.33 kPa and the temperature at 200~220 °C. Sterols and vitamins in rice bran oil are easily lost by entrainment of distillates, so the deodorization production is generally carried out by circulation steam stirring a decolorization tower, or a combination of two or more towers of a packed tower (Pestana-Bauer et al. 2012).

4.5.5 Dewaxing

The wax content in rice bran oil is about 3%~5%, which is easy to coagulate and precipitate, and then affects the transparency and taste of the rice bran oil. The commonly used methods of dewaxing include the conventional methods, solvent method, flocculant method, surfactant method, synthesis method, and electrostatic method. However, their principles all belong to the scope of freezing crystallization separation, that based on the physical properties of bran wax. Haraldsson (1983) obtained high-quality dewaxed rice bran oil by centrifuge separation of rice bran oil that had been stored at low temperatures.

4.5.6 Industrial Refining

In industrial production, rice bran oil is mostly refined by hydration degumming, active clay decolorization, steam deacidification and deodorization, and low-temperature freezing dewaxing. The specific method is to add 2‰~3‰ phosphoric acid and a proper amount of water to treat the crude rice bran oil at 90 °C for degumming. Subsequently, 5% active clay was added and stirred continuously at 100 °C to decolorize. Then, deacidification and deodorization were carried out in the distillation column under the conditions of pressure 0.8~1.3 kPa and temperature 200~220 °C. Finally, the crude oil was treated by static settlement, and the wax was filtered after precipitation to dewax.

4.6 Technologies to Remaining the Bioactive Compounds in Rice Bran Oil

In addition to rich resources and low raw material costs, bioactive compounds such as oryzanol, VE, and plant sterols are also important factors that attract attention (Lai et al. 2019). However, most of these compounds are removed in the refining process of rice bran oil. At present, to increase the nutrient content in the refined rice bran oil, some manufacturers extract these bioactive compounds from the by-products of rice bran oil processing (soap-stock, deodorized distillate, etc.) and then add them back. Although this method can increase the content of bioactive compounds in rice bran oil, the process of extracting these components from the by-products of rice bran refining is quite complicated. The physiologically active functions of these components may also have been affected. Therefore, how to retain the nutrients in the rice bran oil to the greatest extent during the refining process has become one of the key topics in the research of rice bran oil production technology today.

4.6.1 Retention of Oryzin

The oryzanol content in rice bran oil is much higher than that of other plant oils. The difference in the oil process will affect the content of oryzanol in rice bran oil significantly. For example, the content of oryzanol in raw rice bran oil obtained by high-temperature pressing and solvent extraction is much higher than that of low-temperature pressing. And this might be attributed to that the pressing and leaching process destroyed the cell tissue of the rice bran more thoroughly, and then more oryzanol was dissolved in the oil (Punia et al. 2021). Gopala Krishna et al. (2001) showed that during the refining process, the oryzanol loss in the alkali refining section was the largest, which accounted for about 83–95%. The degumming and dewaxing process showed little effect on the loss of oryzanol, which only accounted for about 1.1% and 5.9%, respectively. The process of decolorization, deacidification, and deodorization also has little effect on the content of oryzanol; additionally, the physical refining process seems to remain more oryzanol than the chemical refining process, which accounts for about 1.1–1.7%, and 0.19–0.20%, respectively.

4.6.2 Retention of Vitamin E

As an effective natural antioxidant, vitamin E is easily destroyed, it has been known that both the oil processing and the storage conditions would affect the vitamin E content, and every step of the oil refining process will reduce the vitamin E content, especially in the deodorization process. The adsorption of activated clay in the decolorization process also has a certain effect on the vitamin E content. Hoed et al. (2010) also confirmed that deodorization is the main step that induces the loss of VE from rice bran oil. Therefore, to reduce the loss of vitamin E during the refining process, it is necessary to improve the condition of the refining process, for example, reduce the amount of clay, lower the deodorization temperature, shorten the deodorization time, or adopt a dual-temperature deodorization process, etc.

4.6.3 Retention of Phytosterol

The alkali refining of rice bran oil, decolorization, and steam distillation deodorization sections will all cause the loss of sterols, especially the adsorption decolorization section. It has been reported that most of the sterols flowed to the soap-stock, especially the first alkali refining soap-stock, second alkali refining soap-stock, and deodorizing distillate of rice bran oil. Hoed et al. [51] reported that the loss of phytosterols in rice bran oil is mainly in the decolorization and deodorization process. Therefore, to retain sterols during the oil refining process, the refining degree should be reduced, and the influence of each section on the sterols should be considered comprehensively.

4.7 Application of Rice Bran Oil

4.7.1 Application in Food Processing

Rice bran oil has attracted wide attention because of its aromatic smell, high-temperature resistance, storage resistance, and almost no harmful substances. As fried food oil, it has a synergistic effect on the flavor of fish and snacks. Rice bran oil can also be used in the manufacture of margarine and salad oil, and rice bran oil as a cooking material could also stimulate appetite and improve stomach function.

4.7.2 In Medical Treatment

The unsaturated fatty acids in rice bran oil can improve the sterol hormone of the human body, reduce its excessive deposition in the distribution of the blood vessel walls and prevent or relieve cardiovascular disease, hyperlipidemia, and atherosclerosis. Rice bran oil contains vitamin E, squalene, active fatty acid, sitosterol, ferulic acid, and other components, which play an important role in regulating human physiological functions, brain health, anti-inflammatory, anti-drug, and anti-aging.

4.7.3 In Washing Cosmetics

Rice bran oil is widely used in the chemical industry, surfactants, cosmetics, and spices, such as acetylation of rice bran oil due to easy to form of waterproof protective films, and soft regulating function, thus, can be used as a hand or body moisturizer, softener, etc., and rice bran oil product production can also be used in shampoo and hair care formulas. Polyethylene ether rice bran oil shampoos, washing powders, hand sanitizers, and bath lotions, can reduce the dry feeling of the skin, produce a shiny feeling, and can protect the skin and hair. The quarterly ammoniated rice bran oil conditioner greatly improves dry hair, making it easy to comb.

References

Aprara S, Khare SK, Gupta MN (2001) Enzyme-assisted aqueous extraction of rice bran oil. J Am Oil Chem Soc 78(9):949–951

Bollinedi H, Singh AK, Singh N et al (2020) Genetic and genomic approaches to address rapid rancidity of rice bran. Crit Rev Food Sci Nutr 1718598

Brenner S (1988) The molecular evolution of genes and proteins: a tale of two series. Nature 334(6182):528–530

Chen L (2014) Study on the extraction technology of rice bran oil by enzyme. Food Res Dev 35(17):69–71

Cui FG, Li AP, Xie BX (2012) Effect of different treatment methods on quality stability of rice bran. Food Industry Sci Technol 33(5):141–144

Ding L, Zhou WR, Zhang SY et al (2009) Research on physiological function and extraction technology of rice bran oil. Cereal and Food Industry 16(4):9–17

Dolui AK, Latha M, Vijayaraj P (2020) OsPLB gene expressed during seed germination encodes a phospholipase in rice. Biotechnology 10(1):30

Duansak N, Piyabhan P, Srisawat U, Naowaboot J, Lerdvuthisopon N, Schmid-Schönbein G (2020) The effect of rice bran extract on arterial blood pressure, hepatic steatosis, and inflammation in mice fed with a high-fat det. J Nutr Metab:1–9

Elsaid AF, Shaheen M, Ghoneum M (2018) Biobran/MGN-3, an arabinoxylan rice bran, enhances NK cell activity in geriatric subjects: a randomized, double-blind, placebo-controlled clinical trial. Exp Ther Med 15:2313–2320

FAOSTAT (2019) Database of Food and Agriculture organization of the United Nations. Available from: http://www.fao.org/faostat/en/#data/QC/visualize

Feng GZ, Liu GQ, Zhu CS (2004) Study on the activity of lipophylase in rice bran under microwave irradiation. China Oils 29(2):71–74

Garcia A, De Lucas A, Rincón J, Alvarez A, Gracia I et al (1996) Supercritical carbon dioxide extraction of fatty and waxy material from rice bran. J Am Oil Chem Soc 73(9):1127–1131

Garofalo SF, Tommasi T, Fino D (2021) A short review of green extraction technologies for rice bran oil. Biomass Conver Biorefinery 11:569–587

Ghosh M (2007) Review on recent trends in rice bran oil processing. J Am Oil Chem Soc 84:315–324

Gopala Krishna AG, Sakina Khatoon PM, Shiela CV, Sarmandal TN, Indira AM (2001) Effect of refining of crude rice bran oil on the retention of oryzanol in the refined oil. J Am Oil Chem Soc 78(2):127–131

Gul K, Yousuf B, Singh AK et al (2015) Rice bran: nutritional values and its emerging potential for development of functional food-a review. Bioactive Carbohydrates and Dietary Fiber 6(1): 24–30

Hanmoungjai P, Pyle DL, Niranjan K (2001) Enzymatic process for extracting oil and protein from rice bran. J Am Oil Chem Soc 78(8):817–821

Haraldsson G (1983) Degumming, dewaxing and refining. J Am Oil Chem Soc 60:251–254

He DP (2010) Oil refining and processing technology. Chemical Industry Press, Beijing

Irakli M, Kleisiaris F, Mygdalia A, Katsantonis D (2018) Stabilization of rice bran and its effect on bioactive compounds content, antioxidant activity and storage stability during infrared radiation heating. J Cereal Sci 80:135–142

Jaeger KE, Dijkstra BW, Reetz MT (1999) Bacterial biocatalysts: molecular biology, three-dimensional structures, and biotechnological applications of lipases. Annu Rev Microbiol 53: 315–351

Juliano B, Tuaño APP (2019) Gross structure and composition of the rice grain, *Rice(4th edition)*, 31–53

Kahlon TS (2009) Rice bran: production, composition, functionality and food applications, physiological benefits. In: Cho S, Samuel P (eds) Fiber ingredients food applications and health benefits. CRC Press, pp 305–318

Lai OM, Jacoby JJ, Leong WF, Lai WT (2019) Nutritional studies of rice bran oil. In: rice bran and rice bran oil. *AOCS Press*, 19-54

Li JY, Wei T, Gu HX (2012) Study on extraction of rice bran oil with organic solvent. *Jiangsu*. Agric Sci 40(6):267–268. 362

Liu Y (2005) Study on extraction and deacidification technology of rice bran oil with high acid value. Grain Oil Process 12:48–51

Moraes CA, Fernandes IJ, Calheiro D, Kieling AG, Brehm et al (2014) Review of the rice production cycle: by-products and the main applications focusing on rice husk combustion and ash recycling. Waste Manage Res 32:1034–1048

Nardini W, Dijkstara BW (1999) α/ β hydrolase fold enzymes: the family keeps growing. Curr Opin Struct Biol 9(6):732–737

Noor AF, Tajul AY (2011) Producing high quality edible oil by using eco-friendly technology: a review. Adv J Food Sci Technol 3(4):317–326

Park SY, Kim JT, Kang SG et al (2007) A new esterase showing similarity to putative dienelactone hydrolase from a strict marine bacterium, Vibrio sp. GMD509. Appl Microbiol Biotechnol 77(1):107–115

Pestana-Bauer VR, Zambiazi RC, Mendonça CRB, BeneitoCambra M, Ramis-Ramos G (2012) γ-Oryzanol and tocopherol contents in residues of rice bran oil refining. Food Chem 134(3): 1479–1483

Phongthai S, Rawdkuen S (2020) Fractionation and characterization of antioxidant peptides from rice bran protein hydrolysates stimulated by in vitro gastrointestinal digestion. Cereal Chem 97(2):316–325

Prabhakar JV, Venkatesh KVL (1986) A simple chemical method for stabilization of rice bran. J Am Oil Chem Soc 63(5):644–646

Proctor A, Browan DJ (1996) Ambient–temperature extraction of rice bran oil with hexane and isopropanol. J Am Oil Chem Soc 73:811–813

Punia S, Kumar M, Siroha AK, Purewal SS (2021) Rice bran oil: emerging trends in extraction, health benefit, and its industrial application. Rice Sci 28(3):217–232

Qiao Z, Zhang YH, Zhang MW et al (2012) Effect of different treatment technologies on stabilization of rice bran. Guangdong Agricu Sci 39(15):90–93

Saunders RM (1985) Rice bran: composition and potential food uses. Food Rev Int 1:465–495

Shen WB, Zheng TQ, Zhai HQ (2002) The status and strategy of the research on the relationship between lipoxygenase-3 in rice (*Oryza sativa* L.) embryos and storability. Sci Agric Sin 35(9): 1139–1144

Spaggiari M, Dall'Asta C, Galaverna G, del Castillo Bilbao MD (2021) Rice bran by-product: from valorization strategies to nutritional perspectives. Foods 10:85

Sreenarayanan VV, Chattopadhyay PK (1985) Rice bran stabilization by dielectric heating. J Food Process Preserv:89–98

Umate P (2011) Genome-wide analysis of lipoxygenase gene family in Arabidopsis and rice. Plant Signal Behav 6(3):335–338

Wang P (2006) Fine Mapping of embryo LOX-3 gene in rice (Oryza sativa L.). Master's Thesis. Nanjing: Nanjing Agricultural University: 2-4. (in Chinese)

Wang YH, Yang B, Yang JG (2004) Study on new phospholipase for enzymatic degumming of rice bran oil. China Oils 29(12):24–26

Wu XJ, Li F, Wu W (2020) Effects of rice bran rancidity on the oxidation and structural characteristics of rice bran protein. LWT-Food Sci Technol 120:108943
Xie Y (2013) Study on the preparation and refining process of rice bran oil. Master's Thesis. Central South University of Forestry and Technology, Changsha
Zheng LN, Liu YL, Wang DM (2015) Influence of storage conditions of rice bran on quality and bleaching effect of crude rice bran oil. China Oils Fats 40(4):42–46. (in Chinese)
Zhou T (2014) The fully utilization of rice bran, Master's thesis. Fujian Agriculture and Forestry University, Fuzhou

Chapter 5
Extraction and Identification of Rice Polyphenols and Related Health Benefits

Yu Zhang, Junda Chen, Sen Li, Kai Huang, and Xiao Guan

Abstract Cereal grains are important sources of dietary polyphenols, which contribute greatly to people's health. In this part, after a brief introduction of rice polyphenols, the extraction methods, as well as the components of rice polyphenols, are introduced and then the health benefits such as antioxidant, anti-obesity, and anticancer activity were discussed, after that, the future application of rice polyphenols are looked into the distance.

Keywords Polyphenols · Extraction · Health benefits · Antioxidant activities

Abbreviations

BHT	2, 6-di-tert-butylp-cresol
CVD	Cardiovascular disease
DPPH	2,2-Diphenyl-1-picrylhydrazyl
DW	Dry matter
GAE	Gallic acid equivalents
LDL-C	Low-density lipoprotein-cholesterol
ORAC	Oxygen Radical Absorbance Capacity
P3G	Peonidin-3-glucoside
TE	Trolox equivalents
TPC	Total phenolic acids

Y. Zhang · J. Chen · S. Li · K. Huang · X. Guan (✉)
School of Health Science and Engineering, Shanghai, China

J. Tian et al. (eds.), *Science of Rice Chemistry and Nutrition*,
https://doi.org/10.1007/978-981-99-3224-5_5

5.1 Introduction

Rice (*Oryza sativa* L.) is one of the major staple foods in most countries. Rice grains contain rich nutrients, such as carbohydrate (starch 75–80%), protein (about 8%), lipid (about 3%), and vitamin B complexes, and thus play an important role in providing energy and nutrients to the human body (Verma and Srivastav 2020; Saleh et al. 2019). Fig. 5.1 shows the structure of a rice grain, which mainly composes of rice bran, endosperm, and germ (Juliano and Tuaño 2019). Rice grains usually experience a series of processing steps, including drying, milling, polishing, and packaging before consumption (Zhang et al. 2023). Milling is one of the most important steps of rice processing, which mainly includes two steps. The first step of milling aims to remove the husk of rice grain to obtain brown rice grain, while the second step of milling aims to remove the bran of the brown rice grain to obtain polished or white rice (Saleh et al. 2019). Other than carbohydrates mainly existing in endosperm, a rich source of bioactive components, such as proteins, fatty acids, dietary fiber, and polyphenols exist in rice bran, which represents 5–8% of the total rice grain (Shao and Bao 2015). Though being considered as the byproduct of rice processing, rice bran has caught increasing attention due to its rich bioactive components and health benefits.

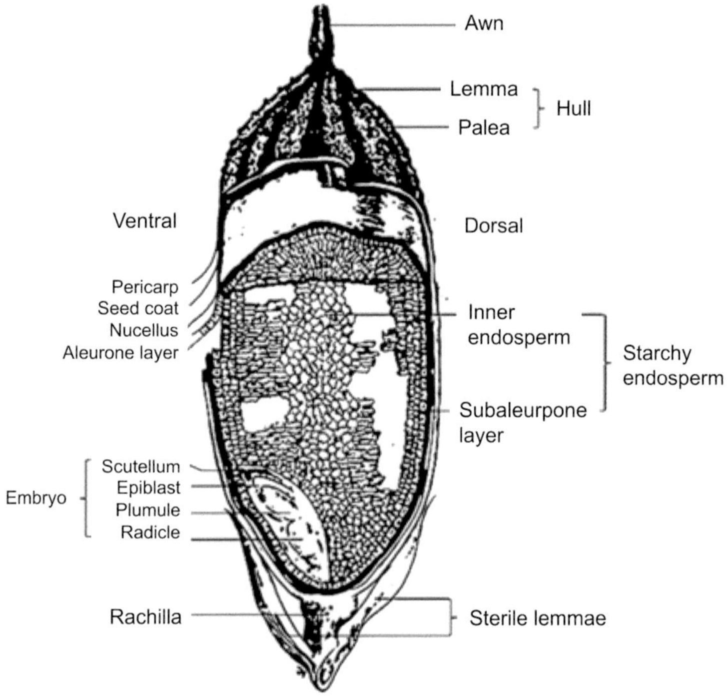

Fig. 5.1 The structure of a rice grain (Reprinted with permission from Juliano and Tuaño 2019)

A number of studies have shown that consuming whole-grain rice exhibited stronger antioxidant, anti-inflammatory, anti-hyperlipidemic effects than de-bran rice (Sohail et al. 2017), which indicated the health benefits of the bioactive components in the rice bran. The US Dietary Guidelines and China Dietary Guidelines also recommend the consumption of whole-grain foods (Curtain and Grafenauer 2019). This brings more attention to investigating the bioactivities of rice bran components, including polyphenols. Thus, this chapter mainly discusses the extraction methods, the composition of polyphenols from different rice varieties, and their related health benefits.

5.2 Extraction and Analysis of Rice Polyphenols

5.2.1 Extraction Methods

Cereal grains are important sources of dietary polyphenols, which contribute greatly to people's health. Different extraction methods have been widely applied in extracting polyphenols from different varieties of rice, including solvent extraction, supercritical fluid extraction, and ultrasound- and microwave-assisted extraction, (Table 5.1). Among these, solvent extraction is the most widely used method due to its effectiveness and ease of use. The yield of rice polyphenol mainly depends on the variety of rice, degree of processing, extraction solvent, time, and temperature. Methanol, ethanol, acetone, water, and their combinations are commonly used in solvent extraction of rice polyphenols. Methanol with a proportion of HCl is the most widely used solvent, followed by acetone and ethanol (Ti et al. 2014a, 2014b). Generally, pigmented rice contains higher contents of polyphenols than non-pigmented rice. Rocchetti et al. (2017) used 0.1% formic acid in 70% methanol assisted with ultrasound to extract polyphenols from rice flour. Their results showed that black rice flour contained the highest level of polyphenols (147.9 mg (gallic acid equivalents) GAE/100 g), followed by brown rice flour (35.9 mg GAE/100 g) and white rice flour (18.3 mg GAE/100 g). Similarly, Callcott et al. (2018) used acetone/water/acetic acid 70:29.5:0.5 (v/v/v) to extract polyphenols in white, brown, red, and purple rice. Results showed that the total anthocyanin content and proanthocyanidins content of purple rice were 25.63 ± 2.06 mg/100 g cyanidin-3-glucoside (C3G) equivalents and 325.2 ± 21.81 mg/100 g catechin equivalents (CE), respectively. While 4 brown rice varieties did not show detectable anthocyanins or proanthocyanidins. Other than the color of rice, different parts of rice grains contain varied levels of polyphenols. Huang & Ng used solvent extraction (80% methanol) to compare the content of polyphenols in 11 commercial rice varieties in Taiwan (Huang and Ng 2012). Their results showed that the order of polyphenols content was as follows: rice bran (1.29–14.28 g GAE/kg) > rice husk (0.61–1.45 g GAE/kg) > brown rice (0.24–1.78 g GAE/kg) > polished rice (0.06–0.34 g GAE/kg). This was in line with a basic finding that polyphenols mainly exist in the bran of rice grains.

Table 5.1 Extraction methods and content of polyphenols from rice (drawn by the author)

Types of rice	Solvent	Extraction methods	Content of polyphenols	Reference
White-milled long-grain rice	Methanol	Solvent extraction	125.7 mg GAE/g dry extract	Igoumenidis et al. (2016))
Six cultivars of Thai germinated brown rice	Methanol	Cold plasma treatment	356–786 μg GAE/g	Yodpitak et al. (2019)
Five indica rice varieties	Methanol/HCl 85:15 (v/v)	Solvent extraction and alkaline hydrolysis	Free phenolic content in rice bran: 647.2–777.9 mg GAE/100 g. bound phenolic content in rice bran: 91.1–126.8 mg GAE/100 g. TPC: 754.9–895.0 mg GAE/100 g.	Ti et al. (2014a)
Fresh brown rice, Tianyou 998	Methanol/HCl 85:15 (v/v)	Solvent extraction and alkaline hydrolysis	TPC: 254.9–284.1 GAE/100 g DW. Free phenolic content: 158.6–187.5 mg GAE/100 g DW during the germination process. Bound phenolic content: 90.3–106.6 mg GAE/100 g DW during the germination process.	Ti et al. (2014b)
Eleven commercial rice varieties in Taiwan	80% methanol	Solvent extraction	TPC: Rice bran (1.29–14.28 g GAE/kg); rice husk (0.61–1.45 g GAE/kg); brown rice (0.24–1.78 g GAE/kg); polished rice (0.06–0.34 g GAE/kg).	Huang and Ng (2012)
Artemide rice (Italy)	80% methanol	Solvent extraction assisted with ultrasound	TPC: 10.9 g of CE/kg rice.	Papillo et al. (2018)
Fifteen rice genotypes	80% methanol	Solvent extraction and alkaline hydrolysis	Pigmented genotypes: Soluble-free TPC (0.79–51.86 mg GAE/100 g); soluble-conjugated TPC (4.52–40.83 mg GAE/100 g); insoluble-bound TPC (33.17–62.55 mg GAE/100 g).	Shao et al. (2018)
Brown, Indian white basmati rice, and Thai jasmine rice	80% methanol	Solvent extraction assisted with ultrasound	Brown rice (630.2 mg/kg DW); Thai jasmine rice (175.6 mg/kg DW); Indian white basmati rice (171.3 mg/kg DW).	Chmiel et al. (2018)

(continued)

Table 5.1 (continued)

Types of rice	Solvent	Extraction methods	Content of polyphenols	Reference
Rice flour	0.1% formic acid in 70% methanol	Solvent extraction assisted with ultrasound	White rice flour (18.3 mg GAE/100 g); brown rice flour (35.9 mg GAE/100 g); black rice flour (147.9 mg GAE/100 g)	Rocchetti et al. (2017)
Spain rice	10–90% ethyl acetate in methanol	Microwave-assisted extraction	Exotic rice grains (68.41 μg/g); aromatic rice grains (6.57–9.39 μg/g); regular rice (13.30–17.62 μg/g); whole short grain rice (34.40 μg/g)	Setyaningsih et al. 2015
Wild rice	Ethanol	Solvent extraction	Jingzhou wild rice (4.8 mg GAE/g); Huai'an wild rice (2.1 mg GAE/g); red rice (1.4 mg GAE/g); white rice (1.3 mg GAE/g)	Chu et al. (2018)
Artemide (one of the Italian pigmented rice)	50% ethanol	Milling, stirring, and ultrasound-assisted extraction	407 mg of CE/g of dry extract, corresponding to 10.9 g/kg of rice	Papillo et al. (2018)
White, brown, red, and purple rice	Acetone/ water/acetic acid 70:29.5: 0.5 (v/v/v)	Solvent extraction	Purple rice: Total anthocyanin content of 25.63 ± 2.06 mg/100 g C3G equivalents; total proanthocyanidins content of 325.2 ± 21.81 mg/100 g CE. Four brown rice varieties: Did not contain detectable anthocyanins or proanthocyanidins.	Callcott et al. (2018)
Seven pigmented rice	70% acetone	Solvent extraction and alkaline hydrolysis	Light black rice: 1.204–1.437 mg GAE/g. black-colored rice: 1.827 mg GAE/g. red-colored rice: 3.743–4.616 mg GAE/g.	Bhat and Riar (2017)
Rice bran (Hangzhou, Zhejiang, China)	Aqueous glycerol	Immersion (140 rpm)	549.6 ± 8.05 mg GAE/100 g DW	Aalim et al. (2019)
Black rice (Hefei, Anhui, China)	NaOH	Water extraction, membrane separation, and resin adsorption	0.86 g anthocyanins/ 100 g black rice	He et al. 2017

(continued)

Table 5.1 (continued)

Types of rice	Solvent	Extraction methods	Content of polyphenols	Reference
Fresh rice (*O. sativa* japonica, Kaohsiung 145)	H_2SO_4 and NaOH pretreatment, ether/ethyl acetate 1:1 (v:v)	Acid-alkaline pretreatment heated under high-pressure, solvent extraction	3428 mg GAE/100 g husk	Lieu et al. (2018)

C3G Cyanidin-3-glucoside, *CE* catechin equivalents, *DW* dry weight, *GAE* gallic acid equivalents, *TPC* total polyphenol content

Assisted extraction methods, such as ultrasound, heat treatment, and high pressure, might help to improve the yield of polyphenols to some extent, however, they are not the decisive factors. Many studies used 80% methanol to extract polyphenols from rice of different varieties in different places (Papillo et al. 2018; Chmiel et al. 2018). Huang and Ng (2012) did not apply ultrasound-assisted extraction, the content of polyphenols in rice bran was about 1.29–14.28 g GAE/kg. Although Papillo et al. and Chmiel et al. both applied ultrasound-assisted extraction, their yields of rice polyphenols showed great variation (TPC of Artemide rice from Italy: 10.9 g of CE/kg rice; TPC of brown rice from Thailand: 630.2 mg/kg DW; TPC of white rice from India: 171.3 mg/kg DW) (Papillo et al. 2018; Chmiel et al. 2018). Similarly, Setyaningsih et al. (2015) optimized microwave-assisted extraction to extract rice polyphenols using 10–90% ethyl acetate in methanol. The contents of rice polyphenols ranged from 6.57 to 68.41 μg/g depending on different varieties. Although using a microwave, such level was not comparable to that of some reported studies (Huang and Ng 2012). Thus, the variety of rice grains plays a key role in the content of rice polyphenols (Table 5.2).

5.2.2 *Components of Rice Polyphenols*

Polyphenols can be mainly classified into free, conjugated, and bound forms depending on their binding with food matrix. Free polyphenols exist in the plant cell vacuole and their contents in rice have been widely investigated (Zhang et al. 2020a). Conjugated polyphenols are covalently bound to sugars, fatty acids, or other components, and can be directly extracted from aqueous solvents (Zhang et al. 2020b). Being different from free and conjugated polyphenols, bound polyphenols are reported to link to macromolecules of cell wall in food matrix via covalent bonding, hydrogen bonding, and hydrophobic interactions. Bound polyphenols include some classes of polyphenols, such as high molecular weight proanthocyanidins, flavonoids, and phenolic acids, and can be released through hydrolysis, acid, alkaline, and enzymatic hydrolysis. Rice bran also contains rich bound polyphenols, and thus, acid or alkaline hydrolysis is widely applied to extract

Table 5.2 Identification and structure of rice polyphenols (drawn by the author)

Types of rice	Analysis and detection	Composition	References
Unprocessed and extruded milled fractions of black rice	HPLC	Gallic acid, chlorogenic acid, vanillic acid, caffeic acid, syringic acid, p-coumaric acid, and ferulic acid	Ti et al. (2015)
Rice grains with different colors	HPLC-DAD	Phenolic acids (gallic acid, protocatechuic acid, 4-hydroxybenzoic acid, vanillic acid, caffeic acid, syringic acid, p-coumaric acid, ferulic acid, sinapic acid and cinnamic acid); flavanols (catechin and epicatechin); flavonols (myricetin, quercetin, kaempferol, quercetin-3-O-glucoside, hyperoside, and rutin); flavones (luteolin and apigenin)	Irakli et al. (2012)
Yellow grain rice mutant	UHPLC-DAD-ESI-Q-TOF-MS	Flavonoids (isoorientin, isoorientin 2″-O-glucoside, vitexin 2″-O-glucoside, isovitexin, isoscoparin 2″-O-glucoside and isoscoparin)	Kim et al. (2018)
Eight varieties of Australian-grown rice	UHPLC online-ABTS system	Phenolic acids (gallic acid, ferulic acid, vanillic acid, o-coumaric acid, p-coumaric acid, syringic acid, protocatechuic acid); flavanols (catechin); anthocyanins (C3G, P3G)	Rao et al. (2018)
Rice bran samples from Thailand	HPLC	Phenolic acids (protocatechuic acid, vanillic acid, p-coumaric acid, ferulic acid, and sinapic acid); flavonoids (rutin, myricetin, and quercetin-3-glucoside), anthocyanins (C3G and P3G)	Peanparkdee et al. (2019)
Jyothi red rice and Burma black rice (India)	HPLC	Phenolic acids (caffeic acid, ferulic acid, vanillic acid, and protocatechuic acid); flavanols (catechin)	Sapna et al. (2019)
Rice bran (Hangzhou)	HPLC	Phenolic acids (ferulic acid, p-coumaric acid, chlorogenic acid, caffeic acid, and syringic acid); flavonols (rutin)	Aalim et al. (2019)
6 varieties of Thai rice	HPLC	Pigmented rice: Ferulic acid, p-coumaric acid, vanillic acid, and protocatechuic acid	Ratseewo et al. (2019)
15 varieties of rice from China	HPLC	Black rice: Protocatechuic acid, vanillic acid, conjugated protocatechuic and 2,5-dihydroxybenzoic acids, bound protocatechuic and vanillic acids; Red rice: Conjugated protocatechuic and 2,5-dihydroxybenzoic acids,	Shao et al. (2018)

(continued)

Table 5.2 (continued)

Types of rice	Analysis and detection	Composition	References
		bound ferulic, p-coumaric, and 2,5-dihydroxybenzoic acids	
Artemide rice (Italy)	HPLC-DAD	Phenolic acids (protocatechuic acid, ferulic acid, vanillic acid, coumaric acid, sinapic acid); anthocyanin (C3G, cyanidin-3-O-rutinoside, cyanidin-3,5-di-O-glucoside, P3G); flavonoids (quercetin and myricetin)	Papillo et al. (2018)
7 different traditional pigmented rice	HPLC-MS	A total of 40 compounds were identified: 6 types of phenolic acids, 6 types of flavonoids, 11 types of hydroxycinnamic acid derivatives, 7 types of hydroxybenzoic acid derivatives, 3 types of anthocyanins and 7 types of flavonoid glucosides.	Bhat and Riar (2017)
Brown, India white basmati rice and Thai jasmine rice	HPLC-DAD	Phenolic acids (protocatechuic acid, p-hydroxybenzoic acid, chlorogenic acid, caffeic acid, vanillic acid, isovanillic acid, syringic acid, p-coumaric acid, ferulic acid, sinapic acid, and isoferulic acid); protocatechuic aldehyde, p-hydroxybenzaldehyde, vanillin	Chmiel et al. (2018)
Fresh brown rice, Tianyou 998	HPLC	Phenolic acids (protocatechuic acid, chlorogenic acid, caffeic acid, syringic acid, ferulic acid, and coumaric acid) in the free and bound fractions of germinated brown rice	Ti et al. 2014b
5 varieties of indica rice	HPLC	Phenolic acids (gallic acid, protocatechuic acid, chlorogenic acid, caffeic acid, syringic acid, coumaric acid and ferulic acid)	Ti et al. (2014a)
Black rice	HPLC-electrospray ionization-mass spectrometry	Anthocyanins (C3G, P3G, cyanidin-3,5-diglucoside, cyanidin-3-rutinoside)	Hou et al. (2013)
11 commercial rice varieties in Taiwan	RP-HPLC	Phenolic acids (ferulic acid, p-coumaric acid, vanillic acid, and caffeic acid); flavanols (catechin)	Huang and Ng (2012)

C3G Cyanidin-3-glucoside, *P3G* peonidin-3-glucoside

bound polyphenols from rice (Zhang et al. 2020a). Lieu et al. (2018) used acid-alkaline pretreatment and ether/ethyl acetate extraction to release bound polyphenols and thus obtained a high yield of TPC (3428 mg GAE/100 g rice husk). Shao et al. (2018) compared the contents of free, conjugated, and bound polyphenols in 15 rice genotypes. Their results revealed that the content of bound polyphenols

(33.17–62.55 mg GAE/100 g) was higher than that of conjugated (4.52–40.83 mg GAE/100 g) and free polyphenols (0.79–51.86 mg GAE/100 g) in pigmented rice. Such results were consistent with some findings that cereal grains contained about ~60% of bound polyphenols, which was much higher than those in fruits and vegetables (~24%) (Zhang et al. 2020a). Various types of polyphenols have been detected in different varieties of rice, including phenolic acids, anthocyanin, flavonoids, and proanthocyanidins.

(a) Phenolic acids: Phenolic acids can be found in many plants, particularly in cereal grains and vegetables. They are types of aromatic acid compounds with a phenolic ring and an organic carboxylic acid function (C6-C1 skeleton) (Horvat et al. 2020). Due to the existence of phenolic rings and several hydroxyl groups, phenolic acids exhibit great antioxidant activities by stabilizing and delocating unpaired electrons. Ferulic acid, protocatechuic acid, vanillic acid, caffeic acid, syringic acid, and p-coumaric acid are the most common phenolic acids in rice and other cereal grains (Horvat et al. 2020). Among these, the levels of ferulic acid and p-coumaric acid in rice are generally higher than those of other phenolic acids (Huang and Ng 2012). Ferulic acid was found to be the most abundant phenolic acid in rice grains and was mainly distributed in rice bran. While, *p*-coumaric acid was present mainly in rice husk (Huang and Ng 2012). Chmiel et al. (2018) showed that the level of ferulic acid ranged from 3.4 to 84.0 mg/kg DW in non-pigmented rice grains, and 12.4–362.0 mg/kg DW in pigmented rice grains. *p*-coumaric acid, *p*-hydroxybenzoic acid, and syringic acid constituted about 15–17% of the phenolic fraction depending on the variety of rice. Also, both free and bound forms of phenolic acids were reported in rice grains (Shao et al. 2018). Maisuthisakul and Changchub (2014) showed that bound phenolic acids contributed to an important part of the cell wall structure of Thai rice grains. Besides, processing methods would affect the contents of phenolic acids in rice. Sapna et al. found that in red rice, plate milling increased the contents of caffeic acid and ferulic acid by 42% and 379%, whereas vanillic acid and protocatechuic acid decreased by 72% and 63%, compared to control (Sapna et al. 2019). Further, toasting caused an increase of protocatechuic acid and ferulic acid by 95% and 7%, respectively. Similar results were also observed in black rice as well. **(b) Anthocyanins:** anthocyanins belong to flavonoids synthesized through the phenylpropanoid pathway and contribute to the color of cereal grains and fruits (Saigo et al. 2020). The glycosides of cyanidin, delphinidin, malvidin, pelargonidin, and peonidin are the most commonly found anthocyanins in cereal grains and fruits. Peanparkdee et al. (2019) found that anthocyanins comprised 30% of the total quantifiable peaks in purple rice, among which C3G and peonidin-3-glucoside (P3G) accounted for approximately 20.29% and 9.45%. While, in red and non-pigmented brown rice, C3G and P3G were not detected. Similar results were also observed in rice bran samples from Thailand. Other than C3G and P3G, Papillo et al. also detected cyanidin-3-O-rutinoside, cyanidin-3,5-di-O-glucoside in 7 different traditional pigmented rice. Hou et al. (2013) also found cyanidin-3,5-diglucoside, cyanidin-3-rutinoside in black rice. **(c) Flavonoids:** Irakli et al. showed that rice grains with different colors also contained rich flavonoids, such as flavanols (catechin and epicatechin), flavonols (myricetin, quercetin, kaempferol, quercetin-3-O-glucoside, hyperoside, and rutin) and flavones (luteolin

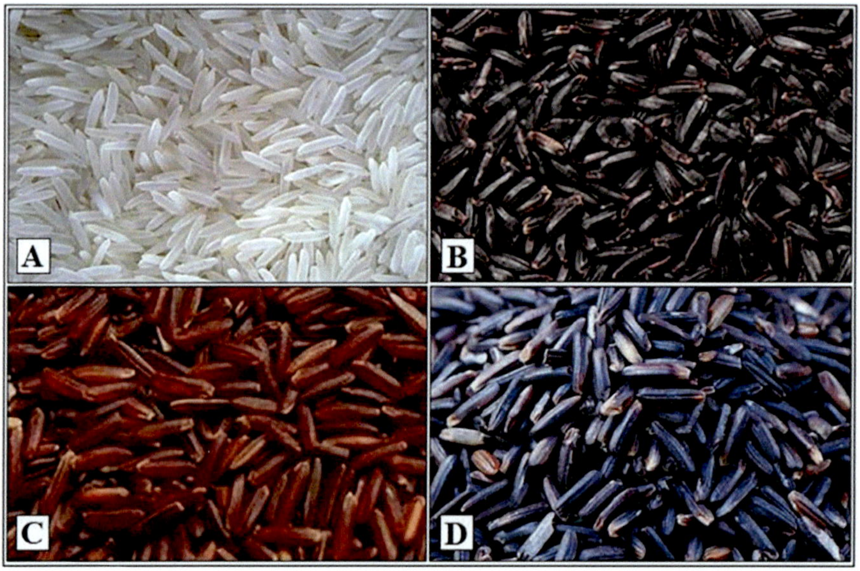

Fig. 5.2 Rice in different colors. (**a**) White rice; (**b**) black rice; (**c**) red rice, and (**d**) purple rice (Reprinted with permission from Verma and Srivastav 2020)

and apigenin) (Fig. 5.2) (Irakli et al. 2012). Similar results were observed in bran samples of four cultivars in a study by Peanparkdee et al. (2019), and in an Italian rice variety in a study by Papillo et al. (2018). Some uncommonly seen flavonoids (isoorientin, isoorientin 2″-O-glucoside, vitexin 2″-O-glucoside, isovitexin, isoscoparin 2″-O-glucoside, and isoscoparin) were detected in yellow grain rice mutant by Kim et al. (2018).

5.3 Health Benefits of Rice Polyphenols

5.3.1 Antioxidant Activities

Excessive free radicals not only accelerate oxidation in food system but also induce oxidative damage to cells. Rice bran or pigmented rice containing a considerable level of polyphenols have been widely investigated and were reported to exhibit strong antioxidant activities (Deng et al. 2013). **(a) in vitro studies-chemical:** many studies have shown that rice polyphenols in different varieties showed strong chemical antioxidant activities by DPPH, ATBS, FRAP, and ORAC assays (Butsat and Siriamornpun 2010). Shao et al. compared 15 varieties of rice and the ABTS values of samples ranged from 0.11 to 13.11 μM trolox equivalents (TE)/g, and the DPPH values ranged from 0.05 to 5.69 μM TE/g (Shao et al. 2018). Butsat and Siriamornpun (2010) showed that all rice fractions including bran, husk, brown rice, and milled rice exhibited the ability to scavenge DPPH radical at a rate higher than BHT at 0.2 mg/mL, but lower than ascorbic acid at 0.2 mg/mL. The FARP of different rice fractions ranged from 4.6 to 32.2 μmol $FeSO_4$/g. The varied radical scavenging activity depended on the amount of total polyphenol contents in different fractions. Also, pigmented rice generally showed higher antioxidant activity than non-pigmented rice, which might be due to the presence of three major polyphenols, proanthocyanidins, phenolic acids, and oryzanol (Verma and Srivastav 2020). Although these assays are quick to evaluate the radical scavenging activity of rice polyphenols, however, they are limited to reflect the antioxidant activities of rice polyphenols comprehensively. Some cellular and animal studies are also required. **(b) In vitro studies—cellular:** other than chemical antioxidant activities, many studies also investigated the antioxidant activities of rice polyphenols at cellular levels. Vargas et al. showed that red and black rice brans extracts exhibited strong free radical scavenging activities in SH-SY5Y cells (Vargas et al. 2018). Chiang et al. (2006) also showed that black rice extracts suppressed the levels of superoxide anions and reactive oxygen species, while increasing the levels of superoxide dismutase and catalase activities by 161.6% and 73.4% in HepG2 cells, respectively. Seo et al. (2013) also found that new red rice exhibited strong nitric oxide scavenging activity on RAW264.7 cells. **(c) In vivo studies:** Imam et al. investigated the antioxidant effects of white rice, brown rice, and germinated brown rice in a type II diabetic rats model and found that brown rice and germinated brown rice improved the total antioxidant status and reduced plasma glucose in rats (Imam et al. 2012). Esa et al. (2013) also showed similar findings and further revealed that germinated brown rice had a higher antioxidant activity (increased antioxidant enzyme activity) than that of white rice and brown rice in high-fat diet-induced rabbits. Brown rice and germinated brown rice generally showed stronger antioxidant activity than white rice in animal models (Wu et al. 2013), which might be due to a higher level of rice polyphenols.

5.3.2 *Anti-Obesity Activities*

The prevalence of obesity or overweight is increasing all over the world, especially in developed countries (Saleh et al. 2019). Obesity and hyperlipidemia are associated with a high risk for the development of several chronic diseases, such as metabolic syndrome and cardiovascular disease. Long-term use of drugs, such as statins, might induce liver function abnormalities and rhabdomyolysis (Zhang et al. 2017). Other than medical treatment, exercise, diet, and food components play important roles in therapeutics for hyperlipidemia and obesity. Many studies have shown that pigmented rice containing rich rice polyphenols exhibited anti-obesity and anti-hyperlipidemic effects. Since pancreatic lipase is a primary lipase enzyme that hydrolyzes dietary fat molecules to monoglycerides and free fatty acids, inhibiting its activity is effective to inhibit the absorption of dietary fat and thus to improve obesity (Liu et al. 2020). Some drugs are invented as pancreatic lipase inhibitors, such as orlistat. Lim et al. (2014) showed that germinated brown rice extracts using hexane showed the highest inhibitory effect (13.58%) at 200 μg/mL. Aalim et al. (2019) also showed similar results that aqueous glycerol extract from rice bran could inhibit the activity of pancreatic lipase (12.4%). Imam et al. further compared the effects of white rice, brown rice, and germinated brown rice on hypercholesterolemic rats, and found that brown rice and germinated brown rice ameliorated cardiovascular disease (CVD) risk by modulating lipid metabolism and oxidative stress, while white rice did not show positive effects. Similar findings were also observed in a study by Lim et al. (2016) which showed that 50% or 100% germinated brown rice significantly decreased body weight gains and improved lipid profile in obese rats. Ho et al. (2013) revealed that germinated brown rice inhibited adipocyte differentiation by reducing lipid accumulation and downregulating the mRNA expression of adipogenic genes. Besides, Wilson et al. compared the cholesterol-lowering effects of different components of rice bran, including trans-ferulic acid, γ-oryzanol, and rice bran oil. Results showed that the levels of low-density lipoprotein-cholesterol (LDL-C) and total plasma cholesterol were significantly lowered in groups fed with diet with trans-ferulic acid (24 and 22% reduction, respectively) (Wilson et al. 2007). Diet with γ-oryzanol (77 and 70%, respectively) and rice bran oil (70 and 64%, respectively) also served positive effects on regulating plasma cholesterol profile. These studies collectively revealed that brown rice and germinated brown rice can be suggested as functional foods for the prevention or treatment of obesity.

5.3.3 Anti-Cancer Activity

Cancer is one of the leading causes of death all over the world, especially in low- and middle-income countries (Siegel et al. 2021). It is estimated that only 5–10% of all cancers are due to genetic factors, while 90–95% are directly or indirectly due to lifestyles and dietary habits (Anand et al. 2008). Healthier diets play a key role in the reduction of cancer risks. Many studies have shown the beneficial effects of whole-grain rice or rice phytochemicals in the prevention of cancer, such as breast cancer, colorectal cancer, liver cancer, and ovarian cancer. Bioactive components from rice prevented different types of cancer in different mechanisms.

Nam et al. (2005) investigated the anti-cancer effects of bran from different pigmented and non-pigmented rice varieties. Their results revealed that 70% ethanol extract of bran from pigmented rice exhibited greater inhibitory effects on lymphoblastoid B cells compared with bran from non-pigmented rice. Li et al. (2011) also investigated the effects of brown rice, rice bran, and white rice on 1,2-dimethylhydrazine-induced colon carcinogenesis rats. Their results showed that brown rice and rice bran significantly reduced the cyclooxygenase-2 expression of the middle colon in rats compared to white rice. Wunjuntuk et al. (2016) further showed that regular intake of a parboiled germinated brown rice-supplemented diet significantly ameliorated liver inflammation and fibrosis in rats compared with brown rice- or white rice-supplemented diets. They explained that the anti-inflammatory and anti-cancer effects of germinated brown rice might be due to the combined effects of rice polyphenols and other phytochemicals, such as ferulic acid, coumaric acid, γ-oryzanol, γ-tocotrienol, and γ-aminobutyric acid. Researchers also reported that a type of rice bran polyphenols-cycloartenyl ferulate exhibited strong inhibitory effects on human colorectal adenocarcinoma SW480. They explained that the health benefits of whole grains might not be completely associated with their fiber content, but also be associated with others bioactive components or their combination effects. Upanan et al. (2019) showed that proanthocyanidin-rich fractions from red rice's germ and bran inhibited the proliferation of HepG2 cells effectively without affecting p53. Pintha et al. also showed that proanthocyanidin-rich fractions from red rice significantly reduced the migration and invasion of MDA-MB-231 human breast cancer cells by reducing the nuclear factor NF-κB. They also found that proanthocyanidin-rich fractions from red rice degraded the extracellular matrix-associated protein and collagenase activity (Pintha et al. 2015). Limtrakul et al. (2015) also showed that 50 μg/mL of proanthocyanidin reduced the activities of AP-1 and NF-κB in LPS activated RAW264.7 cells. The anti-cancer activities of rice extracts varied a lot from different varieties. Forster et al. (2013) investigated the inhibitory effects of 7 varieties of rice bran on the growth of colorectal cancer cells. Their results showed that the inhibitory effects of rice bran extracts ranged from 0 to 99%. A positive correlation was observed between the total polyphenols and the reduction of colorectal cancer cells. Thus, rice polyphenol-enriched fractions exhibited strong anti-cancer effects and can be suggested as a novel dietary supplement for cancer prevention.

5.4 Concluding Remarks and Future

With increasing attention being paid to the consumption of whole grains, bioactive components of rice bran, such as polyphenols worth further and deeper investigation. Some processing methods can increase the content of polyphenols in rice, such as germination, microwave processing, and ultrasound processing. Increasing the content of polyphenols and diversifying the components of polyphenols from rice would also improve the health benefits of whole-grain rice. Future studies should clarify the mechanisms of the health benefits of rice polyphenols and their synergistic effects with other bioactive components, such as dietary fiber, vitamins, and proteins.

References

Aalim H, Belwal T, Jiang L, Huang H, Meng X, Luo Z (2019) Extraction optimization, antidiabetic and antiglycation potentials of aqueous glycerol extract from rice (*Oryza sativa* L.), vol 103. LWT-Food Sci Technol, bran, pp 147–154

Anand P, Kunnumakara AB, Sundaram C, Harikumar KB, Tharakan ST, Lai OS, Sung BY, Aggarwal BB (2008) Cancer is a preventable disease that requires major lifestyle changes. Pharm Res 25:2097–2116

Bhat FM, Riar CS (2017) Characterizing the pigmented traditional rice cultivars grown in temperate regions of Kashmir (India) for free and bound phenolics compounds and in vitro antioxidant properties. J Cereal Sci 76:253–262

Butsat S, Siriamornpun S (2010) Antioxidant capacities and phenolic compounds of the husk, bran and endosperm of Thai rice. Food Chem 119:606–613

Chu MJ, Liu XM, Yan N, Wang FZ, Du YM, Zhang ZF (2018) Partial purification, identification, and quantitation of antioxidants from wild rice (*Zizania latifolia*). Molecules 23(11):2782

Chiang AN, Wu HL, Yeh HI, Chu CS, Lin HC, Lee WC (2006) Antioxidant effects of black rice extract through the induction of superoxide dismutase and catalase activities. Lipids 41:797–803

Curtain F, Grafenauer S (2019) Health star rating in grain foods does it adequately differentiate refined and whole grain foods? Nutrients 11(2):415

Callcott ET, Santhakumar AB, Strappe P, Luo J, Blanchard CL (2018) Polyphenols from Australian-grown pigmented red and purple rice inhibit adipocyte differentiation. J Cereal Sci 81:140–146

Chmiel T, Saputro IE, Kusznierewicz B, Bartoszek A (2018) The impact of cooking method on the phenolic composition, total antioxidant activity and starch digestibility of rice (*Oryza sativa* L.). J Food Proc Preserv 42:e13383

Deng GF, Xu XR, Zhang Y, Li D, Gan RY, Li HB (2013) Phenolic compounds and bioactivities of pigmented rice. Crit Rev Food Sci Nutr 53:296–306

Esa NM, Kadir KKA, Amom Z, Azlan A (2013) Antioxidant activity of white rice, brown rice and germinated brown rice (in vivo and in vitro) and the effects on lipid peroxidation and liver enzymes in hyperlipidemic rabbits. Food Chem 141:1306–1312

Forster GM, Raina K, Kumar A, Kumar S, Agarwal R, Chen MH, Bauer JE, McClung AM, Ryan EP (2013) Rice varietal differences in bioactive bran components for inhibition of colorectal cancer cell growth. Food Chem 141:1545–1552

He S, Lou Q, Shi J, Sun H, Zhang M, Li Q (2017) Water extraction of anthocyanins from black rice and purification using membrane separation and resin adsorption. J Food Proc Preserv 41: e13091

Horvat D, Simic G, Drezner G, Lalic A, Ledencan T, Tucak M, Plavsic H, Andric L, Zdunic Z (2020) Phenolic acid profiles and antioxidant activity of major cereal crops. Antioxidants 9(6): 527

Huang SH, Ng LT (2012) Quantification of polyphenolic content and bioactive constituents of some commercial rice varieties in Taiwan. J Food Compos Anal 26:122–127

Hou Z, Qin P, Zhang Y, Cui S, Ren G (2013) Identification of anthocyanins isolated from black rice (*Oryza sativa* L.) and their degradation kinetics. Food Res Int 50:691–697

Ho JN, Son ME, Lim WC, Lim ST, Cho HY (2013) Germinated brown rice extract inhibits adipogenesis through the down-regulation of adipogenic genes in 3t3-l1 adipocytes. Plant Food Hum Nutr 68:274–278

Irakli MN, Samanidou VF, Biliaderis CG, Papadoyannis IN (2012) Simultaneous determination of phenolic acids and flavonoids in rice using solid-phase extraction and RP-HPLC with photodiode array detection. J Sep Sci 35:1603–1611

Imam MU, Musa SN, Azmi NH, Ismail M (2012) Effects of white rice, brown rice and germinated brown rice on antioxidant status of type 2 diabetic rats. Int J Mol Sci 13:12952–12969

Igoumenidis PE, Lekka EG, Karathanos VT (2016) Fortification of white milled rice with phytochemicals during cooking in aqueous extract of Mentha spicata leaves. An adsorption study. LWT-Food Sci Technol 65:589–596

Juliano BO, Tuaño APP (2019) Gross structure and composition of the rice grain. Rice:31–53

Kim B, Woo S, Kim MJ, Kwon SW, Lee J, Sung SH, Koh HJ (2018) Identification and quantification of flavonoids in yellow grain mutant of rice (*Oryza sativa* L.). Food Chem 241:154–162

Liu TT, Liu XT, Chen QX, Shi Y (2020) Lipase inhibitors for obesity: a review. Biomed Pharmacother 128:110314

Lim SM, Goh YMB, Kuan W, Loh SP (2014) Effect of germinated brown rice extracts on pancreatic lipase, adipogenesis and lipolysis in 3T3-L1 adipocytes. Lipids Health Disease 13: 3379

Lim SM, Goh YM, Mohtarrudin N, Loh SP (2016) Germinated brown rice ameliorates obesity in high-fat diet induced obese rats. BMC Complement Altern Med 16:2371

Lieu YS, Chang YC, Chen HH (2018) Synthesis of silver nanoparticles by using rice husk extracts prepared with acid–alkali pretreatment extraction process. J Cereal Sci 82:106–112

Li SC, Chou TC, Shih CK (2011) Effects of brown rice, rice bran, and polished rice on colon carcinogenesis in rats. Food Res Int 44:209–216

Limtrakul P, Yodkeeree S, Pitchakarn P, Punfa W (2015) Suppression of inflammatory responses by black rice extract in raw 264.7 macrophage cells via downregulation of NF-kB and AP-1 signaling pathways. Asian Pac J Cancer Prev 16:4277–4283

Maisuthisakul P, Changchub L (2014) Effect of extraction on phenolic antioxidant of different Thai rice (*Oryza Sativa* L.) genotypes. Int J Food Prop 17:855–865

Nam SH, Choi SP, Kang MY, Koh HJ, Kozukue N, Friedman M (2005) Bran extracts from pigmented rice seeds inhibit tumor promotion in lymphoblastoid B cells by phorbol ester. Food Chem Toxicol 43:741–745

Papillo VA, Locatelli M, Travaglia F, Bordiga M, Garino C, Arlorio M, Coisson JD (2018) Spray-dried polyphenolic extract from Italian black rice (*Oryza sativa* L., var. Artemide) as new ingredient for bakery products. Food Chem 269:603–609

Peanparkdee M, Patrawart J, Iwamoto S (2019) Effect of extraction conditions on phenolic content, anthocyanin content and antioxidant activity of bran extracts from Thai rice cultivars. J Cereal Sci 86:86–91

Pintha K, Yodkeeree S, Limtrakul P (2015) Proanthocyanidin in red rice inhibits mda-mb-231 breast cancer cell invasion via the expression control of invasive proteins. Biol Pharm Bull 38: 571–581

Rocchetti G, Chiodelli G, Giuberti G, Masoero F, Trevisan M, Lucini L (2017) Evaluation of phenolic profile and antioxidant capacity in gluten-free flours. Food Chem 228:367–373

Rao S, Callcott ET, Santhakumar AB, Chinkwo KA, Vanniasinkam T, Luo J, Blanchard CL (2018) Profiling polyphenol composition and antioxidant activity in Australian-grown rice using UHPLC online-ABTS system. J Cereal Sci 80:174–179

Ratseewo J, Warren FJ, Siriamornpun S (2019) The influence of starch structure and anthocyanin content on the digestibility of Thai pigmented rice. Food Chem 298:124949

Saleh ASM, Wang P, Wang N, Yang L, Xiao Z (2019) Brown rice versus white rice: nutritional quality, potential health benefits, development of food products, and preservation technologies. Compr Rev Food Sci Food Saf 18:1070–1096

Shao Y, Bao J (2015) Polyphenols in whole rice grain: genetic diversity and health benefits. Food Chem 180:86–97

Shao Y, Hu Z, Yu Y, Mou R, Zhu Z, Beta T (2018) Phenolic acids, anthocyanins, proanthocyanidins, antioxidant activity, minerals and their correlations in non-pigmented, red, and black rice. Food Chem 239:733–741

Sohail M, Rakha A, Butt MS, Iqbal MJ, Rashid S (2017) Rice bran nutraceutics: a comprehensive review. Crit Rev Food Sci Nutr 57:3771–3780

Setyaningsih W, Saputro IE, Palma M, Barroso CG (2015) Optimization and validation of the microwave-assisted extraction of phenolic compounds from rice grains. Food Chem 169:141–149

Saigo T, Wang T, Watanabe M, Tohge T (2020) Diversity of anthocyanin and proanthocyanin biosynthesis in land plants. Curr Opin Plant Biol 55:93–99

Sapna I, Kamaljit M, Priya R, Jayadeep PA (2019) Milling and thermal treatment induced changes on phenolic components and antioxidant activities of pigmented rice flours. J Food Sci Technol-Mysore 56:273–280

Seo WD, Kim JY, Song Y-C, Cho J-H, Jang KC, Han S-I, Ra J-E, Oh S-H, Kang H-J, Kim B-J, Baek N-I, Jeong R-H, Nam MH (2013) Comparative analysis of physicochemicals and antioxidative properties in new red rice (*Oryza sativa* L. cv. Gunganghongmi). J Crop Sci Biotechnol 16:63–68

Siegel RL, Miller KD, Fuchs HE, Jemal A (2021) Cancer statistics, 2021. CA Cancer J Clin 71:7–33

Ti H, Zhang R, Zhang M, Wei Z, Chi J, Deng Y, Zhang Y (2015) Effect of extrusion on phytochemical profiles in milled fractions of black rice. Food Chem 178:186–194

Ti H, Li Q, Zhang R, Zhang M, Deng Y, Wei Z, Chi J, Zhang Y (2014a) Free and bound phenolic profiles and antioxidant activity of milled fractions of different indica rice varieties cultivated in southern China. Food Chem 159:166–174

Ti H, Zhang R, Zhang M, Li Q, Wei Z, Zhang Y, Tang X, Deng Y, Liu L, Ma Y (2014b) Dynamic changes in the free and bound phenolic compounds and antioxidant activity of brown rice at different germination stages. Food Chem 161:337–344

Upanan S, Yodkeeree S, Thippraphan P, Punfa W, Wongpoomchai R, Limtrakul P (2019) The proanthocyanidin-rich fraction obtained from red rice germ and bran extract induces hepg2 hepatocellular carcinoma cell apoptosis. Molecules 24(4):813

Vargas CG, da Silva Junior JD, Rabelo TK, Moreira JCF, Gelain DP, Rodrigues E, Augusti PR, Rios AO, Flores SH (2018) Bioactive compounds and protective effect of red and black rice brans extracts in human neuron-like cells (SH-SY5Y). Food Res Int 113:57–64

Verma DK, Srivastav PP (2020) Bioactive compounds of rice (*Oryza sativa* L.): review on paradigm and its potential benefit in human health. Trends Food Sci Technol 97:355–365

Wilson TA, Nicolosi RJ, Woolfrey B, Kritchevsky D (2007) Rice bran oil and oryzanol reduce plasma lipid and lipoprotein cholesterol concentrations and aortic cholesterol ester accumulation to a greater extent than ferulic acid in hypercholesterolemic hamsters. J Nutr Biochem 18:105–112

Wunjuntuk K, Kettawan A, Rungruang T, Charoenkiatkul S (2016) Anti-fibrotic and anti-inflammatory effects of parboiled germinated brown rice (*Oryza sativa* 'KDML 105') in rats with induced liver fibrosis. J Funct Foods 26:363–372

Wu FF, Yang N, Toure A, Jin ZY, Xu XM (2013) Germinated brown rice and its role in human health. Crit Rev Food Sci Nutr 53:451–463

Yodpitak S, Mahatheeranont S, Boonyawan D, Sookwong P, Roytrakul S, Norkaew O (2019) Cold plasma treatment to improve germination and enhance the bioactive phytochemical content of germinated brown rice. Food Chem 289:328–339

Zhang Y, Chen S, Wei C, Chen J, Ye X (2017) Proanthocyanidins from Chinese bayberry (*Myrica rubra* Sieb. Et Zucc.) leaves regulate lipid metabolism and glucose consumption by activating AMPK pathway in HepG2 cells. J Funct Foods 29:217–225

Zhang B, Zhang Y, Li H, Deng Z, Tsao R (2020b) A review on insoluble-bound phenolics in plant-based food matrix and their contribution to human health with future perspectives. Trends Food Sci Technol 105:347–362

Zhang XW, Dong LH, Jia XC, Liu L, Chi JW, Huang F, Ma Q, Zhang MW, Zhang RF (2020a) Bound Phenolics ensure the Antihyperglycemic effect of Rice bran dietary fiber in db/db mice via activating the insulin signaling pathway in skeletal muscle and altering gut microbiota. J Agric Food Chem 68:4387–4398

Zhang Y, Li F, Huang K, Li S, Cao HW, Xie J, Guan X (2023) Structural changes of starch under different milling degrees affect the cooking and textural properties of rice. Food Chem 17:100627

Chapter 6
Dietary Fiber from Rice by-Products and Its Applications

Zhong Zhang

Abstract As a by-product of rice milling, rice bran is mainly composed of dietary fiber, starch, ash, protein, and lipid. In this part, the ways to improve the rice bran dietary fiber functionality are analyzed and then a detailed discussion on the health benefits of rice bran dietary fiber are performed (i.e., hypoglycemic and antioxidant binding with harmful metal ions). After that, the application of rice bran dietary fiber in different foods is discussed and the future directions in the usage of rice bran dietary fiber are illustrated.

Keywords Rice bran · By-products dietary fiber · Hypoglycemic · Applications

Abbreviations

ACU	Area under the curve
APP	Asian pear powder
CEC	Cation-exchange capacities
DF	Dietary fiber
DSC	Differential scanning calorimetry
FTIR	Fourier Transform infrared spectroscopy
IDF	Insoluble dietary fiber
OHC	Oil holding capacities
PBG	Postprandial blood glucose
RB	Rice bran
RBDF	Rice bran dietary fiber
SC	Swelling capacity
SDF	Soluble dietary fiber
SEM	Scanning electron microscopy
SMRBIFs	Sulfuric acid-modified rice bran-insoluble fibers
TPA	Texture profile analysis

Z. Zhang (✉)
Ningxia University, Yinchuan, Ningxia, China

J. Tian et al. (eds.), *Science of Rice Chemistry and Nutrition*,
https://doi.org/10.1007/978-981-99-3224-5_6

WHC	Water holding capacities
XRD	X-ray diffraction

6.1 Dietary Fiber in Rice Bran

As a by-product of rice milling, rice bran (RB) is mainly composed of dietary fiber (20.5%–33.3%), starch (16.1%–26.7%), ash (9.2%–13.9%), protein (13.2–18.6%), and lipid (9.5%–22.9%). The global production of RB was approximately 59 million tons in 2014 (FAO 2016). Hence, it can be seen as a good source of dietary fiber (DF) for food applications. Methods for extracting DF from RB mainly involve the removal of lipid, starch, and protein from the bran, including chemical, physical, enzymatic, and microbial methods (Liu et al. 2021). Rice bran dietary fiber (RBDF) is mainly composed of cellulose, hemicellulose, lignin, and pectic substances. Most DF from rice bran is insoluble and called insoluble dietary fiber (IDF); only a small part is soluble and called soluble dietary fiber (SDF).

The IDF contains cellulose, lignin, some hemicellulose, and chitosan, while SDF contains some hemicellulose, sodium alginate, guar gum, pectin, dextran, and so on. Moreover, SDF and IDF have different physiological functions. IDF can be used to support the growth of intestinal flora, increase the volume of feces, inhibit pancreatic lipase activity, and decrease the intestinal transit. The SDF plays an important role in cutting down the glycemic reaction and plasma cholesterol, immunomodulatory activity, preventing colorectal cancer, and reducing the risk of angiocardiopathy (Jia et al. 2019). Additionally, compared with IDF, SDF has a higher capacity to form gels and act as an emulsifier, enabling it to be readily incorporated into food products. The proportion of SDF in total dietary fiber (TDF) must be at least 10% to be recognized as high-quality DF. However, the content of SDF was much lower than this value in the RBDF.

The content of the TDF in RB is approximately 20–30%, but nearly 90% of that content consists of IDF (Zhao et al. 2018). RBDF can be prepared by many processing methods, such as enzyme treatment, inorganic acid, and alkaline treatments. Different chemical treatments may cause irreversible modifications to the insoluble fiber of RB and affect its original structure to different extents, which may induce changes in its physicochemical properties (e.g., swelling property, water holding capacities (WHC), oil holding capacities (OHC), and cation-exchange capacities (CEC) (Qi et al. 2016).

RB also contains bioactive components such as ferulic acid and γ-oryzanol, both of which contain antioxidant properties. These bioactive components are primarily bound to the DF components. Physical and enzymatic treatments may free up these bound phytochemicals, leading to much better bioavailability for the bioactive (Dang and Vasanthan 2019). Although RB has been mainly used as animal feed, it has tremendous potential for use in human consumption with suitable processing and application technologies. RB has been recognized as an excellent source of IDF and

SDF, both being known for their good physical and bioactive properties (Dang and Vasanthan 2019).

6.2 Improvement in RBDF Functionality

As mentioned before, most extracted RBDF is insoluble, with only a small proportion being soluble components, and SDF and IDF have different physiological functions. Therefore, to enhance the SDF content and functionality of RBDF, RBDF was modified by the following methods. Some modifications may also enhance the sensory quality of RBDF.

6.2.1 Use of Enzymes

To further improve the physicochemical and functional properties of DF extracted from RB, enzymatic modification methods have been used to further modify RBDF. It was reported that cellulase treatment did not change the WHC or improve the OHC, but significantly increased the swelling capacity (SC) and CEC of RBDF. RBDF was obtained using NaOH and α-amylase to remove the starch from RB. Cellulase hydrolysis further reduced the WHC and OHC, but increased the SC and cholesterol absorption capacity (CAC) of the RBDF (Liu et al. 2021). It was reported that complex enzymes, such as α-amylase, glucoamylase, protease, and cellulase, could significantly improve antioxidant activities by increasing the total phenolic and flavonoid content of RB.

RBDF subjected to pre-water-washing and complex enzyme treatment using thermostable α-amylase, alcalase, and glucoamylase had significantly higher ($p < 0.05$) proportions of cellulose, hemicellulose, lignin, and lower proportions of lipid, protein, and starch than RBDF subjected to complex enzyme treatment without pre-water-washing. Cellulase modification of RBDF significantly decreased ($p < 0.05$) cellulose, hemicellulose, starch, and protein content while the relative lignin content increased. Cellulase modification significantly improved ($p < 0.05$) WHC, OHC, SC, cholesterol absorption capacity (CAC), and glucose adsorption capacity, while decreasing the emulsifying capacity and glucose dialysis retardation index. The changes in the physicochemical and functional properties of fiber samples after cellulase modification were attributed to the increased porosity of the fiber surface, greater exposure of binding sites caused by reduced crystallinity, and changes to the chemical composition (Liu et al. 2021).

6.2.2 Use of Fermentation

Defatted RB, the residue of RB extract oil, contains lots of nutrients including DF and other ingredients. Due to the rough taste of defatted RB, the production usage rate is extremely low, resulting in the waste of functional ingredients and resources. Therefore, the use of defatted RB in functional foods has great significance (Jia et al. 2019).

Trichoderma viride was used to ferment the defatted RB, and SDF was prepared before and after fermentation, named as B-SDF and F-SDF, respectively (Jia et al. 2019). Optimum fermentation conditions for the yield of F-SDF were determined using response surface methodology. The yield of F-SDF was significantly increased from 10.5 (w/w, %) to 33.4 (w/w, %) under the optimal conditions (inoculum size = 10%, pH = 5.8 and time = 41 h). Fourier transform infrared spectroscopy demonstrated that both B-SDF and F-SDF had typical characteristics of the infrared spectrum of polysaccharides, and X-ray diffraction revealed that they were mainly substances of amorphous structure. Additionally, F-SDF showed a looser structure by scanning electron microscopy (SEM) analysis, increased molecular weight by high-performance gel permeation chromatography analysis, and more complex monosaccharide composition by high-performance anion-exchange chromatography analysis. Furthermore, functional properties analysis showed that F-SDF exhibited higher WHC, OHC, water solubility (WS), and CAC than B-SDF. In summary, this study suggested that SDF from defatted RB obtained through *Trichoderma viride* fermentation has a higher potential to be used as a functional ingredient in foods (Jia et al. 2019).

6.2.3 Physical Processing

RB-IDF possesses certain hypoglycemic and hypolipidemic properties, such as altering the conformation and inhibiting the activity of porcine pancreatic lipase, binding glucose, and retarding a-amylase action. RB-IDF has been proven to have a hypocholesterolemic effect in rats and certain beneficial effects in mildly hypercholesterolemic men. Furthermore, RB-IDF was demonstrated to decrease both fasting and postprandial glucose levels in patients with diabetes. Despite its various health benefits, RB-IDF, similar to IDFs from other sources, is not favored by people due to its rough texture and bad flavor. Therefore, many efforts have been made to modify IDFs, including reducing their particle size. Reducing the particle size of IDF may result in certain changes in its structure, porosity, surface area, and functional properties (Zhao et al. 2018). The effects of superfine processing on the functional properties of RB-IDF and its phenolic profiles and bioaccessibility were investigated. Superfine RB-IDF powder exhibited higher WHC, SC, and nitrite ion adsorption capacity, and a lower OHC than the other IDF powders. Moreover, the superfine RB-IDF powder exhibited higher extractability in both free and bound phenolics, higher phenolic bioaccessibility, and antioxidant properties than its coarse counterpart. The extractable phenolic profile in superfine RB-IDF was also different from those of other IDFs. These results suggest that superfine grinding can improve the functional properties of RB-IDF and benefit its application in functional foods (Zhao et al. 2018).

There are few studies on the influence of DF from extruded RB on the thermal transition properties of rice starch (Wu et al. 2021). It was studied to investigate the gelatinization and retrogradation behavior of rice starch with the addition of SDF and IDF from extruded RB compared to those from un-extruded RB. DSC, X-ray diffraction (XRD), texture profile analysis (TPA), and SEM were used for investigation. As the amount of DF increased (0%, 4%, 8%, 16%, 20%), the gelatinization temperature increased, while the values of setback and breakdown decreased. Both SDF and IDF from extruded RB showed weaker inhibition on short-term staling of rice starch than those from un-extruded RB, respectively. The results confirmed that the DF from RB exhibited suppression of long-term staling of rice starch, and the inhibition abilities of SDF and IDF from extruded RB at the same concentration were higher than that of SDF and IDF from un-extruded RB, respectively. It may be useful for the usage of DF from extruded RB for preventing rice starch staling and quality improvement of rice products (Wu et al. 2021).

6.2.4 Combination of Enzymes and Physical Processing

It was proved that SDF had health bencfits and the potential to improve the sensory and physicochemical properties of foods. Thus, the conversion of RB-IDF to RB-SDF would be beneficial in value addition to the RB as a functional food ingredient. Soluble pentosan content in RB increases using physical (extrusion) and enzymatic (xylanase) technologies individually and/or in combination on water-washed RB. Water-washing was necessary to remove water-soluble compounds, all to increase the proportion of total dietary fiber (IDF + SDF) that could be converted to SDF (Dang and Vasanthan 2019). Although water washing led to the loss of native SDF along with starch and other solubles, this step substantially increased the content of IDF. Both enzyme treatment and extrusion improved the solubility of RBDF and other soluble components, especially the sequential extrusion-enzyme treatment significantly increased the total soluble pentosan content compared to individual or simultaneous treatments. Warm water (37 °C)-soluble pentosan content of treated RB was 6.5% in the sequential extrusion-enzyme treatment, and 4% by either extrusion alone or xylanase treatment alone or simultaneous extrusion-enzyme treatment. The total hot water (100 °C) solubility was 25% (including 10.5% soluble pentosans) in the sequential extrusion-enzyme treatment, which was approximately a fourfold increase compared to the untreated RB, superior to extrusion alone or xylanase treatment alone (Dang and Vasanthan 2019). Higher SDF content in RB is beneficial in relation to both food formulation and nutrition, due to SDF's superior ability to hydrate and blend the food matrix (Esposito et al.

2005). Additionally, there are added human health benefits such as gut health, diabetic management, and heart health (Chawla and Patil 2010). A combination of cellulase, xylanase, and ball milling increased the SDF content in RB (Wen et al. 2017).

Using cellulase, xylanase, and ball milling, the influences of enzyme and enzyme-micronization treatments on the structural and functional properties of RBDF were investigated. Due to the degradation of IDF, the cellulase, xylanase, micronization, combined enzymes, and enzyme-micronization treatments increased the SDF content by 3.8, 4.7, 3.5, 10.0, and 11.4-fold, respectively. Scanning electron microscopy analysis indicated the enzymatic treatments caused the breakage of RBDF structure, and the enzyme-micronization treatment totally broke the RBDF matrix. Cellulase and xylanase increased RBDF crystallinity because of the hydrolysis of hemicelluloses and the amorphous portion in cellulose, while enzyme-micronization reduced RBDF crystallinity due to the destruction of the crystalline structure. Infrared spectroscopy indicated the breakage of intra-molecular hydrogen bonding and increased oligosaccharides, and differential scanning calorimetry (DSC) analysis showed the oligosaccharides melting and reduced water-evaporating peaks because of the enzymatic and enzyme-micronization treatments. Additionally, enzyme and enzyme-micronization reduced the WHC and OHC, but increased the SC, cholesterol, and sodium taurocholate absorption capacity of RBDF. The results suggest that cellulase and xylanase can modify the structural and functional attributes of RBDF, and the enzymatic treatments assisted with micronization are more effective in modifying the RBDF properties (Ya et al., 2017). Overall, it was indicated that extracting and modifying RBDF using different methods leads to changes in functional and physicochemical properties. The physicochemical and functional properties of DF are closely linked with its chemical composition and structure (Liu et al. 2021).

6.3 Health Benefits

Accumulating evidence favors the view that increased intake of DF in an otherwise low-fiber diet can have beneficial effects in both human and experimental animals. These benefits include the prevention or alleviation of maladies such as cardiovascular disease, diabetes, diverticulosis, and colon cancer. Many studies have repeatedly shown that RB gives interesting health benefits. Compositional analysis reveals that RB consists of almost 27% DF and has positive effects, such as laxative and cholesterol-lowering ability. This suggests that RB is a good fiber source that can be added to various food products (Azizah and Yu 2000).

Studies have shown that defatted oil-bearing materials (such as sesame husk, RB, and flaxseed) give interesting health benefits. Compositional analysis reveals that sesame husk, RB, and flaxseed consist of almost 68%, 27%, and 39% DF, respectively, and have positive health effects, such as in the form of a laxative and cholesterol-lowering agent (Nandi and Ghosh 2015). The components and their contents of RBDF are different from other DF.

6.3.1 Hypoglycemic Effects

Significant effects of both soluble and insoluble fibers prepared from RB in decreasing both fasting and postprandial blood glucose (PBG) levels of patients with diabetes have been demonstrated. The effects of RB fibers in lowering serum glucose levels may be different by different processing and testing methods (Qi et al. 2016).

With the removal of starch and protein, the porosity and specific surface area of the insoluble fiber matrix could be effectively improved and their capacities in binding glucose and lessening the activity of a-amylase would be enhanced. Some authors have previously reported that the porosity and specific surface area of RB-IDF were effectively improved by the severity of acid or alkaline treatment (Qi et al. 2016).

The in vitro hypoglycemic properties of three different sulfuric acid-modified rice bran-insoluble fibers (SMRBIFs) produced using 0.2% (0.2%-SMRBIF), 1.25% (1.25%-SMRBIF), 2.0% (2.0%-SMRBIF) (w/v) H_2SO_4 along with 1.25% potassium hydroxide (KOH), respectively, were investigated. The results revealed that the three RBIFs have strong glucose adsorption capacity and α-amylase inhibition activity, with 1.25%-SMRBIF being the most effective. The glucose adsorption capacity increased with increasing porosity and specific surface area of the SMRBIFs, and the maximum glucose-bound amount increased by 2–3 folds for acid-alkaline-treated RBIFs compared to untreated material. The α-amylase inhibition activity of RBIF was associated with the formation of amylase–RBIF complex as confirmed by fluorescence spectroscopy, with the highest amylase activity inhibitory (24.72%) for 1.25%-SMRBIF. However, the glucose diffusion was retarded a bit stronger by these RBIFs compared to the untreated RB, which might be attributed to their larger specific surface area. Therefore, it was suggested that the concentration of postprandial serum glucose lowered by RBIF was mainly ascribed to binding glucose and retarding a-amylase action (Qi et al. 2016).

The effects of RBDF on PBG among Type II patients with diabetes were investigated. A total of nine volunteering Type II patients with diabetes were enrolled in the study. The patients were given, on two separate days, the same breakfasts (including 60 grams of carbohydrates) that were with RBDF (10 g) or RBDF free. Fasting and postprandial blood were collected at intervals of 30 min for 180 min to determine whole blood glucose. The results indicated that the use of RBDF can significantly reduce the area under the curve (AUC glucose) for the PBG plot (36,360 $\pm$ 6073 vs. 32,560 $\pm$ 5546.7 mg/dL $\times$ min, $P < 0.05$) as well as shortened the PBG peak values and peak time. In conclusion, RBDF can be used to improve PBG for Type II Patients with diabetes (Wu and Chen 2016).

Defatted RB was used as the raw material to extract RBDF by the composite enzymatic hydrolysis method. The extracted RBDF was first modified by the optimal steam explosion scheme, and then subjected to ultrafine treatment by jet milling and nano-impact grinding. Finally, two types of RBDF were obtained: steam explosion-jet pulverization and steam explosion-nano pulverization. In the steam explosion-jet pulverization and steam explosion-nano pulverization groups, the blood glucose levels of female rats at the end of the experiment were 18.3%, 4.9%, and 6.4% lower than before. Male rats were 18.4%, 4.2%, and 13.2% lower than before. Biochemical indicators of mice indicated that RBDF could reduce the levels of total cholesterol, total triglycerides, and low-density lipoprotein in serum ($P<0.05$) and increase high-density lipoprotein. The activity of superoxide dismutase in the liver was enhanced, and the content of malondialdehyde was reduced. The effect of steam explosion-nano pulverization was more significant. The results indicated that the composite-modified steam explosion-nano pulverization had a stronger protective effect on diabetic mice (Xu et al. 2020).

6.3.2 *Blood Pressure and Lipid Metabolism*

The high DF rice (Suwon 464) contained DF two times as much as normal rice (Ilpum) in both polished and brown rice. When five groups of spontaneously hypertensive rats were fed one of five kinds of diets (including corn starch diet as a control (CO), polished (IP) and brown rice (IB) of Ilpum, and polished (SP) and brown rice (SB) of Suwon 464) for 8 weeks, body weight gain was lower in the four rice intake groups though there was no significant difference in dietary intake among the experimental groups. The systolic and diastolic blood pressure levels were decreased by 18% and 16% in the SP group compared with the CO group, respectively. LDL-cholesterol level was significantly lower in the SB group and HDL-cholesterol level was significantly higher in the IB, SP, and SB groups though serum triglyceride and total cholesterol levels had no significant difference among the experimental groups. The total hepatic lipid, triglyceride, and total cholesterol concentrations were significantly lower in the four groups fed rice diets, and these three levels were significantly lower in the SB group than in the IP group. Total lipid, triglyceride, and total cholesterol contents excreted to feces were significantly higher in the IB, SP, and SB groups, and these three levels were significantly higher in the SB group than in the IP and IB groups. These results suggest that the high DF rice (Suwon 464) can decrease blood pressure, serum LDL-cholesterol, total hepatic lipid, triglyceride, and total cholesterol by increasing fecal lipid, and increasing serum HDL-cholesterol. It was expected that high DF rice (Suwon 464) can control blood pressure, and serum and hepatic lipid levels (Lee et al. 2014).

As mentioned before, the Suwon 464 rice contained twice of DF than the Ilpum rice, and the contents of DF in the experimental diets were 1.1% in the Cornstarch group (CO), 3.6% in the Ilpum rice group (IR), and 6.6% in the Suwon 464 rice group (SR). For the diabetic mice, there was no significant difference in diet intake, body weight, organ weights, and the concentration of blood glucose among the experimental groups. However, the levels of triglyceride and total cholesterol in the serum and liver were lower in SR, and the levels of total lipid, triglyceride, and total cholesterol excreted in feces were significantly higher in SR than the others. These results suggest that the supplement of the Suwon 464 rice decreased triglyceride and total cholesterol concentrations of serum and liver by increasing fecal lipid excretion in these diabetic mice (Lee et al. 2013).

It is indicated that extracted TDF from flaxseed, RB, and sesame husk could lower total cholesterol and triglyceride levels in serum and increased the brain and liver phospholipids/cholesterol ratio. The cholesterol-lowering effect of DF and polyphenol-rich products obtained from other sources have already been well studied but these three have not yet been reported yet. In the study, the functional properties and nutritional evaluation of these three DFs were reported. The in vivo study was conducted with five groups of animals. One group of five adult male Wistar rats were fed a purified cholesterol-free diet for 32 days, the second group was fed with 1 g/100 g cholesterol-added diet for 32 days and the other three groups with 1 g/100 g cholesterol-added diets for 32 days, along with 3 g/100 g experimental DF diet as a source of DF for the last 28 days. HDL-cholesterol concentration was significantly higher and serum LDL-cholesterol concentrations were significantly lower in rats fed TDF extracted from these oil seeds than in the control hypercholesterolemic group. The results indicate that the TDF extracted from flaxseed, RB, and sesame husk reduces serum total cholesterol and LDL-cholesterol, as well as the atherogenic index, in hypercholesterolemic rats (Nandi et al. 2019).

6.3.3 Antioxidant Properties

RB is a good fiber source that can be added to various food products. A study was conducted to examine the effectiveness of a DF preparation, derived from these sources, as a functional ingredient with a potential for antioxidant activity. The observations established that DF from defatted RB had higher water binding, oil binding, and emulsifying capacity compared to sesame husk DF. However, RB and sesame husk fiber were found to be less viscous than flaxseed fiber. Whereas in the case of antioxidant activity, RB fiber showed higher activity than flaxseed and sesame husk. This study confirms that the DF preparation from defatted RB has great potential in food applications, especially in the development of functional foods (Nandi and Ghosh 2015).

6.3.4 Effects on Intestinal Microbiota

One RB-IDF was obtained from defatted RB by enzymatic treatment. The structure, composition, physicochemical properties, and functional properties in vivo were evaluated. The composition analysis showed that the content of IDF in the preparation was about 64%, and its surface structures became looser and coarser than that of defatted RB. RB-IDF possessed good WHC, OHC, and adsorption capacities of glucose and cholesterol, which could reach about 6.4 mg/g and 12.8 mg/g, respectively. The RB-IDF also exhibited excellent properties in scavenging hydroxyl and DPPH radicals. Intragastric administration with RB-IDF could lead to improvements in the changes in gut microbiota diversity and composition caused by a high-fat diet. The results suggested that the RB-IDF from defatted RB presented satisfactory activities both in vitro and in vivo, providing a theoretical basis for their usage and mechanism research of it (Liu et al. 2020).

6.3.5 Binding with Harmful Metal Ions

To promote the comprehensive usage of RB resources, the extraction method and adsorption characteristics of Pb^{2+} of RB-IDF were studied. The enzyme-chemical method was used to extract IDF from defatted RB. The single factor and orthogonal experiments were used to study the effects of high-temperature amylase addition, NaOH mass fraction, alkaline hydrolysis time, and alkaline hydrolysis temperature on the extraction rate and purity of the product, and the optimal preparation parameter of RB-IDF was determined. The results showed that when the high-temperature amylase was added at 0.3%, with the 3% NaOH mass fraction and the alkaline hydrolysis at 75 °C for 75 min, the purity and yield of the obtained product was 85.74% and 37.40%, respectively. Furthermore, adsorption experiments revealed that when the extract was shaken in a 20-mg/L Pb^{2+}aqueous solution for 180 minutes, the adsorption rate was 96.67%. The kinetics were in line with the pseudo-second order kinetic equation. It could better describe the binding behavior between RB-IDF and Cd^{2+}, and there existed chemical adsorption during the adsorption process. It was found that 25 °C was more conducive to the adsorption process. This process is entropy driven and spontaneous. Simultaneously, the adsorption ability of Pb^{2+} was weak, predicting that the applied force was mainly physical adsorption. Therefore, RB-IDF had a better ability to absorb Pb^{2+}. This test provides a theoretical basis for the extraction conditions of RBDF material and its Pb^{2+} adsorption characteristics (Liu et al. 2020).

6.4 Applications of RBDF

RB has great potential in food products with high nutritional values (Wen et al. 2017). In view of the therapeutic potential of DF, more fiber-incorporated food products are being developed. The addition of DF to a wide range of products will contribute to the development of value-added or functional foods that currently are in high demand. In addition to the physiological benefits provided by high-fiber foods, studies have shown that fiber components can give texture, gelling, thickening, emulsifying, and stabilizing properties to certain foods. By understanding the functional properties of DF, one can increase its use in food applications and aid in developing food products with high consumer acceptance (Azizah and Yu 2000).

6.4.1 Application of RBDF in Baked Products

Improvements in the viscosity, gelatinization temperature, volume, and porosity were reported in the baked products that are enriched with xylanase-treated RB (Dang and Vasanthan 2019). A DF preparation derived from defatted RB was added to baked products as a functional ingredient, and the results show that DF from defatted RB has a comparable water-binding capacity to commercial fiber from sugar beet. RBDF exhibited higher fat binding and emulsifying capacity compared to commercial fiber from sugar beet. However, RBDF was found to be less viscous (Azizah and Yu 2000).

Application of RBDF in Biscuits

The increasing demand for functional foods has boosted up the food industry to produce fiber-enriched products. A study investigated the effects of SDF from defatted RB obtained through *Trichoderma viride* fermentation (F-SDF) on the physical properties and starch digestion characteristics of biscuits. Rheological analysis results showed that the addition of F-SDF delayed the gelatinization of starch. TPA results demonstrated that the addition of F-SDF reduced the hardness, springiness, cohesiveness, gumminess, chewiness, and resilience of the biscuits. The SEM results showed that F-SDF hindered the formation of the gluten network structure in the biscuits, making the biscuits lose. Meanwhile, XRD revealed that the addition of F-SDF reduced the area of the crystallization zone. Additionally, F-SDF led to a significant reduction in the predicted glycemic index (*pGI*) of reformulated biscuits. Sensory evaluation results indicated that the effects of F-SDF addition on the odor, color, and taste of the biscuits were relatively small, while the biscuits with the highest texture and acceptability score were obtained by incorporating 6% of F-SDF in the formulation. In summary, adding the proper amount of F-SDF could improve the physical properties of the biscuits and exhibit the potential to decrease the PBG level resulting from biscuits (Jia et al. 2020).

Application of RBDF in Pizza

It was studied to use high-content DF (26%) stabilized rice bran flour (SRBF) in the production of pizza dough. The pizza dough was developed by mixing wheat flour with SRBF in a proportion of 5:95 and 10:90, water, salt, and yeast. All the pizzas were stored for 60 days at 18 °C and their proximal composition, functional properties, and sensorial characteristics were evaluated at 0, 30, and 60 days. The results showed that the content of DF increased by 3.8% and 5.3% as the level of enrichment increased. The farinographic curves of the pizza dough show that the development time, water absorption, and stability decreased, while the mixing tolerance index and departure time were not affected by the enrichment level. The sensorial test results indicate that the pizza dough with an enrichment level of 5% SRBF was well accepted by the panel and it was stable during 60 days at 18 °C (Delahaye et al. 2005).

Application of RBDF in Bread

Gluten-free (GF) breads are often characterized by low nutritional quality as they are mainly starch based and contain low amounts of vitamins, minerals, and in particular DF. It was studied to improve the physical, nutritional, and sensory quality and shelf life of rice-based GF bread by adding different fractions of RB, containing different amounts of protein, fat, DF, and different ratios of IDF to SDF. As a first step, recipe parameters (content of hydroxypropyl methylcellulose (HPMC), egg albumen, and emulsifier) of GF bread with 10% RB addition were optimized. An amount of 3% HPMC, 1.5% emulsifier, and 2% egg albumen provided acceptable structural and textural quality of GF bread. The main trials demonstrated that the addition of RB, in particular when containing a high amount of SDF, produced better bread color, higher specific volume, softer crumb firmness, and improved porosity profile. All RB-containing GF bread had higher protein and DF content, and the ratio of IDF to SDF was improved substantially depending on the DF fraction of RB. Sensory acceptance was increased and shelf life was extended when selecting an RB source with high SDF content (Phimolsiripol et al. 2012).

The addition of 5 and 10% DF preparation reduced loaf volume significantly and increased the firmness of the breads. Sensory evaluations revealed that breads with 5% and 10% RB fiber were comparable to high-fiber bread available in the market. This confirms that the DF preparation from defatted RB has great potential in food applications, especially in the development of functional foods (Zhao et al. 2018).

Rice bran high-quality dietary fiber(RB-HQDF)was applied to the bread production. Through sensory evaluation and bread texture analysis with different amounts of RB-HQDF, the maximum addition of RB-HQDF into bread was determined as 9%. Under this condition, the bread had a good flavor. It was soft and resilient and had a slight fragrance of RB. Meanwhile, the water retention and aging degree of RB-HQDF bread were better than the control group, which proved that the addition of RB-HQDF could extend the shelf life of the bread (Liu et al. 2014).

6.4.2 Application of RBDF in Encapsulation

Asian pear powder (APP) encapsulated using RBDF and maltodextrin (MD) was investigated for its physical stability, phenolic bioaccessibility, antioxidant activities, and major phenolic compounds (arbutin, ferulic acid, p-coumaric acid). The results showed that moisture content, the angle of repose, solubility, and color values of APP were all significantly ($P < 0.05$) affected by the addition of RBDF and MD at various levels. Water adsorption and hygroscopicity were lower in RBDF/APP mixtures than MD/APP mixtures. X-ray photoelectron spectroscopy, XRD, and SEM showed that RBDF-added APP was more hydrophobic than MD-added APP with significant flowability differences. RBDF/APP showed higher total and bioaccessible phenolics and antioxidant activities compared to MD/APP. During storage for 150 days, high retention of bioactive compounds was observed in RBDF/APP. Therefore, RBDF improved the handling and functional properties of APP compared to MD (Jiang et al. 2020).

6.4.3 Application of RBDF in Pasta Products

Fourier Transform infrared spectroscopy(FTIR)was used to study the changes in the gluten protein structure when the amount of RBDF powder obtained by Ultra high-pressure homogenization(UHPH)was 0%, 5.0%, 10.0%, 15.0%, and 20.0%. The results showed that the secondary structure of gluten protein supplemented with RBDF powder was still dominated by α-helix and β-sheet. However, as the amount of RBDF powder increased, the content of α-helix structure was significantly reduced ($p<0.05$), the decreasing trend of the total content of β-turn and random coils was the same as the increasing trend of the β-sheet content, both decreased first and then increased, that is, they all showed the tendency of β-turn and random coils transformed to β-sheet. When the amount of RBDF nano-powder was more than 15%, the α-helix content in the secondary structure of gluten decreased sharply, the hydrogen bonds in the space conformation of gluten protein were destroyed, and the protein helix structure was seriously damaged, the balance of maintaining the gluten protein's spatial stability was destroyed, and the structure of the gluten protein tended to loosen. Therefore, when the amount of RBDF powder is controlled within 15%, the pasta products can obtain good processing characteristics while having a healthcare function. It can be seen that UHPH treatment is beneficial to the micronization of RBDF, which can improve the comprehensive usage of RB (Xie et al. 2020).

6.4.4 Application of RBDF in Noodles

It was carried out to study the effect of RBDF on the edible quality of noodles by adding different amounts of DF. The results showed that the addition of DF led to a significant increase in the cooking loss of noodles, and the loss rate showed a wavy rising trend increasing from 6.24% to 7.64%. The texture properties of noodles were significantly affected by the addition of DF. With the increase in the addition, the tensile force fell to 8.3 g from 11.78 g and the tensile distance fell to 11.96 mm from 38.73 mm, the firmness increased to 4684.96 g from 3681.11 g, and the cohesion decreased to 0.45 from 0.57. The addition of DF led to a significant decrease in the sensory quality of noodles. The results showed that the noodles with less than 4% of DF could still maintain relatively good quality (Zhou et al. 2017).

6.4.5 Application of RBDF as a Fat Substitute

Response surface methodology was used to optimize the preparation of fat substitute by carboxymethyl substitution of RBDF. A Box-Behnken design involving the independent variables chosen based on single-factor experiments was used to define an experimental design matrix of reaction conditions for RBDF. For the optimization of the independent variables, quadratic regression analysis was performed using Design-Expert 8.0.6 software. The optimal parameters for the preparation of fat substitute were achieved as alkalization of RBDF at 25.9 °C followed by 4.54 h etherification at 69 °C after the addition of 16.27 g of chloroacetic acid per 15-g DF. Under these conditions, the degree of carboxymethyl substitution of RBDF was 1.266 g. Fourier transform infrared spectroscopy confirmed the occurrence of carboxymethyl substitution in the fat substitute prepared with RBDF (Liu et al. 2016).

6.4.6 Application of RBDF in Mayonnaise

SDF from RB was added to the traditional mayonnaise and its most appropriate amount was decided through the analysis of thixotropy. Finally, the rheological properties of the products were investigated. Results showed that the optimal technology as followed: yolk was 11.0%, oil was 74%, stir rate was the fourth gear (14,400 r/min) and stir time was 20 min, SDF from RB was 2.0%, and the mayonnaise has smaller thixotropy. The content of cholesterol was 1.26 mg/g, which was lower than traditional mayonnaise (2.21 mg/g). Compared with Kewpie mayonnaise and Kewpie sweet mayonnaise, the mayonnaise with low cholesterol had slightly worse rheological properties, but was better than Kewpie thousand island mayonnaise (Guo et al. 2015).

6.4.7 Application of RBDF in Sausage

The incorporation of defatted RBDF into sausage deteriorated texture and sensory quality; however, bologna with 5% addition of all treated DF and 10% addition of micronized DF exhibited acceptable sensory and textural qualities. The physically modified DF showed important improved functional and physiological effects (hypoglycemic, hypocholesteremia effects) and can be incorporated as a low-calorie bulk ingredient in foods to reduce calorie levels (Daou 2012).

6.4.8 Application of RBDF in Fortified Rice

RBDF extracted from early indica rice was added to the rice flour. Different concentrations of fortified rice were transformed into steamed rice after cooking, using a texture analyzer to determine their hardness, elasticity cohesiveness, gumminess, and chewiness. And it was compared with the steamed rice prepared from Early Indica Rice to establish the appropriate amount of DF in rice. The results showed that the fortified rice's indices with a 4% addition level of DF were closest to the best texture properties of raw rice. The additional amount of DF was of the greatest impact on the hardness, gumminess, and chewiness of fortified rice, but the elasticity and cohesiveness were less affected (Wang et al. 2011).

6.5 Conclusion of Future Directions

Although RBDF has many health benefits, the usage of RBDF in the food industry is still poor, especially in developing countries. This is mainly due to the following problems: (1) the taste of RBDF is not good, (2) the preparation cost of RBDF is relatively high, and RBDF has no economic advantage. To promote the application of RBDF in the food industry, the suggestions of future research are as follows: (1) study further the functionality and health benefits of RBDF, and conduct in vivo experiments of the human body. (2) Optimize the preparation technology of RBDF and reduce the cost. (3) Treat the RBDF with physical, chemical, biochemical, and microbial methods and improve the taste of RBDF. (4) Use RB as a whole instead of the preparation and use of RBDF. RB has many functional components; however, during the preparation of RBDF, many functional components are lost and the cost is increased. So, using RB as a whole is an economical method and is good for human health. Because RB has unstable factors such as lipids and enzymes, the stabilization treatment of RB should be studied.

References

Azizah AH, Yu SL (2000) Functional properties of dietary fiber prepared from defatted rice bran. Food Chem 68:15–19

Chawla R, Patil GR (2010) Soluble dietary fiber. Compr Rev Food Sci Food Saf 9(2):178–196

Daou C (2012) Dietary fiber in defatted rice bran and its application in meat products.: [D]. Jiangnan University of China

Dang TT, Vasanthan T (2019) Modification of rice bran dietary fiber concentrates using enzyme and extrusion cooking. Food Hydrocoll 89:773–782

Delahaye EP, Jiménez P, Pérez E (2005) Effect of enrichment with high content dietary fiber stabilized rice bran flour on chemical and functional properties of storage frozen pizzas. J Food Eng 68:1–7

Esposito F, Arlotti G, Maria Bonifati A, Napolitano A, Vitale D, Fogliano V (2005) Antioxidant activity and dietary fibre in durum wheat bran by-products. Food Res Int 38(10):1167–1173

FAO (2016) Rice market monitor. Food Agric Organ United Nations 19(2):1–35

Guo TT, Yang Y, Xie BJ et al (2015) Effect of soluble dietary fiber from rice bran on reducing cholesterol content and rheological property of mayonnaise. Chin J Sci Technol Food Ind 36(20):265–270

Jia MY, Yu Q, Chen JJ et al (2020) Physical quality and in vitro starch digestibility of biscuits as affected by addition of soluble dietary fiber from defatted rice bran. Food Hydrocoll 99:105349

Jia MY, Chen JJ, Liu XZ et al (2019) Structural characteristics and functional properties of soluble dietary fiber from defatted rice bran obtained through Trichoderma viride fermentation. Food Hydrocoll 94:468–474

Jiang GH, Kashif Ameer K, Eun JB (2020) Encapsulation of hot air-dried Asian pear powders using rice bran dietary fiber. Food Biosci 38:100742

Liu YL, Zhang HB, Yi CP et al (2021) Chemical composition, structure, physicochemical and functional properties of rice bran dietary fiber modified by cellulase treatment. Food Chem 342: 1–9

Lee SH, Park H, Cho SY et al (2014) The supplementary effects of high dietary fiber rice (suwon 464) on blood pressure and lipid metabolism in spontaneously hypertensive rats. Poster Abstracts/Cardiovasc Pathol 13:S87

Lee SH, Park HJ, Chun HK et al (2013) The supplementary effects of the rice with high dietary fiber on blood glucose and lipid metabolism in diabetic KK mice. Int Sympos Atherosclerosis Poster Session Nutr 2:278

Liu Q, Fan YC, Liu SS et al (2020) Physicochemical properties of insoluble dietary fibers from defatted rice bran and its effects on high fat rats intestinal microbiota. Chin J Food Sci. https://doi.org/10.7506/spkx.1002-6630-20200730-395

Liu TT, Xu YY, Wang DW (2014) The application of rice bran high quality dietary fiber in bread production. Chin J Food Res Develop 35(10):1–5

Liu Y, Song DD, Fu W et al (2016) Optimization of preparation process for fat substitute from rice bran dietary fiber by response surface methodology. Chin J Food Sci 37(04):37–43

Nandi I, Sengupta A, Ghosh M (2019) Effects of dietary fibers extracted from defatted sesame husk, rice bran & flaxseed on hypercholesteremic rats. Bioact Carbohydr Diet Fibre 17:100176

Nandi I, Ghosh M (2015) Studies on functional and antioxidant property of dietary fibre extracted from defatted sesame husk, rice bran and flaxseed. Bioact Carbohydr Diet Fibre 5:129–136

Phimolsiripol Y, Mukprasirt A, Schoenlechner R (2012) Quality improvement of rice-based gluten-free bread using different dietary fibre fractions of rice bran. J Cereal Sci 56:389–395

Qi J, Li Y, Masamba KM et al (2016) The effect of chemical treatment on the in vitro hypoglycemic properties of rice bran insoluble dietary fiber. Food Hydrocoll 52:699–706

Wang WH, Li H, Liu CM et al (2011) Effect of rice bran dietary fiber on texture properties of fortified rice. Chin J Food Machinery 27(03):16–19

Wen Y, Niu M, Zhang BJ et al (2017) Structural characteristics and functional properties of rice bran dietary fiber modified by enzymatic and enzyme-micronization treatments. LWT Food Sci Technol 75:344–351

Wu NN, Qiao CC, Xiao-Hong Tian XH et al (2021) Retrogradation inhibition of rice starch with dietary fiber from extruded and unextruded rice bran. Food Hydrocoll 113:1–9

Wu TY, Chen WH (2016) The effect of rice-bran dietary fiber on postprandial blood glucose for type II diabetic patients. *Poster presentations/*. Diabetes Res Clin Pract 120S1:S65–S211

Xu TH, Su Y, Liang H et al (2020) Hypoglycemic effect of steam explosion - superfine pulverization rice bran dietary fiber on type 2 diabetic mice. J Chin Cereals Oils Assoc 35(10):9–15

Xie FY, Zhao YY, Lei YC et al (2020) Effect of dietary fiber powder by ultra-high pressure homogenized rice bran on the structure of gluten protein. J Chin Inst Food Sci Technol 20(11): 115–121

Zhao GH, Zhang RF, Dong LH et al (2018) Particle size of insoluble dietary fiber from rice bran affects its phenolic profile, bio accessibility and functional properties. LWT Food Sci Technol 87:450–456

Zhou YP, Zhang LP, Xiang ZQ et al (2017) Effect of dietary fiber from rice bran on the edible quality of noodles. Chin J Food Sci Technol 42(05):147–151

Chapter 7
Postharvest Processing of Rice

Natthawuddhi Donlao

Abstract Postharvest processing of rice involves a series of processes applied to rice after the harvesting (i.e., handling, drying, storage, and milling). In this part, a detailed postharvest processing of rice is discussed, including different drying methods, different storage methods, and milling methods. After that, the effects of postharvest processing (drying, storage, and milling) on the physicochemical and nutritional properties of rice are discussed and might provide insights into potential improvements in rice production with high quality.

Keywords Paddy · Postharvest · Storage · Milling · Drying

Abbreviations

DOM	degree of milling
GI	glycemic index
HRY	head rice yield
IR	Infrared radiation
RH	relative humidity
TAA	total content of amino acids

7.1 Introduction

Rice is one of the most important cereal crops in many countries around the world and it is the staple food for two-thirds of the world's population. Depending on the variety and planting method, the harvest time of rice lies between 100 and 140 days after sowing. Harvesting is the process of collecting the rice crop from the field when

N. Donlao (✉)
Mae Fah Luang University, Chiang Rai, Thailand
e-mail: Natthawuddhi.don@mfu.ac.th

J. Tian et al. (eds.), *Science of Rice Chemistry and Nutrition*,
https://doi.org/10.1007/978-981-99-3224-5_7

the plants have reached full growth. Postharvest processing of rice involves a series of processes applied to rice after the harvesting. For a well-managed crop, rice grains are accounted for 70%, while around 30% is taken as by-products from postharvest. Based on seasons and harvesting time, paddy rice normally contains high moisture content ranging between 18% and 24% (wet basis) after harvesting (ManzoorAhmad et al. 2015). High moisture content of grain after harvesting can lead to spoilage stimulate the development of molds, and reduce grain quality. Therefore, after harvesting, it is important to dry the grain as soon as possible to reduce the moisture content to a safe level for storage and reduce the handling cost. This means that postharvest treatment is one of the important factors affecting rice yield and quality. Normally, it requires various postharvest processing to reduce postharvest losses (Hamzah et al. 2019) and maintain the qualities of rice, such as eating quality, storage stability, and functional and physicochemical properties. Rice is normally consumed as white rice, also called milled or polished rice, which is the most common form of consumption. Therefore, milling is a crucial step in the postharvest of rice. The primary objective of a rice milling system is to remove the husk and the bran layers. The final product of the milling process is an edible white rice kernel. The by-products from the milling process are mainly 8–9% of rice bran, 1–2% of rice germ, and 20% of rice husks.

Rice grains are edible parts which contain mainly carbohydrates and a trace amount of protein while rice bran is the inedible part located in the outer layer of rice grain; it is composed of pericarp aleurone, sub-aleurone layer, and germ. However, rice bran contains a high nutritional value with health-promoting benefits such as dietary fiber, vitamins, minerals, and unsaturated fatty acids. Thus, the purpose of postharvest processing is adding value, prolonging the shelflife, and transforming into edible through transforming rice into a suitable form, condition, or composition. Generally, postharvest processing of rice includes threshing, drying, cleaning, dehulling, milling, grading, and storage. However, some postharvest processing of rice significantly impacts the physicochemical, functional, and nutritional properties of rice, including drying, storage, and dehulling/milling. In this chapter, the current postharvest processing and the effects of postharvest processing on the physicochemical and national properties of rice grain will be discussed in the following sections.

7.2 Postharvest Processing of Rice

Suitable postharvest processing is necessary to achieve more stability as well as the desirable quality of rice grains. Postharvest operations begin when the rice panicle is separated from the plant at harvest. Before consumption, the rice grain undergoes various handling, drying, storage, and milling, respectively (Fig. 7.1) (Sonu Poudel et al. 2021).

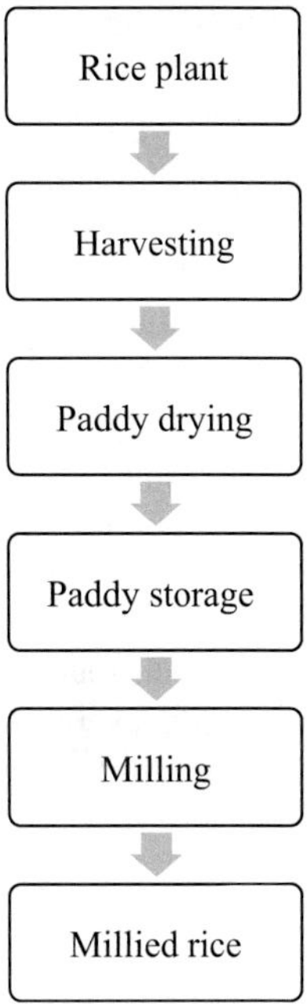

Fig. 7.1 Postharvest processing of rice (drawn by the author)

7.2.1 Drying

Drying is one of the most important postharvest processing, which reduces the moisture content of rice paddy to a safe level during storage. Drying is operated after rice paddy is harvested from the field immediately through traditional, technological, or mechanical methods, e.g., sun drying, natural air drying, low-temperature drying, high-temperature drying, and fluidized bed drying (Atungulu and Sadaka 2019). Drying is an energy-intensive process that significantly affects the quality of milled rice (Firouzi et al. 2017). A suitable drying can prolong the storage life and relieves the deterioration of the product. However, the quality of rice grains may deteriorate during drying due to overheating at high temperatures and rapid moisture loss. Therefore, the postharvest drying of rice is often considered to include drying methods and drying temperature. Various studies have been conducted to investigate the conditions for the drying of paddy as well as the quality of the final product. These studies cover a wide range of topics such as the type of dryer, operating conditions, pre- and post-treatments, and drying rates. Since it is the most important processing after harvesting rice paddy. The moisture content of rice paddy after harvesting ranged from 14 to 22; consequently, the moisture content should be reduced below 13% for appropriate storage (Mukhopadhyay et al. 2019). Numerous studies have investigated the effects of the paddy drying method on the quality of rice. Additionally, the following are various studies relevant to the study of the postharvest drying method of rice.

Sun Drying

Sun drying is a traditional drying method for reducing moisture content by spreading the product under direct sunlight. It is the most common drying method in tropical and subtropical countries, where solar radiation is convenient. Solar energy is an important alternative source of energy because it is abundant, inexhaustible, non-pollutant, renewable, cheap, and environmentally friendly. The main advantages of this traditional drying method are low capital and operating cost, and not much expertise is required. However, the main disadvantages of this method are long drying time, limitations of control of the process, weather uncertainties, high labor costs, required large drying area, insect infestation, mixing with dust, and other foreign materials. Sun drying is one of the traditional methods for paddy drying and it is widely used in many developing countries. The most common sun drying practice in many countries is spreading rice paddy on a flat surface such as mats, nets, canvas, or pavements. The paddy is exposed to sunlight for 2–3 days approximately until the moisture content reaches a safe storage level (Donlao and Ogawa 2017).

Natural Air Drying

Currently, natural air drying or in-bin drying of rough rice has become more attention in paddy drying as it is a low-temperature, relatively slow drying process. It generally prevents kernel cracking and maintains a high rice milling yield. This method uses unconditioned ambient air to dry rice grains in which the moisture content of grain in a bin relies on air equilibrium moisture content. Drying duration depends on various parameters such as initial moisture of grain, grain depth, weather, and airflow rate.

Low-Temperature Drying

This method is normally performed in a perforated bin with an electricity supply. Hence, it is also called electric drying instead of low-temperature drying. Liquefied petroleum and solar energy are often used as thermal sources. Low-temperature drying is capable of drying moisture from rice grains into safe storage moisture content, which is appreciated for long-term storage. During drying, heated air reduces relative humidity inside the bins and then increases the mass transfer of moisture from rice grains to the air until reaching equilibrium moisture content between air and rice grains. Additionally, this method has a low operating cost since rice grain is dried, and it can be stored in the same bin. Hence, this obviously minimizes handling and labor costs.

High-Temperature Drying

High-temperature drying is done either in the bin or a dryer. There are four methods in high-temperature drying, including in-bin batch drying, recirculating bin drying, continuous flow bin drying, and pass drying. In-bin batch drying is quite different from low-temperature drying where air temperatures ranged from 49 to 71 °C. Thus, the drying duration is significantly shortened as a high temperature is applied during drying. However, uniformity drying can occur where the grain near the floor often becomes extremely dry, while the top layer of the grain often remains moist. Hence, stirring devices should be provided to generate more uniform drying. Recirculating bin dryers are bins in which coupled with a sweep auger at the bottom and top of the bins. Once the bin is filled with grains, then the fan and heater are turned on. The sweep auger in the bottom of the bins is then activated by moisture sensors when the system reaches the target condition; the sweep auger makes one full pass and then stops until reaches the target conditions again. Finally, the rice grain is discharged by the sweep auger is placed onto the top of the grain in a similar bin is circulated until obtaining the target condition. Continuous flow bin dryers have a similar configuration of bins as recirculating bin dryers except a cooling bin collects dried grains separately. Pass drying is usually considered as the most rapid drying method for grains. A grain elevator uses some form of pass dryer to dry large amounts of grain rapidly. Hence, this method requires the highest energy inputs compared to all drying methods. The advantage of using pass dryers is that higher capacity can be operated as a higher temperature input, thereby more rapid drying duration. However, Tohidi et al. (2017) reported that the number of fissured kernels was related to the drying rate, where higher drying rates resulted in more fissured kernels.

Fluidized-Bed Drying

Fluidized bed drying is one of the hot air drying processes that is widely used during the last few years as it generates high heat and mass transfer rates between the hot air and rice grains (Dondee et al. 2011). During fluidization, grains are suspended in a gas stream; therefore, hence it allows individual fluidized particles to be exposed to the hot gas stream uniformly, and then the drying rate is improved. However, there are some limitations in the industrial-scale use of the fluidized beds in drying grains including excessive cracking and breakage of grains. Wiriyaumpaiwong et al. (2001) describes that the damage is generated from high heat transfer rates during drying. The quality of rice grain may deteriorate during drying due to overheating at high temperatures and rapid moisture loss (Inprasit and Noomhorm 2001).

Microwave Drying

Olatunde et al. (2017) assessed the feasibility of using an industrial-type microwave heating system operated at 915 MHz to achieve one-pass rough rice drying in a continuous operation. Their results indicated that the microwave drying with the specific energy ranging between 450 and 750 kJ/kg rough rice (for rice bed thicknesses of 0.01 and 0.05 m) provided uniform drying. However, increasing rice bed thickness and specific energy reduced HRY and increased the final viscosity of milled rice, whereas slightly affected rice peak viscosity and surface lipid and protein contents.

Superheated Steam Drying

Srisang et al. (2011) indicated that superheated steam drying could reduce the number of fissured germinated brown rice kernels, compared with hot air drying. However, the superheated steam drying treatment caused an agglomeration of the kernels as opposed to the hot air-drying treatment. Moreover, faster digestion (the higher GI value) of rice starch obtained from SHS drying was observed. Rumruaytum et al. (2014) evaluated the effect of the fluidization drying technique combined with SHS on the physicochemical properties and antioxidant properties of rice. They found that high-temperature drying of paddy can change the physicochemical and antioxidant properties of the rice. The longer time of superheated steam drying also resulted in changes in the physicochemical and antioxidant properties of the rice.

Infrared Radiation (IR) Drying

Pan et al. (2008) studied the feasibility of simultaneous rough rice drying and disinfestations by IR heating. They suggested that simultaneous drying and disinfestation with high rice milling quality can be achieved using a catalytic IR emitter to heat rough rice to 60 °C, followed by tempering and slow cooling. Das et al. (2009) found that the drying rate of paddy dried with a vibration-assisted IR dryer was found to depend on both radiation intensity and grain bed depth. Wang et al. (2017) found that high heating rate, drying efficiency, and milling quality for rough rice can be achieved using IR heating followed by tempering treatment and natural cooling under one- and two-pass drying processes. Moreover, IR heating did not generate adverse effects on the milling quality of rough rice.

Intermittent Tempering Drying

Intermittent tempering drying is a non-continuous drying process with tempering periods. During drying, moisture and temperature gradients are often created inside the rice kernel (Dong et al. 2010; Dong et al. 2009). These gradients induce stresses that can cause fissuring within rice kernels when the failure strength of the kernel is exceeded. However, the tempering process allows the transfer of moisture from the center to the surface of the kernel. This decreases the moisture gradients inside the kernel imposed during the previous drying stage (Dong et al. 2009; Golmohammadi et al. 2012). Dong et al. (2010) reported that drying and tempering durations had significant influences on the moisture gradient in kernels and rice fissuring. The ratio of fissured kernels increased with increasing drying time and decreasing tempering time. The study by Aquerreta et al. (2007) indicated that post-drying tempering at high temperatures (~60 °C) resulted in greater moisture removals and lower drying time. The drying in two or three steps showed a lower percentage of fissured kernels compared with one-step drying. Moreover, tempering at high temperatures reduced the percentage of fissured kernels and enhanced head rice yield (HRY). Cnossen et al. (2003) determined the effect of drying and tempering treatments on rice kernel fissuring and found that the percentage of fissured kernels decreased with increasing tempering duration. They also concluded that the tempering duration required for preventing kernel fissuring might be longer than the tempering duration required for maintaining a high HRY.

Limitation of the Use of a Novel Drying Technique for Paddy Drying

Although new drying methods such as combined Microwave or IR – hot air drying, fluidized bed drying, superheated steam drying, and spouted bed drying have been reported as energy and quality-efficient drying methods for paddy, their use in industrial scales is still limited. Nowadays, the most common drying method for paddy is a fixed bed dryer (Tohidi et al. 2017).

7.2.2 Storage

After rice grains are dried into safe storage moisture content, grains are stored to minimize the loss. Remarkably, stored rice is widely preferred against raw rice as improved flavor, texture, and improved milling yield were observed after storage. Thereby, rice storage can be called "rice aging" that is also one of the important steps for postharvest processing of paddy (Saikrishna et al. 2018). Rice storage plays an important step in the postharvest processing of grains due to storing rice under safe conditions prevents the loss of grains from undesirable changes such as excessive moisture, insects, and microorganisms. Storage conditions are strongly associated with storage temperature and duration. Storage temperature is normally ranging from 4 to 37 °C depending on the type of grain while storage duration is ranged from 3 to 6 months (Parnsakhorn and Noomhorm 2012). Hermetic storage is also widely used for rice preservation which avoids any interactions with the surrounding environment, for example, relative humidity, oxygen (O_2), and carbon dioxide (CO_2). However, under optimum conditions for rice storage extends the shelflife of grains in the prevention of insect survival, such as storage in the hermetic bag (85% RH, with low O_2 and high CO_2 contents) at 24 °C (Carvalho et al. 2019). Thus, facilities for rice storage are varied based on the number of grains, storage purposes, and storage location. Storage containers are widely used such as bag, bulk, or hermetic containers.

Natural Aging of Rice

Natural aging of rice occurs during postharvest drying of paddy. Since this method is naturally operated, it is considered a time-consuming method, requiring large space in the warehouse and this leads to higher operational and maintenance costs. However, quality deterioration of rice can also occur, which leads to the risks of insect infestation, rodents attack, and bird or animal attack. Hence, proper safe storage conditions, as well as careful monitoring, should be taken into account (Saikrishna et al. 2018).

Artificial Aging of Rice

Gujral and Kumar (2003) have studied the artificial aging of rice, which is a resemble with natural aging. It can also be accomplished by heating rough or milled rice to high temperatures also called parboiling of rice. As natural aging requires a longer aging duration, which generally takes 3–4 months minimally. Thereby, this method is an alternative approach to shorten the aging duration. Artificial aging of rice also known as accelerated aging of rice, which induces desired physicochemical changes rapidly.

7.2.3 Milling

Milling is a series of processes that converts rice grains into edible forms for human consumption. The basic objective of a rice milling system is to remove the husk and the bran layers and produce edible rice grains. The rice grain (rough rice or paddy) consists of an outer protective covering, hull (16–28% dry mass basis), and the rice caryopsis. After removing the hull, the rice caryopsis consists of the pericarp (1–2%), aleurone plus seed coat and nucellus (4–6%), the embryo (~2–3%), and starchy endosperm (89–94%) (Zhou et al. 2002). The rice grain (rough rice or paddy) consists of 20% rice hull or husk, 11% bran layers, and 69% starchy endosperm approximately (Dhankhar and Hissar 2014). To reveal the starchy endosperm, the rice should be milled through a series of milling processes such as husk removal, brans layer removal, and foreign matter removal. This process provides whole grains or head rice and broken rice grains. Thereby, grading and shorting must separate broken rice grains from whole grains or head rice.

A rice milling system can be classified into three systems including one-step, two-step, and multi-stage processes, depending on consumer requirements and degree of whiteness after milling. In an ideal milling process, the rice should have a minimum number of broken rice after milling (typically ranging from 12% to 15%) (Dhankhar and Hissar 2014).

One-Step Milling System

This milling system removes the bran and husk layers in a single-pass machine. The machine can produce white rice from paddy directly (Prabhakaran et al. 2017). This machine is also known as an iron huller, which is developed from the Engelberg coffee huller. Although it is easy to operate and a less time-consuming machine, it significantly reduces the HRY through breaking. G. G. Atungulu and Pan (2014) states that this machine continues to be highly considered in many rice-growing developing countries and it is also widely used in custom milling of rice at the household level. Thereby, a single pass is not widely used in the commercial milling of rice.

Two-Step Milling System

A two-stage milling system was developed from an Engelberg coffee huller or single-pass machine. It is also known as a compact rice mill. This milling system has better performance through the separation of hulling and polishing processes. The milling process was first operated using rubber rollers to remove the husk. Before a milling process, the husk and bran layers are removed separately which can be performed either in the same machine or different machines (Atungulu and Pan 2014). Consequently, the two-stage milling system produces a higher yield of head rice or a better milling yield compared to the single-pass milling system. Moreover, it is widely used for custom milling and in agricultural regions of rice production.

Multi-Stage Milling System

Most rice milling at the commercial level is commonly operated through a multi-stage milling system. This milling system involves a more complicated system to reduce grain breakage or increase milling yield. The series of multi-stage milling systems are listed in Table 7.1. Additionally, this milling system generates

Table 7.1 Keys of various unit operations in multi-stage rice milling (drawn by the author)

Unit operation	Purpose
Cleaning	Remove foreign matter such as stones, straw, weed seeds, soil, tree stump, and other materials from paddy rice (Dhankhar and Hissar 2014); "Odior and Oyawale (2011). Application of Time Study Model in Rice Milling Firm: A Case Study. *Journal of Applied Sciences and Environmental Management, 15*, 501–505.").
Dehulling/ dehusking	Remove husks from paddy through mechanical equipment. This method separates brown rice from the husk, which is subsequently available for the milling process.
Milling	Remove bran layer from brown rice through abrasion and friction into milled white kernels. Then, the bran layer is removed by air ventilation. This method typically takes 2–3 cycles based on the required milling degree
Polishing	Produce smooth and bright surface of rice grain through a rolling. This method further removes the bran layer from the germ normally between 8% and 10% of paddy based on the degree of whiteness (Bodie et al. 2019; "Odior and Oyawale (2011). Application of Time Study Model in Rice Milling Firm: A Case Study. *Journal of Applied Sciences and Environmental Management, 15*, 501–505.,").
Grading and sorting	Milled rice is mixed with different sizes and grades such as head rice and broken rice. The purpose of rice grading is to separate broken rice which is not up to the standard quality requirements. The common equipment is the rice grader, which can separate rice into different grades (Bodie et al. 2019; "Odior and Oyawale (2011) Application of Time Study Model in Rice Milling Firm: A Case Study. *Journal of Applied Sciences and Environmental Management, 15*, 501–505.,").

by-products of 31% approximately such as fine broken rice, rice germ, husks, and bran (Kennedy and Burlingame 2003). However, these by-products have been used in various means currently such as industrial and bioactive applications in both humans and animals instead of being discarded as waste.

7.3 The Effects of Postharvest Processing on the Physicochemical and Nutritional Properties of Rice

7.3.1 Effects of Drying on Rice Qualities

Due to the complex chemical and biochemical processes that can occur during the postharvest drying of food and agricultural materials (Chua et al. 2000), the quality attributes of dehydrated products such as color, texture, flavor, and nutritional value changes, are depending on the drying method and process conditions (Krokida et al. 2001; Antonio Vega-Gálvez et al. 2009; A. Vega-Gálvez et al. 2008). Therefore, the selection of reasonable drying technology and conditions is important to provide the appropriate drying time, good final product quality, and economical operation cost

(Beaudry et al. 2004; Antonio Vega-Gálvez et al. 2009). Several methods have been used for drying wet paddy; two common methods are natural drying and artificial drying. Even today, natural drying, i.e., sun drying of agricultural products, remains the most common drying method in many parts of the world. In this method, moist rice grains are spread on a flat surface under sunlight (approximately 2–3 days) until the moisture content reaches the expected value. Although the energy cost for the removal of water is very low for sun drying, it is labor-intensive, unreliable, and rewetting during a sudden downpour degrades the quality of the product (Madamba and Yabes 2005). While in artificial drying, mechanical dryers are used to remove water from wet grains by forcing either ambient air or heated air through the grains, e.g., fixed bed batch dryer, recirculating batch dryer, and continuous flow dryer. Artificial drying is commonly used to dry grains in the commercial rice milling industry. Drying is the most critical operation after harvesting rice crops. Thus, delays in drying, incomplete drying, or ineffective drying will reduce rice grain quality, which includes milling, cooking, and amylograph pasting properties of rice (Pearce et al. 2001). Previous research has shown that postharvest processing, i.e., drying method and storage time, have significant effects on the aroma and milling quality of rice (Wongpornchai et al. 2004). Drying temperature, drying time, and rate of water removal of rough rice affected the physical and physicochemical properties of milled and cooked rice, e.g., milling yield, color, cooking, and eating quality (Inprasit and Noomhorm 2001). Furthermore, drying treatments affected the milling yield, cooking properties, peak viscosity, and functional properties of rice (Pearce et al. 2001).

Crack and Fissure Formation during Paddy Drying

During the postharvest drying process, moisture and temperature gradients created inside the kernel induce stress that can cause the kernel to fissure (Cnossen et al. 2003; Cnossen et al. 2001; Dong et al. 2010). Fissured kernels usually break during the milling process and reduce head rice yield (HRY) (Cnossen et al. 2003; Dong et al. 2010; Iguaz et al. 2006). HRY is defined as the mass of head rice by the original mass of rough rice, and this value is used as the current standard for commercial rice milling quality (Cnossen et al. 2003; Iguaz et al. 2006). Fissured kernels are not only sensible to breakage during the milling process, they also influence the functional properties of milled rice (Cnossen et al. 2001). Generally, there are two different types of rice fissures: one is surface cracks, and the other is a large internal fissure that is usually found to be perpendicular to the long axis of the rice grain. Normally, a rice kernel with two or three cross-sectional fissures will lose its commercial price (Cnossen et al. 2003; Dong et al. 2010). Additionally, the value of broken grain is typically worth half the value of whole grain as the cooking quality of broken rice is very poor (Iguaz et al. 2006). Severe drying conditions can increase the number of fissured kernels after the drying process. This means that a higher drying temperature results in a higher percentage of fissured kernel (Dong et al. 2010). However, several authors have reported that proper tempering using high temperatures is able to

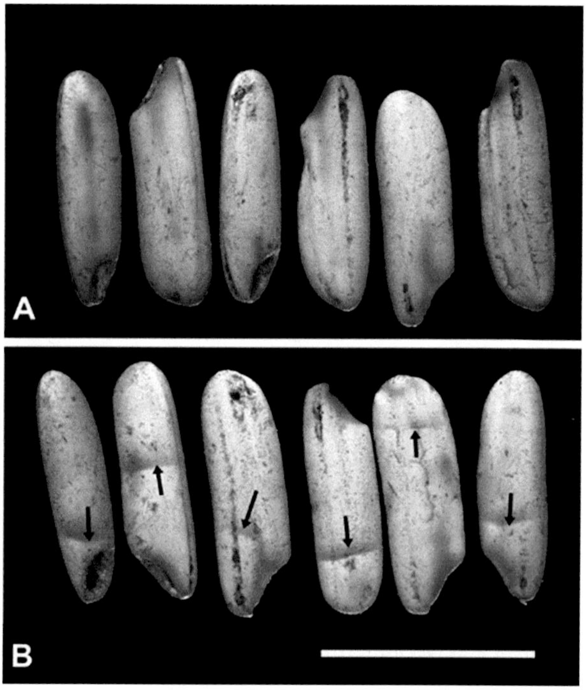

Fig. 7.2 Microscopy images on lateral side of uncooked rice kernels obtained from different drying conditions of unfissured kernels (**a**) and fissured kernels, particularly obtained from hot air drying at 115 °C (**b**). Scale bar shows 5 mm (Reprinted with permission from Donlao and Ogawa 2017)

decrease the amount of fissured kernel from severe drying conditions (Cnossen et al. 2003; Dong et al. 2010).

Breakage of the rice grain is one of the quality deteriorations that can occur due to grain fissuring before milling (Alhendi et al. 2019; Zhang et al. 2005). Previous studies analyze the effects of postharvest drying conditions on milling, textural, pasting, and digestive properties of non-waxy long-grain rice. Their results revealed that hot air drying (40 °C, 65 °C, 90 °C) and sun drying cause very few surface cracks and internal fissures in polished rice kernel (Fig. 7.2a) while internal fissures were found in high-temperature hot air drying at 115 °C (Fig. 7.2b). Hence, a factor that influences rice fissuring is drying temperature due to the increase in drying temperature increase internal fissuring of the kernel. Additionally, internal fissuring could induce void formation which is associated with material leaching material into the cooking water. Thus, the fissured kernel was more susceptible to breakage and head rice yield reduction which subsequently decreases the value of rice produced.

Physicochemical Properties

Numerous studies have investigated the effects of paddy drying on the quality of rice. Zheng et al. (2011) found that the hardness of cooked rice was significantly decreased as increasing in drying parameters (i.e., drying temperature, surface velocity, and initial moisture content), while increased drying temperature and surface velocity raised the cohesiveness of cooked rice. Chewiness and stickiness were significantly decreased when increasing drying temperature, while air velocity and initial moisture content showed no impact on these two parameters. Moreover, drying at higher temperatures is one of the critical factors that influenced the change in fatty acid content and the development of undesirable sensory characteristics in cooked rice (Zheng and Lan 2007). A study by Scariot et al. (2020) demonstrated that lipid content was significantly decreased with increasing drying temperature. Jaiboon et al. (2009) further reported that fluidized bed drying of waxy rice at higher temperatures led to only a small drop in the HRY, whereas the color of rice after drying was slightly darker and pale yellow. Moreover, the translucent kernels were observed when drying at 130 °C and the percentage of translucent kernels increased after the tempering process.

Nutritional Properties

Rice is a relatively good source of starch, protein, lipid, calcium, iron, thiamin, folate, and vitamin E. However, the chemical composition of rice grains varies widely, depending on the growing conditions, soil, and rice variety (Zhou et al. 2002). The major composition of rice is starch, which makes up about 70% of the whole grain. In particular, in the rice endosperm, the starch content ranged from 87.2% to 93.5%. Rice digestibility becomes a particular interest in the nutritional aspect of rice. The glycemic index (GI) is one important parameter in which well-established in vivo and in vitro to measure postprandial blood glucose-raising potential of carbohydrate-containing foods. Srisang et al. (2011) indicated that superheated steam drying could reduce the number of fissured germinated brown rice kernels, compared with hot air drying. However, the superheated steam drying treatment caused an agglomeration of the kernels as opposed to the hot air-drying treatment. Moreover, faster digestion (the higher GI value) of rice starch was obtained from superheated steam drying. A study by Donlao and Ogawa (2017) also indicated that the cooked rice obtained from the hot air-drying treatment showed a relatively lower GI than the rice sample treated with sun-drying treatment. In the hot air-drying treatment, the GI of cooked rice decreased with increasing drying temperature; this behavior is due to the changes in resistant starch content.

7.3.2 *Effects of Storage on Rice Qualities*

Physicochemical Properties

The darkening of rice generally occurs during long-term storage. Many published studies (Hayashi and Yanase 2016; Jungtheerapanich et al. 2017; Ziegler et al. 2017) revealed that the color of rice undergoes a darkening phenomenon during storage that normally associates with a decrease in lightness or L^* values. This phenomenon

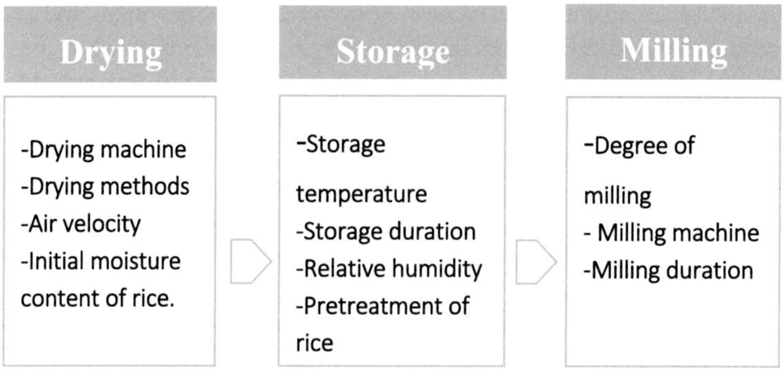

Fig. 7.3 List of postharvest processing factors affecting the physicochemical and nutritional properties of rice (drawn by the author)

resulted from various chemical reactions such as the reduction of polyphenol content, oxidative degradation of proanthocyanidins, and the Millard reaction. Moreover, the lower lightness (*L**) value of isolated starch was obtained from whole rice grains stored at 40 °C for 6 months (Ziegler et al. 2017) (Fig. 7.3).

The morphology of rice starch granules is an important parameter that impacts rice qualities, especially the pasting properties of rice. This parameter is generally measured through the size and shape and surface features (e.g., pores, smooth, coarse, and channels) of rice starch granules under a microscope. A study published by Zhou et al. (2016) revealed that the difference in the morphology between the two cooked rice samples was observed after storage in different conditions. The cooked rice obtained from paddy stored at lower temperatures has a smoother surface than those stored at higher temperatures.

Nutritional Properties

Pigmented rice has been considered high nutritional rice as it contains high content of bioactive compounds such as polyphenol and anthocyanin. According to published research by Hayashi and Yanase (2016), they investigated the changes in total phenolic content in red rice at different storage durations (i.e., fleshly harvested, and after 1, 2, and 5 years). Their results demonstrated that increasing storage duration decreased polyphenol content. The results revealed that the total polyphenol content of red rice dropped to 25% when the storage duration reached 5 years. Moreover, Zhou et al. (2016) investigated the effects of aging-induced changes in white rice grains during storage at 4 °C and 37 °C. The results indicated that aged rice stored at 37 °C has a lower starch digestion rate compared to rice stored at 4 °C. They concluded that storing at higher temperatures induced structural and starch digestibility changes.

7.3.3 *Effects of Milling on Rice Qualities*

Milled rice quality is largely affected by the degree of milling (DOM). It can be characterized by measuring the quantity of bran left on the surface of milled rice kernels or by assessing the level of removal of germ or bran layers from brown rice kernels after various milling procedures (Puri et al. 2014). Thus, there are various approaches in DOM determination, such as reflectance instruments (e.g., milling meter and color meter), light transmittance, Nano-image resonance technology, and image analysis.

Physicochemical Properties

Textural properties (e.g., stickiness, cohesiveness, and hardness) are one of the important physical qualities of milled or polished rice. These properties are also associated with the sensory characteristics of rice after cooking (Okabe 1979). The changes in textural properties were influenced by the milling process (i.e., milling machine and abrasive instrument) of rice paddy. The results revealed that under the same conditions, the milling method has a significant impact on the hardness value due to different milling methods generating variation on DOM. Higher DOM is associated with a higher capacity of water absorption in cooked rice, which led to a softer texture in cooked rice (Sasaki et al. 2015).

Nutritional Properties

According to the study by Lim et al. (2017), they studied the impact of the degree of milling (DOM) on rice constituents. Their results indicated that the reduction of sugar (e.g., D-fructose, D-galactose, and D-glucose) and sugar alcohol (e.g., D-arabitol, D-mannitol, and D-glucitol) was observed when increasing DOM. Additionally, the impacts of milling durations (i.e., 0, 15, 30, 45, and 60 s) on rice constituents, especially protein and lipid content, have been investigated by Olatunde and Atungulu (2018). The results obtained from their study demonstrated that protein content steadily decreased as the milling time increased. (Wang et al. 2021) found that the lipid content of rice obtained from different DOM varied from 0% to 12%. The same behavior was also observed by Li et al. (2021). Normally, most lipid contents are found in the outer layer of brown rice (e.g., aleurone and bran layer) (Bhattacharya 2013). Therefore, the removal of the aleurone and bran layer during the milling process significantly decreased the lipid contents. The changes in amino acids were also observed by Lim et al. (2017). They found that different DOM levels (i.e., 0, 5, 7, 9, and 11) impact the concentration of L-proline. The lowest concentration of L-proline was observed at higher DOM levels. Liu et al. (2017) also reported the change in the total content of amino acids (TAA) and various minerals in milled Chinese rice due to the impacts of DOM. Moreover, Wang et al. (2021)

reported that increasing DOM (0–12%) steadily decreased the content of vitamin E (E), vitamin B1 (B1), vitamin B2 (B2), and niacin (Na). Polyphenols are considered a high nutritional value constituents in rice, especially in pigmented rice. However, the polyphenols could be decreased during milling at different DOM (Choi et al. 2017). Moreover, DOM was also found to be important factors affecting the starch digestibility of rice (Wang et al. 2021). Sasaki et al. (2015) studied the effects of different milling of Japanese rice on in vitro starch digestibility and in vivo blood glucose response. The results showed that starch digestibility was influenced highly by the milling method.

7.4 Summary and Future Direction

Significant improvement has been made over the last few decades in terms of increasing rice productivity at the farm level and designing postharvest processing to minimize losses and retain the high nutritional value of rice. However, several published research has been demonstrated that postharvest processing of rice still significantly influenced various changes in rice, especially the physicochemical and nutritional properties of rice. Moreover, studies related to the impact of postharvest processing on human health benefit-related traits are still limited. This information has presented a huge opportunity for researchers to improve rice production and reduction in rice postharvest losses. Further investigation may contribute to a better understanding of the impacts of postharvest processing on rice grain quality and provide insights into potential improvements in rice production.

References

Alhendi A, Al-Rawi S, Jasim A (2019) Effect of moisture content of two paddy varieties on the physical and cooked properties of produced rice. Braz J Food Technol 22

Aquerreta J, Iguaz A, Arroqui C, Vírseda P (2007) Effect of high temperature intermittent drying and tempering on rough rice quality. J Food Eng 80(2):611–618

Atungulu GG, Pan Z (2014) Rice industrial processing worldwide and impact on macro- and micronutrient content, stability, and retention. Ann N Y Acad Sci 1324:15–28

Atungulu GG, Sadaka S (2019) 15 - postharvest technology: Rice drying. In: Bao J (ed) Rice, 4th edn. AACC International Press, pp 473–515

Beaudry C, Raghavan GSV, Ratti C, Rennie TJ (2004) Effect of four drying methods on the quality of osmotically dehydrated cranberries. Dry Technol 22(3):521–539

Bhattacharya K (2013) Degree of milling (DM) of rice and its effect. In: Rice quality, pp 100–115

Bodie AR, Micciche AC, Atungulu GG, Rothrock MJ, Ricke SC (2019) Current trends of rice milling byproducts for agricultural applications and alternative food production systems. Front Sustain Food Syst 3:47

Carvalho MO, Fradinho P, Martins MJ, Magro A, Raymundo A, de Sousa I (2019) Paddy rice stored under hermetic conditions: the effect of relative humidity, temperature and storage time in suppressing Sitophilus zeamais and impact on rice quality. J Stored Prod Res 80:21–27

Choi S, Seo H-S, Lee K, Lee S, Lee J (2017) Effect of cultivars and milling degrees on free and bound phenolic profiles and antioxidant activity of black rice. Applied. Biol Chem 61

Chua KJ, Mujumdar AS, Chou SK, Hawlader MNA, Ho JC (2000) Convective drying of banana, guava and potato pieces : effect of cyclical variations of air temperature on drying kinetics and color change. Dry Technol 18(4–5):907–936

Cnossen AG, Jimenez MJ, Siebenmorgen TJ (2003) Rice fissuring response to high drying and tempering temperatures. J Food Eng 59(1):61–69

Cnossen AG, Siebenmorgen TJ, Yang W, Bautista RC (2001) An application of glass transition temperature to explain rice kernel fissure occurrence during the drying process*. Dry Technol 19(8):1661–1682

Das I, Das SK, Bal S (2009) Drying kinetics of high moisture paddy undergoing vibration-assisted infrared (IR) drying. J Food Eng 95(1):166–171

Dhankhar P, Hissar T (2014) Rice milling. IOSR J Eng 4(5):34–42

Dondee S, Meeso N, Soponronnarit S, Siriamornpun S (2011) Reducing cracking and breakage of soybean grains under combined near-infrared radiation and fluidized-bed drying. J Food Eng 104(1):6–13

Dong R, Lu Z, Liu Z, Koide S, Cao W (2010) Effect of drying and tempering on rice fissuring analysed by integrating intra-kernel moisture distribution. J Food Eng 97(2):161–167

Dong R, Lu Z, Liu Z, Nishiyama Y, Cao W (2009) Moisture distribution in a rice kernel during tempering drying. J Food Eng 91(1):126–132

Donlao N, Ogawa Y (2017) Impact of postharvest drying conditions on in vitro starch digestibility and estimated glycemic index of cooked non-waxy long-grain rice (Oryza sativa L.). J Sci Food Agric 97(3):896–901

Firouzi S, Alizadeh MR, Haghtalab D (2017) Energy consumption and rice milling quality upon drying paddy with a newly-designed horizontal rotary dryer. Energy 119(Supplement C):629–636

Golmohammadi M, Rajabi-Hamane M, Hashemi SJ (2012) Optimization of drying–tempering periods in a paddy rice dryer. Dry Technol 30(1):106–113

Gujral HS, Kumar V (2003) Effect of accelerated aging on the physicochemical and textural properties of brown and milled rice. J Food Eng 59(2):117–121

Hamzah A, Ahmad MT, Shahar A, Sahari Y, Ahmad R (2019) Post-harvest management of rice for sustainable food security in Malaysia. FFTC-AP, Taipei, Taiwan

Hayashi S, Yanase E (2016) A study on the color deepening in red rice during storage. Food Chem 199:457–462

Iguaz A, Rodriguez M, Virseda P (2006) Influence of handling and processing of rough rice on fissures and head rice yields. J Food Eng 77(4):803–809

Inprasit C, Noomhorm A (2001) Effect of drying air temperature and grain temperature of different types of dryer and operation on rice quality. Dry Technol 19(2):389–404

Jaiboon P, Prachayawarakorn S, Devahastin S, Soponronnarit S (2009) Effects of fluidized bed drying temperature and tempering time on quality of waxy rice. J Food Eng 95(3):517–524

Jungtheerapanich S, Tananuwong K, Anuntagool J (2017) Aging kinetics of low amylose rice during storage at ambient and chilled temperatures. Int J Food Prop 20

Kennedy G, Burlingame B (2003) Analysis of food composition data on rice from a plant genetic resources perspective. Food Chem 80(4):589–596

Krokida MK, Maroulis ZB, Saravacos GD (2001) The effect of the method of drying on the colour of dehydrated products. Int J Food Sci Technol 36(1):53–59

Li H, Xu M, Chen Z, Li J, Wen Y, Liu Y, Wang J (2021) Effects of the degree of milling on starch leaching characteristics and its relation to rice stickiness. J Cereal Sci 98:103163

Lim D, Phuoc Long N, Choo S, Mo C, Dong Z, Kim G, Kwon S (2017) Impact of milling on rice constituents (Oryza Sativa L.): a metabolomic approach. Anal Lett 50

Liu K-L, Zheng J-B, Chen F-S (2017) Relationships between degree of milling and loss of vitamin B, minerals, and change in amino acid composition of brown rice. LWT Food Sci Technol 82:429–436

Madamba PS, Yabes RP (2005) Determination of the optimum intermittent drying conditions for rough rice (Oryza sativa, L.). LWT Food Sci Technol 38(2):157–165
ManzoorAhmad S, Munir A, Wajid A (2015) Design, development and performance evaluation of a small-scale solar assisted paddy dryer for on farm processing. J Food Process Technol 06
Mukhopadhyay S, Siebenmorgen TJ, Mauromoustakos A (2019) Effect of tempering approach following cross-flow drying on rice milling yields. Dry Technol 37(16):2137–2151
Odior A, Oyawale F (2011) Application of time study model in Rice milling firm: a case study. J Appl Sci Environ Manag 15:501–505
Okabe M (1979) Texture measurement of cooked rice and its relationship to the eating quality. J Texture Stud 10(2):131–152
Olatunde GA, Atungulu GG (2018) Milling behavior and microstructure of rice dried using microwave set at 915 MHz frequency. J Cereal Sci 80:167–173
Olatunde GA, Atungulu GG, Smith DL (2017) One-pass drying of rough rice with an industrial 915 MHz microwave dryer: quality and energy use consideration. Biosyst Eng 155(Supplement C):33–43
Pan Z, Khir R, Godfrey LD, Lewis R, Thompson JF, Salim A (2008) Feasibility of simultaneous rough rice drying and disinfestations by infrared radiation heating and rice milling quality. J Food Eng 84(3):469–479
Parnsakhorn S, Noomhorm A (2012) Effects of storage temperature on physical and chemical properties of brown rice, parboiled brown rice and parboiled paddy. Thai J Agric Sci 45:221–231
Pearce MD, Marks BP, Meullenet JF (2001) Effects of postharvest parameters on functional changes during rough rice storage. Cereal Chem 78(3):354–357
Poudel S, Poudel P, Poudel S, Pokhrel S, Aryal S (2021) Impact of postharvest operations on Rice grain quality: a review. Rev Food Agric 2(1):20–23
Prabhakaran P, Ranganathan R, Kumar V, Rathanasamy R, Devakumar L, Pal SK (2017) Review on parameters influencing the rice breakage and rubber roll wear in sheller. Arch Metall Mater 62
Puri S, Dhillon B, Sodhi N (2014) Effect of degree of milling (Dom) on overall quality of rice—a review. Int J Adv Biotechnol Res 5
Rumruaytum P, Borompichaichartkul C, Kongpensook V (2014) Effect of drying involving fluidisation in superheated steam on physicochemical and antioxidant properties of Thai native rice cultivars. J Food Eng 123(Supplement C):143–147
Saikrishna A, Dutta S, Subramanian V, Moses JA, Anandharamakrishnan C (2018) Ageing of rice: a review. J Cereal Sci 81:161–170
Sasaki T, Okunishi T, Sotome I, Okadome H (2015) Effects of milling and cooking conditions of rice on in vitro starch digestibility and blood glucose response. Cereal Chem J 93
Scariot MA, Karlinski L, Dionello RG, Radünz AL, Radünz LL (2020) Effect of drying air temperature and storage on industrial and chemical quality of rice grains. J Stored Prod Res 89:101717
Srisang N, Varanyanond W, Soponronnarit S, Prachayawarakorn S (2011) Effects of heating media and operating conditions on drying kinetics and quality of germinated brown rice. J Food Eng 107(3):385–392
Tohidi M, Sadeghi M, Torki-Harchegani M (2017) Energy and quality aspects for fixed deep bed drying of paddy. Renew Sust Energ Rev 70(Supplement C):519–528
Vega-Gálvez A, Di Scala K, Rodríguez K, Lemus-Mondaca R, Miranda M, López J, Perez-Won M (2009) Effect of air-drying temperature on physico-chemical properties, antioxidant capacity, colour and total phenolic content of red pepper (Capsicum annuum, L. var. Hungarian). Food Chem 117(4):647–653
Vega-Gálvez A, Lemus-Mondaca R, Bilbao-Sáinz C, Fito P, Andrés A (2008) Effect of air drying temperature on the quality of rehydrated dried red bell pepper (var. Lamuyo). J Food Eng 85(1):42–50

Wang T, Khir R, Pan Z, Yuan Q (2017) Simultaneous rough rice drying and rice bran stabilization using infrared radiation heating. LWT Food Sci Technol 78(Supplement C):281–288
Wang Z, Zhang M, Liu G, Deng Y, Zhang Y, Tang X et al (2021) Effect of the degree of milling on the physicochemical properties, pasting properties and in vitro digestibility of Simiao rice. Grain Oil Sci Technol 4(2):45–53
Wiriyaumpaiwong S, Soponronnarit S, Prachayawarakorn S (2001) Effect of temperature on soybean quality using spouted bed technique. Agric Nat Res 35(4):460–466
Wongpornchai S, Dumri K, Jongkaewwattana S, Sirri B (2004) Effects of drying methods and storage time on the aroma and milling quality of rice (Oryza sativa L.) cv. Khao Dawk Mali 105. Food Chem 87(3):407–414
Zhang Q, Yang W, Sun Z (2005) Mechanical properties of sound and fissured rice kernels and their implications for rice breakage. J Food Eng 68(1):65–72
Zheng X & Lan Y (2007) Effects of drying temperature and moisture content on rice taste quality. In The Proceedings of the 5th Asia-Pacific Drying Conference (pp. 1112–1117): World Scientific Publishing Company
Zheng X-Z, Liu C-H, Chen Z-Y, Ding N-Y, Jin C-J (2011) Effect of drying conditions on the texture and taste characteristics of rough Rice. Dry Technol 29(11):1297–1305
Zhou Z, Robards K, Helliwell S, Blanchard C (2002) Composition and functional properties of rice. Int J Food Sci Technol 37(8):849–868
Zhou Z, Yang X, Su Z, Bu D (2016) Effect of ageing-induced changes in rice physicochemical properties on digestion behaviour following storage. J Stored Prod Res 67:13–18
Ziegler V, Ferreira CD, Goebel JT, El Halal SL, Santetti GS, Gutkoski LC et al (2017) Changes in properties of starch isolated from whole rice grains with brown, black, and red pericarp after storage at different temperatures. Food Chem 216:194–200

Chapter 8
Cooking of Rice

Masatsugu Tamura

Abstract The cooking methods of rice are indeed diverse and vary across different cultures, histories, and dietary habits. In this section, the preference and cooking methods of rice in the world are introduced in brief, followed by a detailed exploration of the history of rice cooking and development of rice cookers in Japan. Subsequently, the cooking process of rice with limited water is discussed and the changes of tissue structure, moisture, starch as well as the texture of rice grains during cooking are explained. Finally, delve into potential future advancement in cooking methods aimed at improving both the palatability and functionality of rice.

Keywords Rice · Cooking methods · Gelatinization · Microstructure · Quality · Digestibility

Abbreviations

DOP	Degree of polishing
eGI	Estimated glycemic index
EWM	Excess water method
GABA	γ-aminobutyric acid
GI	Glycemic index
HP	High-pressure
HR	Head rice
IH	Induction heating
LWM	Limited water method
RDS	Rapidly digested starch
RS	Resistant starch
SDS	Slowly digested starch

M. Tamura (✉)
Utsunomiya University, Utsunomiya, Japan
e-mail: m-tamura@cc.utsunomiya-u.ac.jp

J. Tian et al. (eds.), *Science of Rice Chemistry and Nutrition*,
https://doi.org/10.1007/978-981-99-3224-5_8

VIP Vacuum Insulation Panel
W/R Water-to-rice

8.1 Introduction

Rice is one of the world's three largest cereal crops along with wheat and maize. In 2019, 504 Mt. of rice (polished rice) was produced, of which 452 Mt. was produced in Asia, making it a major food resource (FAOSTAT 2021). Only two types of rice are cultivated worldwide: Asian rice (*Oryza sativa*) and African rice (*Oryza glaberrima*). Asian rice is widely cultivated worldwide and is broadly referred to as rice. Asian rice is often classified into two subspecies or ecotypes, Japonica (*Oryza sativa ssp. japonica*) and Indica (*Oryza sativa* ssp. *indica*), based on morphological, physiological, and genetic characteristics. This broadly defined japonica species are sometimes divided into temperate japonica (japonica in the narrow sense) and tropical japonica (javanica). The world's predominant type of rice is indica, a long-grained variety with high-amylose content that is grown at low latitudes from southwestern China to Southeast Asia and South Asia. Japonica rice, on the other hand, is grown at high latitudes, such as in Japan and the Korean Peninsula, and is either a medium- or short-grain variety with relatively low amylose content.

Wheat and corn are milled and prepared by adding secondary ingredients and seasonings. In contrast, rice is typically prepared and consumed as whole grains. In other words, rice is prepared after the bran is separated from the rice grains through the process of polishing, then cooked by applying heat and adding water. This cooking method has been eveloved over time in various countries and regions, taking into account different preferences, rice varieties, cultivation techniques, harvesting practices, and storage conditions. This chapter presents the preferences for rice, cooking methods, and effects of rice cooking on rice.

8.2 Preferences and Cooking Methods of Rice

Although there are countless articles and books on the spread of rice as a plant, few books discuss the relationship between rice as a food source and civilization (Sato 2003). Among ancient civilizations, Mesopotamian civilization is considered to have been the earliest to have practiced agriculture. Mesopotamian civilization was thought to have cultivated and consumed wheat and barley in a primitive manner. Egyptian civilization was directly influenced by Mesopotamian civilization, and wheat and barley were considered to be the main cereals. In the Yellow River civilization, millet and other minor grains appeared in the early stages but were replaced by wheat by the time of the Shang and Zhou dynasties. In the Americas, the Mayan and Inca civilizations were supported by corn and potatoes. As described above, rice has never been seen in the formation of world civilization. Although

Table 8.1 Essential amino acid composition of main cereals (mg/g protein by amino acid composition) (Ministry of Education, Culture, Sports, Science and Technology 2020) (reprinted from the Ministry of Education, Culture, Sports, Science and Technology 2020)

	Brown rice (*japonica*)	Non-waxy polished rice (*japonica*)	Waxy polished rice (*japonica*)	Polished rice (*indica*)	Wheat	Corn
Isoleucine	46	47	48	47	41	43
Leucine	93	96	95	95	79	170
Lysine	45	42	41	42	24	20
Methionine	27	28	28	33	19	27
Phenylalanine	60	62	61	61	57	60
Threonine	45	44	43	45	33	38
Tryptophan	17	16	16	17	13	5.8
Valine	70	69	70	69	49	53
Histidine	32	31	30	29	26	33

there are many theories, rice is thought to have originated in the middle and lower reaches of the Yangtze River 6000–7000 years ago (Takano and Taniguchi 2021). Table 8.1 shows the essential amino acid composition of the main cereals (Ministry of Education, Culture, Sports, Science, and Technology 2020). Rice contains more lysine, an essential amino acid, than wheat or corn; therefore, it is less likely to cause malnutrition even if eaten alone, and it contains no gluten or allergens. For this reason, the number of cultures that have chosen rice as their staple grain has never decreased, and it is said that "no ethnic group has ever stopped eating rice after learning to do so." From the rice and processed rice products eaten today, it can be inferred that basic rice cooking and manufacturing techniques have developed over the years and are closely related to the types and characteristics of raw rice in each region and people's preferences. Currently, a variety of rice dishes around the world are adapted to the local rice varieties, characteristics, and food cultures. In this section, the preferences and cooking methods for rice in the world are introduced, as well as the history of rice cooking in Japan and the birth and development of rice cookers.

8.2.1 Rice Preferences

Consumer preferences for rice differ greatly among nationalities. Several studies classified preferences and clarified their trends. Rice grain preferences in the worldwide can be broadly classified into regions that prefer sticky rice grains, such as Japan and Korea; regions that prefer non-sticky rice grains, such as South Asia; and regions that prefer rice grains with ungelatinized starch in the core, such as Italy. The rice preferences of several countries can be classified into five categories by discriminant analysis: (1) long-grain preference (Southeast Asian and South Chinese) (2) short grain preference (Japanese/Korean, North Chinese/Taiwanese), (3) Basmati preference (South Asian/Middle East); (4) nonspecific preference (American/Canadian, European, Australian/New Zealander, and other nationalities), and (5) Jasmine preference (Thailand) (Suwannaporn and Linnemann 2008). Suwannaporn and Linnemann (2008) concluded that eating quality, especially hardness, stickiness, and aroma, were the best discriminators for each country. According to Custodio et al. (2019), consumers in Southeast Asian countries, namely, Thailand, the Philippines, Vietnam, and Cambodia, view the characteristics of premium rice quality as nutritional benefits, soft texture, long and slender grains, whiteness, translucency, aroma, and high-head rice (HR) (i.e., minimal or no broken grains), while consumers in South Asian countries, namely, Eastern India and Bangladesh, consider the appearance of cooked rice grains (uniformity of size and shape, whiteness, and fineness) and the feeling of fullness to be important (Table 8.2) (Custodio et al. 2019). Regarding the characteristics of low-quality rice in Southeast Asia, hardening due to cooling is always mentioned, whereas in South Asia, sticky is always listed. Furthermore, differences between countries are also shown; for example, consumers in Thailand, Cambodia, and Eastern India consider aroma to be a characteristic of good- and premium-quality rice, but consumers in the Philippines and Bangladesh consider good-quality rice to be non-aromatic and have no bad smell, respectively. The grain size and shape requirements for high-quality rice also appeared to differ across the countries surveyed. In Cambodia and Bangladesh, rice with medium-sized grains is considered to be of good quality, whereas in Thailand, the Philippines, Vietnam, and eastern India, good-quality rice is long and slender. A study of the perception and acceptance of the texture of cooked rice (2 short grain, 2 medium grain, 2 long grain, and 2 waxy grain) between Koreans and Americans indicates the following (Choi et al. 2018). Americans give high marks to cooked rice grain for visual stickiness, stickiness, and chewiness, while Koreans give high marks for hardness. For Americans, chewiness is considered a negative response, whereas for Koreans, visual stickiness and stickiness are identified as positive responses to the overall liking for cooked rice grains. For both Americans and Koreans, familiarity with cooked rice was a prominent factor that increased their overall positive response. In Japan, sticky and soft rice varieties have been widely accepted by consumers a long time; however, in recent years, there has been a demand for more sticky, soft, and grainy rice, suggesting that rice preferences are becoming more diverse (Machida et al. 2020).

8.2.2 Cooking Diversity and Preferences

Ishitani (2009) classified rice cooking methods of rice worldwide according to the amount of water added, the method of watering and heating (cooking method 1), and the cooking method after heating (cooking method 2). Table 8.3 shows the cooking method for rice grains (Ishitani 2009) with minor modifications to this classification method. The properties of rice-based dishes vary significantly depending on the amount of water added. Porridge, zousui (porridge of rice and vegetables containing

Table 8.2 Perception of premium-, good-, and low-quality rice of value chain actors in selected locations in South and Southeast Asia in 2013–2014 (Custodio et al. 2019)

Location	Attribute	Premium quality	Good/Medium quality	Low quality
Thailand	Texture	Soft[d]; Slightly sticky (in the Northeast region)	Soft; Becomes hard when cooled; Slightly sticky	Becomes hard when cooled
	Size and shape	Long[d]	Long[d]	Short[d]
	Color	White; Translucent; Not chalky	White; Translucent but less than premium; Not chalky[d]	Chalky[d]
	Aroma	With fragrance[d]	With fragrance	No fragrance
	Purity	No impurities[a]		
	Homogeneity	High HR; Uniform size, shape and color	With variations in size and shape; With 5% broken grains and color variation	Higher % of broken grains; With variations in size, shape and color
	Others	Shiny; Good cooking quality; High nutritional value[d]; High volume expansion	High-volume expansion	Poor cooking quality; Not tasty
Philippines	Texture	Soft[d]	Moderately soft after cooking but hard when cooled	Becomes hard when cooled
	Size and shape	Long[d]; Slender	Long[d]; Slender	
	Color	White[d]		Yellowish; Discolored[d]
	Aroma	With fragrance	No fragrance	With bad smell[d]
	Purity			
	Homogeneity	High HR[d]	With 5–15% broken grains; High HR[d]	With 25% broken grains[d]
	Others	High volume expansion; Tasty[d]		Not shiny
Vietnam	Texture	Soft[d]; Sticky[d]		Non-sticky; Becomes hard when cooled[d]; Dry[d]
	Size and shape	Slender	Long	Short[d]
	Color	White; Translucent		Yellowish; Chalky[d]
	Aroma	With fragrance[d]		
	Purity			With impurities[a]
	Homogeneity	High HR	With 10–15% broken grains	High % of broken grains; With

(continued)

Table 8.2 (continued)

Location	Attribute	Premium quality	Good/Medium quality	Low quality
				variations in size, shape and color
	Others	High nutritional value[d]; Tasty; Low volume expansion	Sweet; Medium volume expansion	Not sweet; High volume expansion[d]
Cambodia	Texture	Soft; soft and sticky	Soft but not as soft as premium; Soft and sticky; Not too sticky	Becomes hard when cooled
	Size and shape	Slender; Long	Slender; Medium size	
	Color	Translucent	Less chalky grains	Chalky
	Aroma	With fragrance	With fragrance	No fragrance
	Purity	No impurities[a]		
	Homogeneity	High HR	High HR; Uniform size, shape and color	High % of broken grains
	Others	Shiny; Tasty	Shiny; Tasty; Longevity[b]	Not shiny; Not tasty
Eastern India	Texture	Soft[d]; Non-sticky[d]		Sticky[d]; Very soft
	Size and shape	Long; Slender[d]; Very fine to fine[c,d]	Less slender than premium; Fine[c]	Short[d]; Bold[d]
	Color	White		Discolored
	Aroma	With fragrance	With fragrance; Less aromatic than premium	No fragrance
	Purity	No impurities[a]	No impurities[a]	With impurities[a]
	Homogeneity	High HR; Uniform size, shape and color[d]	High HR; With 5% broken grains, color variation and admixture	High % of broken grains; With variations in size, shape and color[d]
	Others	Shiny; Less cooking time[d]; Very tasty[d]; Easy to digest	Polished but not very shiny; Tasty	Long cooking time; Not tasty[d]
Bangladesh	Texture	Non-sticky	Non-sticky	Sticky; Soft/ becomes too soft if cooked rice is soaked in water overnight
	Size and shape	Long[d]; Slender; Very fine to fine[c,d]	Fine to medium fine[c,d]; Medium size	Coarse[c,d]; Bold
	Color	White	White but not as white as premium; White even if parboiled	Not very white
	Aroma	With fragrance[d]	No bad smell	With bad smell

(continued)

Table 8.2 (continued)

Location	Attribute	Premium quality	Good/Medium quality	Low quality
	Purity		With 5% impurities[a, d]	With impurities[a]
	Homogeneity		With 5% broken grains	Higher % of broken grains
	Others	Tasty[d]; Longevity[b]	Tasty[d]; Longevity[b]	Not tasty[d]

[a]Impurities refer to black grains, stones, or husks
[b]Longevity refers to cooked rice which does not easily spoil
[c]There are different classifications of grains in South Asia Very fine to fine grains refer to extra-long and long slender grains, medium fine grains refer to medium slender grains, coarse grains refer to short bold grains
[d]Refer to attribute, which was also mentioned by farmers in the focus group discussions

meat or fish), and risotto are dishes that require large amounts of water. However, the amount of water added to the steamed rice is small. In addition to water, other liquids can be added to rice, such as milk, tea, soup, and fruit juice. There are two common types of rice cooking methods: limited water method (fixed water-to-rice ratio, LWM) and excess water method (excess water/minimum cooking time, EWM) (Tamura et al. 2021). Other methods include frying in oil, cooking with water, steaming, baking, or frying after cooking rice.

Specific research has also been conducted on rice-cooking methods, eating habits, and the interests of consumers in various countries. Suwannaporn et al. (2008) showed that the preferences of several consumers of different nationalities decreased as the degree of processing increases. In other words, natural, unprocessed rice grains are most preferred, followed by semi-processed products, such as portioned rice and nutritionally enhanced rice. This is related to the healthy image of consumers, inferred from the degree of processing and additives. Son et al. (2013) showed consumers' attituded toward rice cooking processes in Korea, Japan, Thailand, and France. The most important motivation for choosing a cooking method is obtaining the desired taste and texture, such as pressure cooking for Korean consumers, soaking for Japanese consumers, and rapid boiling for French consumers. Korean and Japanese consumers have highly positive attitudes toward both brown rice germination and the addition of other types of rice, cereals, and beans cooking methods in terms of nutrition and health compared to Thai and French consumers. It is also found that positive attitudes toward the safety aspects of rinsing or washing rice are common in all countries. The authors concluded that the more popular a cooking process is in a country, the more people tend to develop positive beliefs. Murakami and Aida (2010) surveyed the consumption and cooking of rice in the Tohoku and Hokkaido regions, which were the major rice-producing areas in Japan. The most preferred rice dishes were cooked white rice, Japanese seasoned rice, and nigiri-zushi (hand-rolled sushi), whereas risotto, porridge, and oshi-zushi (pressed sushi) were less preferred. The frequency of rice cooking is 1–3 times a week for rice balls, 1–2 times a month for fried rice, and several times a year for

Table 8.3 Cooking method for rice grains (Ishitani, 2009)

Cooking method 1		Cooking method 2	Name of dish	Typical countries and regions
Cooking with appropriate amount of water.	Cooking with appropriate amount of water.	Using as-is (japonica rice)	Cooked white rice	Japan, Korea, North area of China
		Adding the ingredients and mixing (japonica rice)	Mixed rice	Japan
			Bibimbap	Korea
		Forming with the ingredients (japonica rice)	*Onigiri*, *Omusubi* (Rice ball)	Japan
		Putting the ingredients on top	Com dia	Vietnam
			Katsudon, *Tendon*, *Oyakodon*, etc.	Japan
		Pouring the soup with the ingredients	Curry and rice	South Asia and Southeast Asia
			Hashed meat with rice	Japan
			Feijoada	Brazil
		Pouring hot water or tea over it.	*Ochazuke*	Japan
		Add hot soup and ingredients	Bokbap, Kalbi Bokbap, etc.	Korea
		Add the vinegar and top with the filling or mixing. (japonica rice)	*Sushi*	Japan
		Adding broth and ingredients and cooking (japonica rice).	*Zousui* (Porridge), *Ojiya* (Gruel)	Japan
		Adding soup and ingredients and cooking.	Risotto	Italy
			Squid ink risotto	Mediterranean region

(continued)

Table 8.3 (continued)

Cooking method 1		Cooking method 2	Name of dish	Typical countries and regions
		Adding the ingredients and frying	Fried rice	China
			Nasi goreng	Indonesia and Malaysia
		Stir-frying with pineapple	Khao phat, Thai pineapple steamed rice	Thai
		Covering with white sauce and baking in the oven	Doria, Rice gratin	Europe and the United States
		Stir-frying chicken, onions, etc. with ketchup	Chicken rice, Ketchup rice, Omelet rice	the United States and Japan
		Adding cream, eggs, milk and sugar, and steaming	Rice pudding	Europe and the United States
		Chilling with gelatin to harden	Rice bavarois	
		Stuffing the squid and cooking (japonica rice)	*Ikameshi*	Japan
		Mashing grains well, shaping and baking (Japonica rice)	*Kiritanpo*	Japan
	Cooking with appropriate amount of water, pressurizing a little, and steaming.		Cooked white rice	Japan
	Pressure cooking with appropriate amount of water (brown rice).		Cooked brown rice	Japan
	Repeating cooking with appropriate amount of water (brown rice).		Cooked brown rice	Japan
Cooking with various ingredients	Cooking with a small amount of black rice or millet (japonica rice)		Multigrain rice	Japan

(continued)

Table 8.3 (continued)

Cooking method 1		Cooking method 2	Name of dish	Typical countries and regions
	Cooking with red beans, green beans, chestnuts, etc. (japonica rice)		*Sekihan* (Red rice), *Kurigohan*	Japan
	Cooking the ingredients and glutinous rice with seasonings (japonica rice)		*Okowa*	Japan
	Rolling meat, rice, and chopped vegetables in grape leaves or cabbage and stewing.		Sarmasi (Sarma)	Balkan Peninsula
	Stuffing the rice with green pepper and tomatoes and cooking.		*Dolma*	Turkey and Balkan Peninsula
	Stewing together with chicken and avocado slices.		*Seco de pollo*	Balkan Peninsula
	Stewing with tomatoes and vegetables		Jollof rice	West Africa
Boiling with water	Discarding the hot water in the middle of cooking and let the rice steam (indica rice)		Cooked rice	North Asia and Southeast Asia
	Boiling in a large amount of water, draining the water, and placing in a basket to steam (indica rice)		Cooked rice	
	Boiling in a large amount of water (indica rice)		Boiled rice	
Cooking with more water (japonica rice and indica rice)			Rice porridge	China, Southeast Asia, Europe and the United States
Steaming (Waxy rice)	Steaming in a basket		Thai sticky rice	Thailand
	Soaking and steaming (Japonica rice)	Adding cooked wild vegetables.	Steamed rice with wild vegetables	Japan
	Steaming with cooked red beans, green beans, chestnuts, etc. (japonica rice)		*Sekihan* (Red rice), *Kuriokowa*	Japan
	Soaked glutinous rice is filled with ingredients,			

(continued)

Table 8.3 (continued)

Cooking method 1		Cooking method 2	Name of dish	Typical countries and regions
	wrapped in bamboo leaves or bamboo grass leaves, and steamed		Waxy rice dumpling, *Chimaki*	China, Korea, Japan, and Southeast Asia
	Wrapping the lotus leaves and ingredients together and steaming		Waxy rice dumpling	Yunnan, China
	Steam in a bamboo tube	Adding coconut milk.	Sweet waxy rice in bamboo	Thailand
Cooking with something other than water	Cooking with milk.		Milk rice	Europe and the United States
			Arroz con leche	Spain, Central and South America
	Cooking porridge in broth with vegetables.		Rice soup	Europe and the United States
	Cooking in broth with butter		Butter rice	
	Cooking with bacon and vegetables		Minestrone	Italy
	Making a ball with raw rice, vegetables and minced meat, boiling in broth, wrapping in grape leaves, and boiling		Dolmadakia	Greece
	Boiling and steaming the ingredients and raw rice in a small amount of broth		*Paella*, Squid ink rice	Spain and Philippines
Boiling or cooking rice using the hot water method.	Boiling or cooking rice using the hot water method	Using as is	Rice salad	
		Pour curry on top	Curry rice	South Asia, Southeast Asia, South area of China, etc.
Frying raw rice	Frying in butter	Cooking with broth	Butter rice	
		Cooking with fried onions	Onion rice	
		Adding saffron and cooking	Saffron rice	India

(continued)

Table 8.3 (continued)

Cooking method 1		Cooking method 2	Name of dish	Typical countries and regions
		Cooking with turmeric	Turmeric rice	
	Frying in oil	Cooking with orange juice	*Pulau*	Afghanistan
		Cooking with tomato juice	*Biryani*	Pakistan
		Cooking with ingredients and soup	Pilaf	Turkey, Europe and the United States, India, Iran, Russia
		Cooking with chicken and broth	*Arroz con pollo*	Peru
		Cooking bacon, ham and vegetables together in the oven.	*Jambalaya*	South America
Cooking rice, shaping, and baking or frying.	Cooking rice, filling with ingredients, forming into a ball, and frying.		Rice balls with cheese	Italy
			Rice croquette	
	Cooking rice, making hamburger steak with ground meat and vegetables, and baking		Rice Hamburg	
	Cooking rice, forming into a thin layer, and baking or frying	Pouring into a soup	Scorched rice in soup	Korea, Madagascar, China, Iran
Cooking rice and fermentation.		Lactic acid fermentation	*Narezushi* (Fermented sushi)	Japan, China, and Southeast Asia
		Lactic acid fermentation, alcoholic fermentation	*Tapai, Tape*	Indonesia, Brunei

mochi (rice cake). The frequency of rice consumption is particularly high: 46.6% of consumers consume three servings of rice per day, 38.4% consume two servings, and 11.6% consume one serving. In particular, 91.1% of the consumers eat rice as a staple food for dinner. Hatae et al. (1997) also showed that rice is consumed an average of 12.1 times per week in Japan, while the frequency of eating rice in China, Spain, and France is 8.2, 0.9, and 0.9 times per week, respectively. Cooked plain rice is the most popular menu item in China. It is sometimes prepared from dried fruit or nuts. Porridge, made of green beans, peanuts, and other ingredients and prepared in a variety of ways, is as popular as regular rice and is often eaten for breakfast. Paella, now world famous, is the most popular rice dish in Spain. There are countless paella recipes that use rice as the main ingredient and are cooked with meat, vegetables, and fish in a broth flavored with saffron. After paella, boiled rice, particularly tomato sauce, is the most popular. Since rice itself has no flavor, it seems essential for the Spanish to use olive oil, broth, and garlic to flavor rice. In France, rice is considered neither a staple food nor a main dish. Although boiled rice and Creole-style rice are quite popular, their role is as an accompaniment to that of meat or fish. This simply cooked rice is usually seasoned with salt and flavored with butter. Rice pudding is cooked in milk with sugar and eaten as dessert or breakfast. According to Karizaki (2016), Iranians usually consume rice-based food once or twice a day. The most famous and common rice-based food products in Iran are *chelow kebab* (pilaf-kebab), *chelow gheimeh* (pilaf-gheimeh), *ghormeh sabzi* (herbs stew-pilaf), and *joojeh kebab* (grilled chicken pilaf). The list of ethnic and traditional Iranian rice-based foods can be categorized into main meals, soups/ pottages, and sweets/desserts. In addition, the rice-based main dish can be subdivided into six categories (cooked rice, mixed pilaf, rice and stew, rice and high-protein dishes, stuffed vegetables and kofta, and tahchin).

8.2.3 History of Rice Cooking and Rice Cookers in Japan

The beginning of agriculture in Japan was considered to have occurred after the Yayoi period (1000 BC–mid to 300 AD), but recently it has been suggested that it was during the Jomon period (14,000 BC–1000 BC). Paddy rice cultivation technology is considered to have been introduced from the Korean Peninsula around 950 BC and spread to the Japanese archipelago, except for Hokkaido and Okinawa (Sato 2020). Since then, the Japanese food culture centered on rice has begun, and this food culture has had a great influence on the history of Japanese food. In fact, little is known about how rice was eaten in ancient Japan. Various methods have been used to address this question. One method involves inferring the cooking method from tools excavated from archaeological sites, and consensus has been reached (Yokoo 1995; Sato 2020; Nagai 1959).

In ancient Japan, it was believed that fresh or dried unhulled rice was baked or roasted, and the husks were removed before eating. However, unhulled rice was not always ripe completely. Subsequently, porridge, made by boiling rice in a pot or kettle with water, and *kowaii* (*Okowa,* a type of steamed rice) appeared. Subsequently, the husks on the surface of the rice grains were removed before steaming, and the bran was gradually removed using a mortar and pestle to produce rice in a state similar to that of white rice. In the Heian period (AD 794–1185), *kowaii* was made by pouring water over the rice and steaming, so it was not very sticky. On the other hand, porridge was added with potatoes, azuki beans, chestnuts, fish, wakame seaweed, abalones, and minor grains such as wheat and millet have become popular in various cooking methods. *Kowaii* has been eaten for a long time, but around the beginning of the fifteenth century, it was replaced by porridge, which is similar to today's rice, and porridge with less water (*Himeii*). However, these rice-based meals were regular diets with only the upper classes. The middle class and below, including the farmers who produced rice, did not have easy access to these meals and ate rice mixed with various minor grains. It was not until the end of the Edo period (AD 1603–1868) that everyone from the ruling class to the working class could enjoy rice equally.

After the end of the war in 1945, the demand for rice increased rapidly in Japan owing to the rapid increase in the population caused by marriage and childbirth booms. By 1960, the rice shortage had been resolved, and a surplus gradually became the norm. This excess of rice is also increasing in Japan today, as people are losing interest in rice. One of the reasons for this was the large amount of wheat flour brought in from the United States in the name of support and the accompanying spread of Western food. The second was related to the shift to nuclear families and the amount of time spent on household work. Compared to bread, which has the characteristics of home-meal replacement, rice tends to be avoided because of the time and effort required for domestic labor, that is, cooking. In those days, cooking rice was a huge task that required housewives to wake up early, light a fire in the furnace, and watch out for spills and scorching, often resulting in a failure. In 1955, Tokyo Shibaura Electric Co., Ltd. (Toshiba Corp.) launched the first automatic rice cooker for indirect cooking. This rice cooker, which cooks rice by simply turning on the switch and automatically turning it off after cooking, greatly reduced housewives' household work. Table 8.4 presents the history of the development of rice cookers in Japan (Hirata 2002; Matsumoto 1982; Nakamura 2011). After the sale of automatic rice cookers, there was a need to improve food quality as society became more affluent. Even though rice cookers make it easy to cook rice and the warm function allows people to eat hot rice, the rice cooked in rice cookers up to that point did not taste good. By contrast, the introduction of the microcomputer-controlled jar rice cooker in 1979 indicated the direction of making rice more delicious. The use of microcomputers has two important implications (Hirata 2002). First, the complex cooking processes of boiling, steaming, and baking rice can be controlled by an algorithm in the same way by anyone at any time. The other is that the various functions of the microcomputer have made it possible to select various cooking methods such as brown rice and rice cooked with barley with the desired degree of cooking, leading to labor savings in household work and a wider variety of rice dishes. In addition, in the midst of an affluent society, the introduction of the Induction Heating (IH) jar rice cooker in 1988 further intensified the pursuit of deliciousness (Hirata 2002). Magnetic field lines are generated when high-frequency current is supplied to the magnetic force-generating coils around the bottom and sides of the IH jar rice cooker. When these magnetic lines of force pass through the stainless steel part of the pot, eddy currents are generated, and the pot generates heat. This heat generation was efficient with little heating loss, making it possible to achieve high heat and temperature control, which was not previously possible. The

Table 8.4 History of the development of rice cookers in Japan (Hirata 2002; Matsumoto 1982; Nakamura 2011) (drawn by the author)

Year of sale	Rice cooker	Features
1955	Automatic rice cooker with indirect cooking	The first automatic rice cooker was produced in Japan. Water is placed between the outer pot and the inner pot, and the inner pot is heated indirectly to cook rice
1956	Automatic rice cooker with direct cooking	The inner pot is placed directly on the heating plate for heating.
1972	Electronic jar rice cooker	Electric rice cooker with electronic jar to keep the cooked rice warm for a long time
1979	Microcomputer-controlled jar rice cooker	Equipped with a microcomputer that precisely controls the temperature of rice according to the cooking menu, making it possible for anyone to cook rice without failure
1988	IH jar rice cooker	IH heating technology is introduced to rice cookers, enabling the pot itself to generate heat through electromagnetic induction heating, resulting in uniform heating
1992	Pressure IH jar rice cooker	When cooking rice, the inside of the pot is raised to 1.2–1.4 atm and the boiling temperature is raised to 105–110 °C
2000	Rice cooker with built-in copper pot	A rice cooker that uses a combination of copper, which has an advantage in uniform heating, and stainless steel, which has good IH heating characteristics
2006	Rice cooker with a VIP	A rice cooker that reduces power consumption while keeping rice warm
2008	Steam-less rice cooker	A rice cooker that collects the steam generated during rice cooking in a water tank to prevent the generation of steam and odor during cooking
2014	Small capacity IH jar rice cooker	The development of rice cookers for small households, such as elderly people, is progressing

power consumption of conventional rice cookers was 500 W in the 1980s and approximately 800 W in the 1990s owing to problems such as melting of the heating plate; however, the IH jar rice cooker can increase the power consumption to approximately 1400 W owing to less radiation. Since the introduction of IH technology, advancement has been made to achieving uniform heating. Significant emphasis has been placed on the selection of kettle materials, leading to the development of copper, charcoal, and ceramic kettles. The pressure rice cooker was introduced in 1992. The pressure cooker increased the boiling temperature to 105–110 °C by increasing the pressure inside the cooker to 1.2–1.4 atm during the cooking process, thereby promoting the starch gelatinization of the rice grains. In 2006, a rice cooker with a VIP (Vaccum Insulation Panel) which reduces power consumption during heat retention, was introduced. Since then, rice cookers has been developed for energy conservation and environmental consciousness. Since 2014, small capacity IH jar rice cookers for small households have been adapted to modern Japanese society, and high-end rice cookers that pursue the deliciousness of each rice variety have been manufactured. It is interesting to note that the ideal rice for each rice cooker manufacturer is always cooked in a traditional furnace. In the future, rice cookers will be sold make rice taste as good as if it were cooked in a furnace.

8.3 Cooking Process of LWM

In the LWM, the water-to-rice (W/R) ratio is constant, and the rice is boiled when there is a lot of water and steamed when there is less water. All the water is absorbed by the rice, except for evaporation. Originally, this method of cooking rice is used only in a few areas of Southeast Asia, including Japan. However, LWM has the advantage that no excess cooking water is produced after cooking rice, so that the rice components can be consumed in their entirety and environmental pollution caused by disposal can be prevented. Recently, LWM has been widely used in rice cookers. In this section, the LWM processes are discussed, and the effect of each process on rice quality is explained.

8.3.1 Washing

Washing rice removes bran and debris from the surface of the rice grains. Rice varieties with high-amylose content, such as Basmati, are washed 3–5 times before cooking with the excess water method in Pakistan, India, or Iran, whereas rice varieties with medium-amylose content are washed 2–3 times before cooking with the LWM in Thailand, China, Philippines, Japan, and Australia (Champagne et al. 2010). According to taste testers, rice grains that undergo three rounds of washing exhibit fewer signs of yellowing, less unpleasant odor, and milder taste deterioration during the warming process for 24 h after cooking compared to rice grains washed only once (Fukai and Tukada 2006). This is because increasing the washing times of rice removes a large amount of lipids and proteins contained in the bran on the surface of the rice grain, which are the main factors in quality deterioration and delay deterioration during warming. Li et al. (2019) showed that washing had no significant effect on either the firmness or stickiness of cooked rice grain and did not affect the starch leaching behavior and molecular structure of leached starch. However, excessive washing of rice is a source of environmental pollution because it can break up rice, causing starch and surface nutrients to leak. For this reason, polished rice that does not require a washing process is commercially available. The rice grains begin to absorb water during washing.

8.3.2 Watering

In the batch-type LWM, no operations other than heating are performed until after the rice is cooked. Therefore, to produce rice with a proper texture and a sufficient degree of starch, water addition, and soaking are necessary to supply sufficient water to the intracellular starch in the center of the rice grain before heating. In other words, in polished rice, watering and soaking increase the moisture content from 13% to 15% to approximately 30%. In general, when rice is cooked with less water, no water remains during the boiling phase, and the boiling phase with water convection cannot be created; therefore, the rice does not expand sufficiently to form small grains, which are hard and less sticky, and retrogradation progresses rapidly. However, rice cooked with a large amount of water causes excessive damage to the rice grain structure, which results in penetration of the cooking water and leaching of starch and other substances, leading to rice that is hard, less sticky, and less desirable. Therefore, the amount of water added to the rice grain (i.e., the W/R ratio) is one of the main factors affecting the texture of the cooked rice. For example, in short grain Japonica rice, firmness at 80% compression of cooked rice grain thickness decreases with increasing W/R ratio, while adhesiveness increases between 1.0 and 1.3 W/R ratios and decreases between 1.3 and 1.7 W/R ratios (Ogawa et al. 2012). Adhesiveness of cooked rice grains at 25% compression tends to remain the same between 1.0 and 2.0 W/R ratios (Tamura and Ogawa 2012). These changes in texture are closely related to cell expansion, cell wall rupture, and disintegration of cell tissues at the periphery, as the W/R ratio increases. Based on experiments on rice cooking, rice with a W/R ratio of 1.3–1.5, that of polished japonica rice, has a higher overall sensory evaluation (Kainuma and Ema 1987; Takano and Taniguchi 2021). Maleki et al. (2020) investigated the attributes that drive consumer preference for fragrant white jasmine rice cooked at different W/R ratios and, found that W/R ratios of 1.125 and 1.875 were optimal for producing fluffy and sticky rice, respectively. They also showed that a W/R ratio of 1.5 was not dominated by any attribute and appeals to consumers who consume rice frequently.

The moisture content of the cooked rice increases with an increase in the W/R ratio. In brown rice, the higher the W/R ratio, the longer the boiling time, which promotes the swelling of cellulose, hemicellulose, and lignin contained in the pericarp and water absorption by the endosperm, resulting in the uptake of more water (Kainuma and Ema 1987; Kuwada et al. 2011; Yamazaki and Fukuba 1982). Brown rice grains that attained 20% moisture content during soaking has higher firmness, T_0, T_c, and ΔH, and lower springiness and adhesiveness when cooked compared with brown rice grain that attained 30% moisture content (Han and Lim 2009). The W/R ratio has been reported to affect the degree of gelatinization (Yamazaki and Fukuba 1982), but it has also been reported that the degree of gelatinization does not change at any water addition ratio (Kainuma and Ema 1987), which may be affected by the variety of rice, temperature history of the cooked rice, and method of measuring the degree of gelatinization. In a sensory test of 14 long-grain indica rice varieties with high to low amylose content by 12 trained panelists (Srisawas and Jindal 2007), the firmness score is affected by variety followed by the W/R ratio, whereas the stickiness score is mostly affected by variety, with little effect on the W/R ratio. There is no difference in the flavors of the long and medium rice grains depending on the W/R ratio, except for aromatic rice (cv. Dellmont) where the water-like/metallic flavor is higher in intensity at the recommended W/R ratio than at a low ratio (Bett-Garber et al. 2007). The amount of water required to achieve the peak overall acceptability score of cooked rice increases with increasing amylose content. Consequently, the W/R ratio and rice variety play a significant role in influencing the overall acceptability of cooked rice. Therefore, adjusting the W/R ratio is effective in improving high-amylose rice to a stickier and softer preference, such as low- or medium-amylose rice (Enoki et al. 2018).

8.3.3 Soaking

Water absorption of rice grains during soaking is affected by factors such as time, water temperature, and pressure. These factors are discussed and explained in the following section.

Effect of Time

Soaking of polished rice for approximately 30 min and brown rice for approximately 2 h is usually required. Brown rice grains reach the same water absorption as the maximum water absorption of polished rice grains after soaking at 20 °C for approximately 17 h, 30 °C for approximately 7 h, and 60 °C for approximately 2 h (Kuwada et al. 2011). Soaked rice grains are noticeably softer than unsoaked (raw) rice grain (Seki and Kainuma 1982). In general, rice grains with insufficient water absorption result in cooked rice grains with ungelatinized starch in the center of the endosperm, and rice grains soaked for too long tend to break up and have a poor gloss and texture. Tomita et al. (2015) showed that cooked rice grains with soaking at 10 °C has higher moisture content than cooked rice grain without soaking, but there were no significant differences in moisture content between cooked rice with soaking for 20–120 min. Prolonged soaking for 0–60 min caused the disappearance of amyloplasts in cooked rice grains, increased area of pores in the porous structure, and decreases the rupture energy, elastic modulus, and viscous modulus. In contrast, soaking the polished rice grains for 60–120 min increases the breaking energy and stickiness of the cooked rice grains. Seki and Kainuma (1982) showed that the amounts of reducing sugars and total sugars in soaked rice grains tended to increase as soaking time increaseing (0.5, 2, 6, and 15 h). This was observed not only during the soaking process but also during the temperature rise and steaming periods, suggesting an effect of the enzymatic reduction of starch.

Effect of Heating

Warming during soaking is a simple method for increasing the water absorption rate. Soaking rice in warm water above room temperature and below the pasteurization temperature produces rice with greater water absorption, higher total sugar, more reducing sugar, more amino acids than rice soaked at room temperature, and better taste, especially in terms of elasticity, stickiness, and sweetness (Arai et al. 1997; Han and Lim 2009; Maruyama and Sakamoto 1992; Yamakura et al. 2005b). The increase in sweetness of cooked rice due to heated soaking is due to the activation of starch-degrading enzymes, increase in free sugars derived from starch, and degradation of starch from outside the rice grain by starch-degrading enzymes eluted from the cells (Arai et al. 1997). Similarly, the increase in stickiness of rice is related to the action of endogenous cell wall-degrading enzymes that partially break down the endosperm cell wall and promote the release of starch outside the rice grains. Yu et al. (2015) reported that the higher the soaking temperature (25, 30, 35, and 40 °C), the higher the water absorption capacity and the shorter the cooking time for brown rice to gelatinize. The firmness of brown rice after soaking decreases with increasing temperature and duration, except between soaking at 25 and 30 °C. Soaking rice grains in warm water (30 °C) for a long time (12 h) has been also reported to degrade the estimation of taste by a taste tester, the b* (yellow) of the color tone, and the occurrence of a strange odor in the process of keeping rice warm after cooking (Fukai et al. 2006). High soaking temperatures lead to kinetically high water absorption rates of the rice grains (Cheevitsopon and Noomhorm 2011; Kashaninejad et al. 2007; Muramatsu et al. 2006), but not necessarily to high equilibrium moisture content. Sakamoto et al. (2015) observed a temperature-dependent trend, in which the water absorption of polished rice at 30–50 °C was higher than that at 5–20 °C up to 30 min of soaking time, but the water absorption reversed after 30 min of soaking. Morii et al. (2019) explained that the reverse phenomenon occured because of water absorption disorder caused by the deterioration of the crystal structure of starch and subsequent swelling of the starch grains on the surface of the rice. It should be noted that if the starch is gelatinized before the rice is cooked by hot soaking, the water is not absorbed to the center of the rice grains, and the ungelatinized portions remain in the grain. Since the effect of hot water soaking has been clarified, many electric rice cookers have incorporated hot water soaking into their programs to shorten the soaking time.

Effect of Pressurization

High-pressure (HP) treatment is an effective method for evenly absorbing water into rice grains, which have complex structures. During soaking, HP treatment replaces the voids inside the rice grain with water and forces starch and protein to hydrate and denature. HP treatment at 400 MPa for 10 min during the soaking process increases the amount of total sugar, reducing sugar, some free amino acids, degree of gelatinization, and starch digestibility of cooked polished rice, while decreasing α-amylase activity and protease activity (Yamakura et al. 2005a, b). HP treatment also affects the microstructure of rice grains, accelerating starch absorption, softening, swelling, and gelatinization, thereby closing the voids inside the grains that are normally found in cooked rice grains. HP treatment at 500 MPa for 5 and 20 min during the soaking process reduces the cooking time until the brown rice gelatinizes to 13 min, which is 60% less than the control. HP treatment at 200 and 400 MPa for 20 min also reduces the cooking time until the polished rice gelatinizes to 10–7.5 and 3.5 min, which is 35 and 75% less than that of the control (Yu et al. 2015). Severe HP treatment (i.e., high-pressure and long-term treatment) during soaking drastically reduces the firmness of brown rice after cooking and the enthalpy (ΔH) of starch gelatinization. HP treatment of brown rice during the soaking process accelerates the enzymatic reaction between glutamic acid and glutamate decarboxylase in cells, which produces γ-aminobutyric acid (GABA). Sasagawa et al. (2006) investigated the amount of GABA in brown rice after pressure treatment (0.1, 50, 100, 200, and 400 MPa at 25 °C for 10 min), draining, and incubating at 25 °C for 18 h. This study revealed that the amount of GABA increases with the acceleration of the enzymatic reaction due to the increase in the degree of cell wall disruption up to 200 MPa, but decreases at 400 MPa due to the effect of HP on the enzyme. Meng et al. (2018) showed that the higher the HP to be treated (200–400 MPa), the greater the amount of leached amylose, firmness decreases, and springiness increases. Additionally, the degree of retrogradation and relative crystallinity of cooked rice after HP soaking decreases compared with those of the control group. Two-cycle HP treatment (100–500 MPa; single or two cycles; total holding time 10 min) has been reported to result in higher L* and degree of damage to the bran layer in brown rice and lower firmness and gumminess than single-cycle HP treatment (Yu et al. 2017). The use of lower pressure levels, similar to two-cycle HP treatment, would be more economical for commercial production.

8.3.4 Heating

The main purpose of the heating process is to allow the starch in the rice grains to gelatinize while allowing the remaining water from the soaking process to be absorbed by the rice. The heating process is divided into three stages: temperature increase, boiling, and braising.

Temperature Rise Stage

The rate of heating during the temperature rise period affects the amount of water remaining unabsorbed, the degree of gelatinization, and the taste of cooked rice. The degree of gelatinization depends on the amount of heat and heating time; however, it takes 7–10 min (Takano and Taniguchi 2021). The process of slow heating, which entails a prolonged boiling time, not only results in a decreased water level and hinders the maintenance of the desired boiling time but also causes a reorganization of the decomposed starch structure, leading to incomplete gelatinization of the rice grains at the center. However, rapid heating requires a short time to boil, the balance between temperature rise and water absorption is poor, and the gelatinization and softening associated with water absorption are slow. Mabashi et al. (2007) investigated the effects of different temperature histories during the cooking process on the free sugar and amino acid content of rice by cooking at different heating rates (1.82, 2.22, 4.17, 9.09, and 16.67 °C/min) and holding temperatures (40, 60, and 80 °C). According to the report, the amount of total sugar, reducing sugar, and glucose in rice increases at a slower heating rate, whereas the amounts of free fructose and sucrose remain almost constant regardless of the heating rate. The amount of free amino acids, such as free aspartic and glutamic acids, tends to increase slightly, but not as much as that of sugars. The total sugar, reducing sugar, and glucose contents increase with temperature holding at 60 °C during the temperature rise period, whereas the total free amino acid content tended to increase with temperature holding at 40 and 60 °C. The occurrence of these taste components in rice and their changes with temperature are closely influenced by endogenous enzymes in rice (Tsuyukubo et al. 2010, 2012, 2013).

Boiling Stage and Braising Stage

The boiling stage is a period of intense boiling of the remaining unabsorbed water. When cooking water boils, the rice grains absorb most of the water. At this time, however, the rice grain is hard, with ungelatinized portions remaining in the center. Therefore, in the subsequent braising stage, the heat is turned down to maintain boiling with the remaining water, and the rice continued to cook to evaporate excess water and spread the water inside the rice grains to promote gelatinization. To cook good tasting rice with starch gelatinization, it is necessary to process the rice at 98 °C or higher for 20 min during both the boiling and braising stages, regardless of the time it takes to reach boiling (Kainuma and Seki 1986; Seki and Kainuma 1986).

The application of HP treatment to rice cooking can make the water distribution in the rice grain uniform and improve the gelatinization degree of rice grain by heating above 100 °C. HP treatment produces softer, sticker, less expansive, and slightly yellowish-cooked rice. The decrease in firmness is related to microstructural characteristics, such as an increase in the pore size and thickness of the sponge-like texture of the inner layer of the rice endosperm as the cooking temperature increases (Leelayuthsoontorn and Thipayarat 2006), and an increase in the moisture diffusion rate in the rice grain, which distributes water sufficiently and evenly to rice grains (Xu et al. 2019). According to Xu et al. (2019), the increase in adhesiveness is influenced by the leaching of amylose from the rice grain, damage to the starch–protein network, and dispersion of porosity during HP cooking. An increase in steam pressure and temperature can also promote the abundance of rice flavor substances, such as 2-acetyl-1-pyrroline, which is the main aroma contributor to fragrant rice through the Maillard reaction. Since the Maillard reaction also affects the yellowish of cooked rice, it has been proposed that an HP of 0.10 MPa is better overall.

8.3.5 Steaming

Rice grains immediately after extinguishing have a watery and hard surface and are not significantly preferred in sensory evaluation (Kainuma et al. 1988). Therefore, during the steaming period, the rice grains are left in a temperature-controlled rice cooker for approximately 10–15 min to avoid burning, and excess water is evaporated or absorbed into the rice grains. Kainuma and Ema (1989) reported that steaming ends at 93 °C for the center of the rice in the cooking pot and 86–88 °C for the air, regardless of the heat retention capability of the cooking pot. Steaming homogenizes the moisture content and size of the rice inside the cooking pot and increases the degree of gelatinization. However, rice in the upper part of the cooking pot absorbs water that cannot be evaporated as the steaming time is prolonged, resulting in high moisture content and low firmness, adhesiveness, and cohesiveness, which deteriorates eating quality (Kainuma and Ema 1990). This change in texture due to steaming is more likely to occur 24 h after cooking than immediately after cooking (Kainuma et al. 1988). Therefore, it is important to finish the steaming operation at an appropriate time, open the lid, and gently stir the rice to reduce the difference in moisture content between the upper and lower layers of the rice cooker.

8.4 Changes in the Properties of Rice Grains During Cooking

In this section, the key factors that affect the quality of cooked rice, such as structure, moisture, starch, and texture, are discussed.

8.4.1 Tissue Structure

Brown rice is an edible part of rice that botanically corresponds to the caryopsis. Brown rice is mainly composed of pericarps, seed coats, embryos, and endosperms (Fig. 8.1). Their mass distribution is 1–2% pericarp, 4–6% seed coat, and aleurone layer, 2–3% embryo, and 89–94% endosperm (Hinton and Shaw 1954; Zhou et al. (2002). Short -and medium-grained brown rice is high in dietary fiber and has poor digestibility and taste. Therefore, it is generally eaten as polished rice after the pericarp, seed coat, embryo, and part of the aleurone layer are removed by polishing. The aleurone layer on the surface of polished rice grains has a slight effect on cooking time, water absorption, volume expansion rate, and amount of dissolved solids, and has a significant effect on texture and viscosity (Wu et al. 2016).

Figure 8.2 shows the morphological and histological structural changes in the rice grains at different cooking stages (Tamura et al. 2014a). In the early stages of washing and soaking, small cracks appear across multiple cells in the center of the rice grain. The cracks elongate at 50–70 °C during the cooking process. At temperatures above 85 °C, the cracks widen and become voids as starch gelatinizes, expands, and water vapor is generated, and the morphology of the rice grain elongates. On the periphery of rice grains, the tissue cells are regularly arranged immediately after cooking, but they swell and some of them are destroyed during steaming (Kainuma et al. 1988). These cracks and voids are formed by the breakdown of the cell wall rather than by the separation of individual cells at the intercellular boundary (see arrows in Fig. 8.2), which facilitates the movement of water and affects the gelatinization and texture of starch.

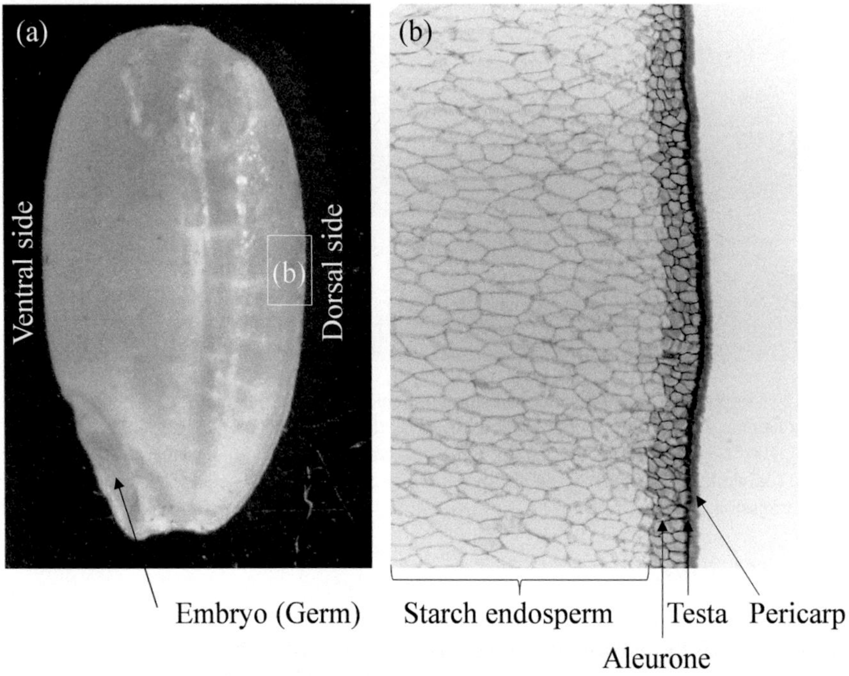

Fig. 8.1 Morphology and histological structure of brown rice grain. Morphology structure (**a**) and histological structure (**b**) are shown (Drawn by the author). This photo (Fig. 8.1b) are provided by Yukiharu Ogawa

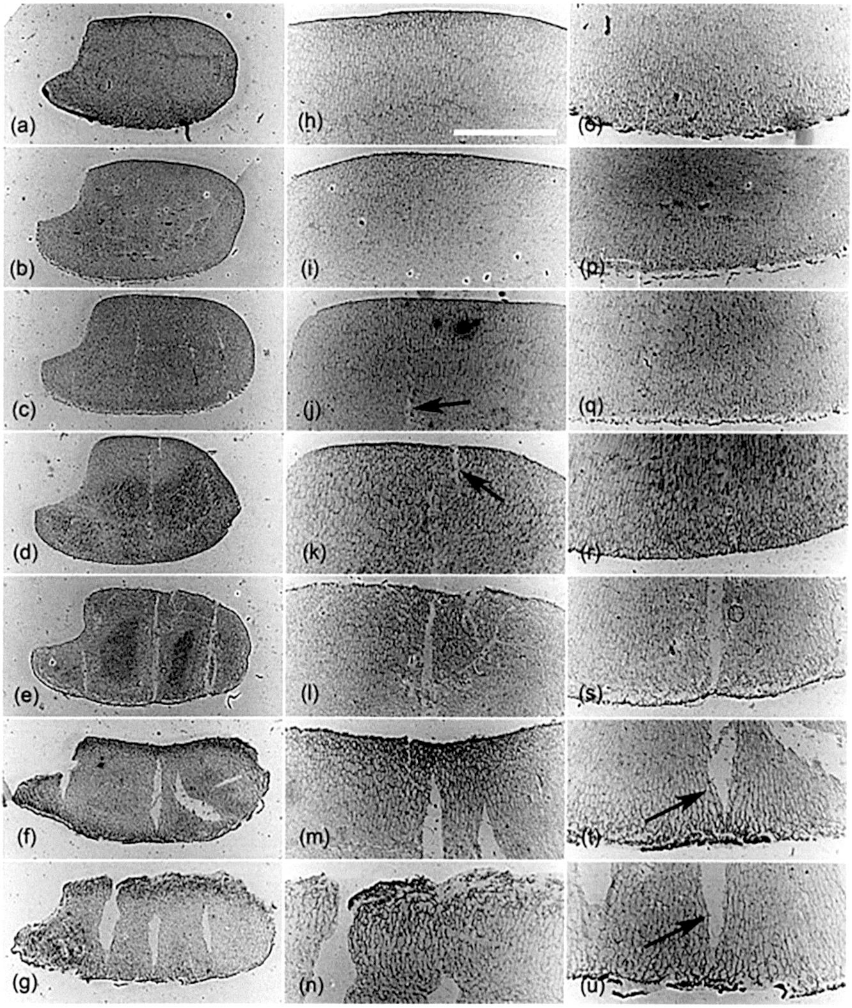

Fig. 8.2 Morphological and histological structural changes in the rice grains at different cooking stages (Tamura et al. 2014a). Polished (a, h, o), 30 °C (b, i, p), 50 °C (c, j, q), 70 °C (d, k, r), 85 °C (e, l, s), 100 °C (f, m, t), cooked grain (g, n, u), and grain dorsal side (h–n), ventral side (o–u) are shown. (drawn by the author). Scale bar shows 1 mm. Scale of h–u is equal to h. Copyright 2013 with permission from Elsevier

8.4.2 Moisture

There is a two- to four-fold difference in water absorption between rice grains and rice starch until equilibrium is reached, which is due to the fact that the grain structure inhibits water absorption (Briffaz et al. 2013). To produce good eating quality cooked rice, it is essential to absorb water to the center of the rice grain, which has a tissue structure, and for this purpose, it is important to understand the amount and distribution of water contained in the rice grain. Water absorption in polished rice grains begins on the ventral side and in areas where the embryo is missing, and is accelerated throughout the central line of the endosperm (Horigane et al. 2006). Water penetrates more rapidly on the ventral side than on the dorsal side and tends to accumulate in the center of the endosperm, where the cells are roughly aligned. Finally, the moisture content of the rice grains increases until equilibrium is reached. The water absorption rate or equilibrium moisture content of rice grains increases with temperature (Muramatsu et al. 2006), and varies with variety and morphology. For example, the trend in relative moisture absorption during the soaking process at 25 °C is as follows: japonica waxy rice > indica waxy rice > japonica non-waxy rice > indica non-waxy rice (Gong et al. 2020). During soaking at 30 °C, the highest equilibrium moisture content is determined for medium-grain rice, followed by short- and long-grain rice, while the moisture content of rice grains from different varieties after soaking at 98 °C is different from that after soaking at 30 °C in the order of long- > short- > medium-grain rice at almost all times (Tamura et al. 2021). During the cooking process with temperature change in a rice cooker, the water absorption of low and medium amylose japonica rice grains is almost the same from 30 to 50 °C, and increases rapidly at 50–70 °C (Tamura et al. 2014b; Yamamoto et al. 1990). NMR imaging revealed the water absorption properties of japonica (cv. Nipponbare) and indica rice (Khao Dawk Mali 105), with similar amylose content (Kasai et al. 2007). Khao Dawk Mali 105 has a brittle microstructure owing to the many cracks that occur during the temperature rise process, resulting in rapid water absorption and rapid homogenization of the moisture distribution in the rice grains. In contrast, indica rice, which has a higher amylose content, has fewer cracks, slower water absorption, and an uneven moisture distribution for the same heating time. Among the components of polished rice grain, amylose content has been reported to be negatively correlated with saturated moisture content (Horigane et al. 2013; Ohisa et al. 2007) and positively correlated with the rate constant of water absorption (Tamura et al. 2021). Miwa et al. (2002) considered that water absorption depended on the crude fat content in the early stage of water absorption and was related to the composition of starch, the main component of rice, as the soaking time increases. The water absorption rate of rice grains increases with increasing degree of polishing (DOP) (Muramatsu et al. 2006). Namely, when brown rice grains are 0% (DOP 0) and the rice grains with mass fractions of 2, 4, 6, and 8% removed by polishing are defined as DOP 2, DOP 4, DOP 6, and DOP 8, respectively, the rate constants of water absorption during soaking at 5–40 °C are shown as follows: 1.313×10^{-3}–9.159×10^{-3} min^{-1} for DOP 0, 1.178×10^{-2}–4.701×10^{-2} min^{-1} for DOP 2, 1.638×10^{-2}–5.676×10^{-2} min^{-1} for DOP 4, 1.928×10^{-2}–5.991×10^{-2} min^{-1} for DOP 6, 2.035×10^{-2}–6.553×10^{-2} min^{-1} for DOP 8. Water uptake in rice is a reversible process in which water molecules enter the amorphous portion of starch granules through cracks in the rice grains and bind to free hydrophilic groups. Amylopectin, the main starch component of waxy rice, has a highly branched structure that easily binds to water molecules and maintains a highly ordered state in the hydrated starch granules. In other words, the higher the amount of amylopectin in the rice grain, the greater the water absorption. In addition, the degree of starch gelatinization, which varies among varieties, the degree of microstructural damage, such as cracks and voids; and the distribution of proteins that inhibit water penetration can also be factors that affect water absorption. In contrast, the initial moisture content of rice grains does not affect the water absorption rate or grain expansion when soaked in boiling water (Yadav and Jindal 2007a, b).

8.4.3 Starch

Gelatinization

Rice starch granules are composed of amylose, an amorphous chain polymer, and amylopectin, a clustered polymer with a mixture of double helical (crystalline) and amorphous parts. Rice starch granules are the smallest among cereal grains and are polygonal with a smooth surface and are 3–8μm in diameter. A few dozen starch monolayers accumulate in the amyloplasts of the organelles. Gelatinization is an endothermic process in which the molecular order within the starch granules is disrupted. In other words, during the cooking process, the starch, particularly amorphous amylopectin, changes from a strong hydrogen bond with the starch to a hydrogen bond with water through water absorption and gelatinization, and the degree of gelatinization reaches almost 100%. Starch gelatinization is an important physicochemical process that occurs during rice cooking. The swelling and gelatinization properties of starch are influenced by the proteins and lipids localized around rice starch (Siaw et al. 2021). In addition, the limited amount of water due to the LWM and the cellular structure of rice grains also inhibit swelling and gelatinization. Under the LWM cooking conditions, the degree of gelatinization of rice grains increases in a stepwise manner, i.e., approximately 30% between 40 and 70 °C, approximately 3% between 70 and 80 °C, about 26% between 80 and 98 °C, about 16% from 98 °C to the end of the boiling period (when the rice cooker is turned off), and approximately 10% between the end of the boiling period and the end of steaming (Ikeda 2001). Seki and Kainuma (1982) investigated the relationship between soaking time and the degree of gelatinization of rice grains. The results showed that, although the degree of gelatinization at 85 °C tended to be lower with a shorter soaking time, the degree of gelatinization immediately after cooking was the same for samples soaked for more than 30 min, and there was no difference in the degree of gelatinization at the end of steaming and after 24 h with or without soaking. Tomita et al. (2015) also reported that the degree of gelatinization of cooked rice after soaking at 10 °C for 20–120 min is more than 90%, with no significant difference. However, with the increase of heating rate (2–15 °C/min) the onset, peak, and end temperatures of starch gelatinization of rice grain gradually shift to higher values, while the enthalpy increased firstly (2.22–4.76 J/g) and then decreased (4.76–2.96 J/g) (Zhu et al. 2020). Thus, different heating rates affect the gelatinization behavior of starch in rice, leading to different textures of the cooked rice grains.

Leached Solids

The starch granules inside rice grains expand by water absorption and gelatinization during the soaking process (Seki and Kainuma 1982) and cooking process (Tamura et al. 2014a, b), physically destroying the cell membrane and cell wall, and leaching to the outside of the cell. On the other hand, hydrolysis of the cell wall by endogenous enzymes at 20–80 °C (Arai et al. 1997; Tsujii et al. 2009) also accelerates the leaching of solids from the inside of the rice grain. The leached solids increase as the amount of water added and the heating time increase, and decrease in the later stages of rice cooking (Ogawa et al. 2012; Tamura et al. 2014a, b; Zhu et al. 2020). The decrease in leached solids in the late stage of rice cooking is not only due to the return of low-molecular-weight starch to the interior of the rice grain but also due to the formation of a surface adhesion layer (coated layer) of high-molecular-weight starch and fragments of cellular tissue on the surface of the rice grains as rice cooking progresses. Figure 8.3 shows the coated layer of cooked rice grain sections (Tamura and Ogawa 2012). The coated layer is 1–7μm thick on the surface of cooked rice grains, depending on the location and amount of water added during cooking (see arrow in Fig. 8.3c), and affects the firmness, stickiness, and gloss of the grain surface.

Non-starch and protein solids account for the majority (40 55%) of the leached solids at the beginning of cooking (30–60 °C), but this percentage decreases with increasing starch and protein leaching at 70–100 °C (Zhu et al. 2021). Hanashiro et al. (2004) reported that the stickiness of cooked rice grains was affected by the properties of amylose and amylopectin, and the higher the amylose content in the leachate, the weaker the stickiness of the grain. Patindol et al. (2010) showed that the amylose–amylopectin ratio of the material that leached out of the grains during cooking was positively correlated with hardness and negatively correlated with stickiness. Wada et al. (2011) studied the relationships between sensory tests using Japanese panelists and leached starch during cooking. The results reveal that the amount of eluted starch is significantly positively correlated with the overall eating quality, glossiness, taste, and stickiness. The amylose ratio of leached starch is significantly negatively correlated with overall eating quality. There are two interesting reports on the modification of rice using leachates in cooking and processing processes. Han and Lim (2009) investigated brown rice in which the cooking water containing leached solids during the soaking process was replaced before heating, and found that the leached solids, 71.3–82.4% of which consists of amylose, increases the hardness of brown rice. It has also been proposed that the content of cooked rice can be controlled by arranging the cooking procedure to remove the appropriate amount of cooking water with leached materials or by replacing it with cooking water during cooking (Tamura and Ogawa 2010). The viscosity of cooking water containing leached solids is also an important factor related to the properties of

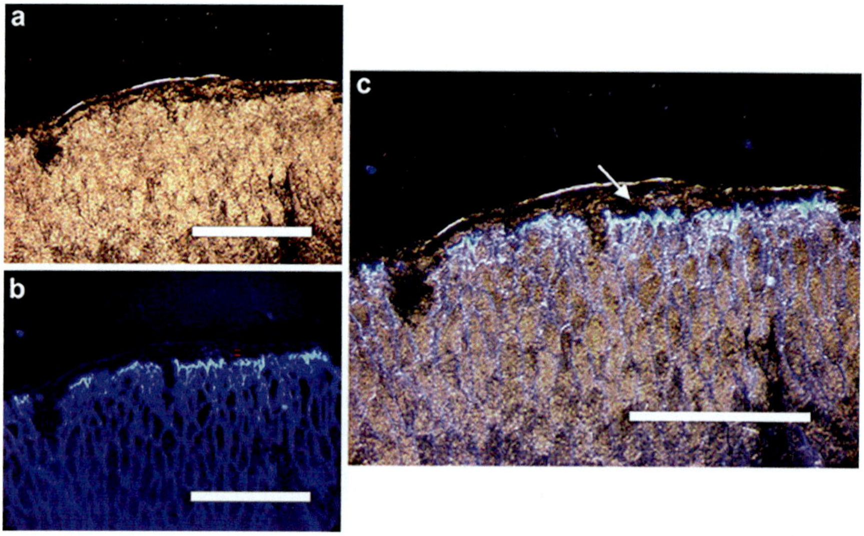

Fig. 8.3 Coated layer of cooked rice grain sections (Tamura and Ogawa 2012). Transmission image (**a**), fluorescence image (**b**), and composite image (**c**). Scale bar shows 0.5 mm. The coated layer can be visualized from a composite image of transmission and fluorescence images of cooked rice grain sections taken at the same location. (Reprinted with permission from Elsevier)

cooked rice. The viscosity of the cooking water increases with increasing heating time (Chen et al. 2019; Nagai et al. 2013). Nagai et al. (2013) pointed out that the apparent viscosity at 70 °C during the rice cooking process was positively correlated with the amount of leached solids at 70 °C, negatively correlated with the firmness when the whole rice grain was compressed, and positively correlated with the adhesiveness when the rice grain surface was compressed. Because there is a positive correlation between the eating quality of rice and the amount of leached solids in rice grains soaked at 70 °C, Taira and Shoji (1999) studied a simple estimation method to measure that rice varieties that absorbed water easily and had more leached solids with better eating quality.

Digestibility

Cooked polished rice has a significantly higher starch digestibility than uncooked (polished) samples in in vivo study (Jung et al. 2009; Wolever et al. 1986) and in vitro studies (Sagum and Arcot 2000; Tamura et al. 2016). Tamura et al. (2016) investigated the impact of the degree of cooking on starch digestibility using uncooked rice grain, partially cooked rice grain (cooked in boiling water for 10 min), and fully cooked rice grain (cooked in boiling water for 20 min). Fully cooked rice grains have a higher estimated glycemic index (eGI) than partially cooked rice grains, but no significant difference is observed in the kinetic constant or equilibrium starch hydrolysis percentage. In addition, significant differences are observed in equilibrium starch hydrolysis percentage and eGI between uncooked and cooked rice grains. Insufficient water absorption during the soaking process results in higher resistant starch (RS) and lower rapidly digested starch (RDS) contents in cooked brown rice (Han and Lim 2009). Furthermore, Han and Lim (2009) showed that the leached solids generated during the soaking process reduced starch digestibility (lower RDS and glycemic index and higher slowly digested starch (SDS) and RS) by inhibiting enzymatic reactions. He et al. (2018) investigated the effects of cooking temperature and time on the starch digestibility of cooked rice. The results shows that the RDS increases significantly, whereas the SDS and RS shows a degree of reduction with an increase in cooking temperature from 50 °C to 90 °C. In contrast, the cooking time (15, 30, and 45 min) has little effect on starch digestibility. Low-temperature cooking (95 °C for 10 min) decreases equilibrium starch hydrolysis, kinetic constant, and eGI of indica long grains and japonica short grains compared to high cooking temperature (100 °C for 8 min) (Guillén et al. 2018). Wang et al. (2020) prepared rice in two rice cookers with different temperature histories and compared the relationships between the starch structure and starch digestibility. The characteristics of the rice cooker with decreased starch digestibility and increased RS and SDS are as follows: (1) Soaking at higher temperatures enhances the diffusion of water into the rice matrix, thus strengthening the starch lamellae without changing the rice morphology and ordered structures, such as starch polymorphs and helices. (2) Shorter heating results in a more fully disrupted rice structure and less prominent nonperiodic molecular organization. (3) At the final cooking stage, braising with a faster temperature drop and longer heating resulted in similar rice morphology, reduced molecular order (e.g., helices), and possibly a higher density of amorphous regions in nonperiodic organizations. These features tend to increase the packing density of molecular chains (mainly starch) in the rice matrix and slow the diffusion of enzymes into the rice matrix and their associated binding and catalysis. In contrast, Wang et al. (2017) concluded that starch hydrolysis of non-waxy cooked rice was not affected by pressure, soaking, or W/R ratio during cooking.

As described above, these studies did not reveal consistent factors regarding the effect of cooking on starch digestibility of rice. The reasons for this include the complex effects of cooking methods, rice varieties, analysis methods and techniques of starch digestibility, as well as the fact that rice samples were prepared in different forms, such as slurry, dry powder, and grains. A study using cooked rice grains whose tissue structure is maintained, shows that the presence of tissue structure influences starch digestibility rather than starch gelatinization (Tamura et al. 2016). Tamura et al. (2021) investigated the relationship between starch digestibility and tissue structure of rice grains (short-, medium-, and long-grain rice) prepared using the LWM and EWM (Fig. 8.4). Starch hydrolysis is significantly higher for rice cooked using LWM than that cooked using EWM, regardless of variety. The rice grains cooked by LWM has larger voids and more cell wall damage from the dorsal side to the ventral side than rice grains cooked by EWM, regardless of the variety. Internal structural damage is highly correlated with the kinetic constant of starch

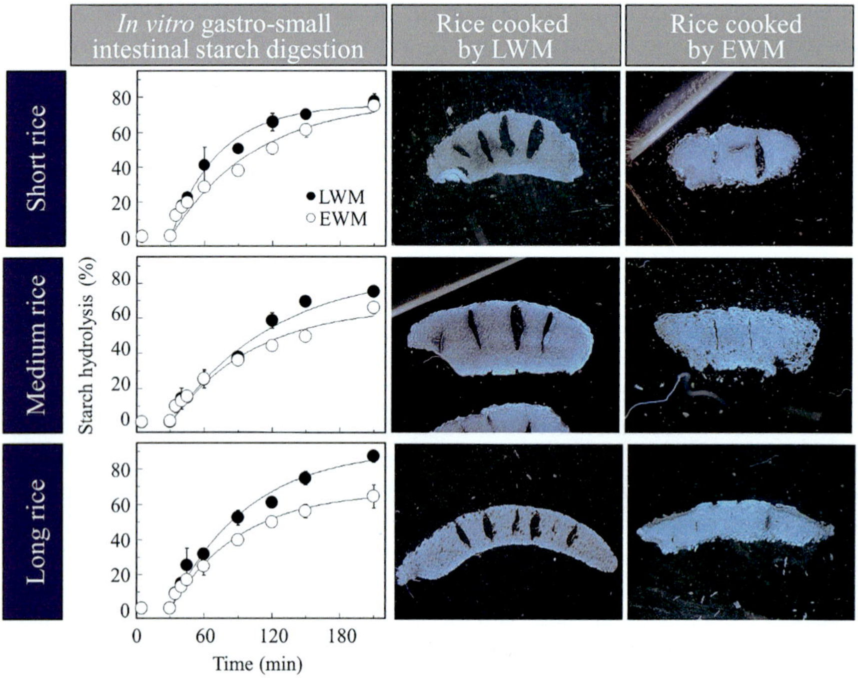

Fig. 8.4 Relationship between tissue structure and starch digestibility of cooked rice grain (Tamura et al. 2021). (Reprinted with permission from Elsevier)

hydrolysis and eGI. This means that internal structural damage dependent on the cooking method may have facilitated the diffusion of digestive juices into the grain core and might have determined starch hydrolysis.

8.4.4 Texture

It is no exaggeration to say that texture, expressed as firmness and stickiness, plays an important role in the eating quality of rice in any country or region (Choi et al. 2018; Custodio et al. 2019; Machida et al. 2020; Rousset et al., 1999; Tani et al. 1969). A test of classification of 13 texture attributes by 10 trained panelists using 18 types of rice with different amylose contents revealed that all texture attributes can be classified into two groups: hardness and stickiness group (Li et al. 2016). The hardness group contains roughness and residual loose particles, whereas the stickiness group contains cohesiveness of mass, uniformity of bite, and toothpacks. Texture is affected not only by the starch structure, which varies among varieties (Cameron and Wang 2005), but also by cooking method.

During the soaking process maintained at temperatures between 30 and 70 °C, a rapid decrease in hardness is observed between 5 and 30 min, and the decrease

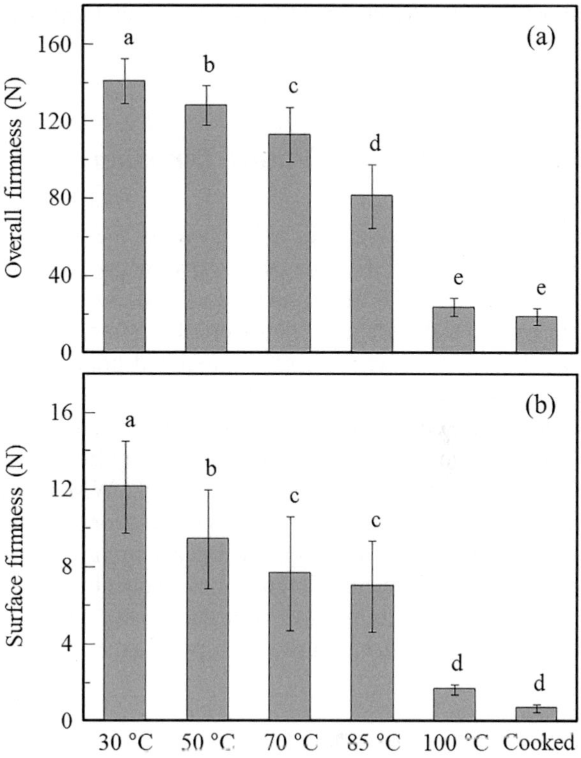

Fig. 8.5 Changes in the overall firmness and surface firmness at different cooking stages (Tamura et al. 2014a). Overall firmness (**a**) and surface firmness (**b**) are defined as the force at 90% and 25% compressions, respectively. Error bars represent the standard deviation ($n = 25$–34). Different letters indicate significant differences ($p < 0.05$) (drawn by the author)

becomes more pronounced as the soaking temperature increases (Zhu et al. 2019). During soaking in boiling water, firmness decreases with soaking time and is higher for rice grains with higher amounts of amylose (Jinorose et al. 2014). Rice grains cooked by soaking followed by the LWM are softer than those without soaking, and the difference in hardness is greater at 85 and 98 °C during the cooking and after 24 h of cooking (Seki and Kainuma 1982). Figure 8.5 shows the changes in the overall firmness and surface firmness at different cooking stages (Tamura et al. 2014a). During the cooking process of LWM, the overall firmness and surface firmness decrease with cooking time. On the surface of the grains, which are easily exposed to hot water during cooking, a decrease in surface firmness due to gelatinization occurs between 50 and 70 °C. Contrary, the decrease in overall firmness from 50 to 70 °C should be influenced by the water penetration into the grain. Overall surface firmness decreased sharply from 85 to 100 °C (30 min after heating), although significant differences are not revealed between 100 °C and cooked grains. It has also been reported that the firmness at 100 °C is not the final value of cooked rice; the decrease continues with increasing time of cooking up to 20 min (Zhu et al. 2021). The textural changes in rice grains during the cooking process are closely related to the effects of cooking, such as the temperature history and amount of water added, and the properties of the rice grain, such as the amount and distribution of water, structure of the starch, properties of the leached solids, and composition and damage of the tissue structure.

8.5 Summary and Future Direction

Rice is mainly cultivated in Asia; however, the food culture and cooking characteristics of rice, which have developed over many years, vary greatly among countries and regions. The cooking method of rice has a significant effect on the overall preference, mainly in terms of texture. Therefore, the cooking method was optimized based on the rice variety and the required preference. The LWM is a simple yet environmentally and nutritionally beneficial method for cooking rice, and its application in rice cookers makes it possible to cook rice in the same way anywhere in the world. Previous studies have examined the effects of the rice washing, watering, soaking, heating, and steaming processes of LWM on the quality of cooked rice. The relationship between the different cooking methods and cooked rice quality and preference was also investigated. Recent studies have explored cooked rice as an external factor affecting functional components and starch digestibility. However, cooking methods to improve both palatability and functionality have not yet been sufficiently studied, and further prospects are expected in the future.

Acknowledgments Photographs of Fig. 8.1b was provided by Yukiharu Ogawa (Chiba University). Takasuke Ishitani (Japan Food Packaging Association) and Elsevier gave permission to reprint. Nwaeche Chiemerie Famous (Tokyo University of Agriculture and Technology) took care of the grammar proofreading. Author would like to express gratitude to them.

References

Arai E, Shimizu C, Watanabe M (1997) Quality improvement of cooked rice by hot-water soaking. J Home Econ Japan 48(9):789–795

Bett-Garber KL, Champagne ET, Ingram DA, McClung AM (2007) Influence of water-to-rice ratio on cooked rice flavor and texture. Cereal Chem 84(6):614–619

Briffaz A, Mestres C, Matencio F, Pons B, Dornier M (2013) Modelling starch phase transitions and water uptake of rice kernels during cooking. J Cereal Sci 58(3):387–392

Cameron DK, Wang YJ (2005) A better understanding of factors that affect the hardness and stickiness of long-grain rice. Cereal Chem 82(2):113–119

Champagne ET, Bett-Garber KL, Fitzgerald MA, Grimm CC, Lea J, Ohtsubo KI et al (2010) Important sensory properties differentiating premium rice varieties. Rice 3(4):270–281

Cheevitsopon E, Noomhorm A (2011) Kinetics of hydration and dimensional changes of brown rice. J Food Process Preserv 35(6):840–849

Chen X, Zhang X, Wang B, Chen P, Xu Y, Du X (2019) Investigation of water migration and its impacts on eating qualities of black rice during cooking process. J Cereal Sci 89:102810

Choi WS, Jarma Arroyo SE, Seo HS (2018) Cross-cultural comparisons between Korean and US adults with respect to texture perception and acceptance of cooked milled rice. Int J Food Sci Technol 53(9):2181–2194

Custodio MC, Cuevas RP, Ynion J, Laborte AG, Velasco ML, Demont M (2019) Rice quality: How is it defined by consumers, industry, food scientists, and geneticists? Trends Food Sci Technol 92:122–137. https://doi.org/10.1016/j.tifs.2019.07.039

Enoki Y, Sasagawa K, Fujimura S, Yamaguchi T (2018) Investigation of optimal water addition ratio for cooking of high-amylose rice "Koshinokaori". J Jpn Soc Food Sci Technol 65(6): 313–317

FAOSTAT (Statistics division of food and agriculture organization of the united nations). https://www.fao.org/faostat/en/#home (accessed 6 December 2021)

Fukai Y, Okamura O, Tukada K (2006) Influence of period and temperature of soaking on quality of cooked rice (Influence of cooking conditions on quality of cooked rice — Part 2). J Jpn Soc Food Sci Technol 53(11):592–595

Fukai Y, Tukada K (2006) Influence of pre-washing on quality of cooked rice maintained at a constant temterature (Influence of cooking conditions on quality of cooked rice — Part 1). J Jpn Soc Food Sci Technol 53(11):587–591

Gong M, Zhou Z, Jin J, Yu Y, Liu S, Han X et al (2020) Effects of soaking on physicochemical properties of four kinds of rice used in Huangjiu brewing. J Cereal Sci 91:102855

Guillén S, Oria R, Salvador ML (2018) Impact of cooking temperature on *in vitro* starch digestibility of rice varieties with different amylose contents. Polish J Food Nutr Sci 68(4):319–325

Han JA, Lim ST (2009) Effect of presoaking on textural, thermal, and digestive properties of cooked brown rice. Cereal Chem 86(1):100–105

Hanashiro I, Ohta K, Takeda C, Mizukami H, Takeda Y (2004) Leaching of amylose and amylopectin during cooking of rice grains and their effect on adhesiveness of cooked rice. J Appl Glycosci 51(4):349–354

Hatae K, Kasahara M, Tanisawa Y, Shimada A, Nakatani K, Zhuorong Y et al (1997) The habit of eating rice in four countries: Japan, China, Spain and France. J Consum Stud Home Econ 21(4): 349–361

He M, Qiu C, Liao Z, Sui Z, Corke H (2018) Impact of cooking conditions on the properties of rice: combined temperature and cooking time. Int J Biol Macromol 117:87–94

Hinton JJC, Shaw B (1954) The distribution of nicotinic acid in the rice grain. Br J Nutr 8(1):65–71

Hirata Y (2002) The recent trend about the electric rice cooker. J Integr Study Dietary Habits 13(3): 147–155

Horigane AK, Suzuki K, Yoshida M (2013) Moisture distribution of soaked rice grains observed by magnetic resonance imaging and physicochemical properties of cooked rice grains. J Cereal Sci 57(1):47–55

Horigane AK, Takahashi H, Maruyama S, Ohtsubo KI, Yoshida M (2006) Water penetration into rice grains during soaking observed by gradient echo magnetic resonance imaging. J Cereal Sci 44(3):307–316

Ikeda H (2001) Relationship between the saccharides extracted from rice grains during cooking and the sensory taste of cooked rice. J Home Econ Japan 52(5):401–409

Ishitani T (ed) (2009) Encyclopedia of rice: From rice cultivation to genomics. Saiwai shobo Co., Ltd., Tokyo, Japan

Jinorose M, Prachayawarakorn S, Soponronnarit S (2014) A novel image-analysis based approach to evaluate some physicochemical and cooking properties of rice kernels. J Food Eng 124:184–190

Jung EY, Suh HJ, Hong WS, Kim DG, Hong YH, Hong IS, Chang UJ (2009) Uncooked rice of relatively low gelatinization degree resulted in lower metabolic glucose and insulin responses compared with cooked rice in female college students. Nutr Res 29(7):457–461

Kainuma Y, Ema S (1987) The effect of ratio of water to rice on cooking. J Home Econ Japan 38(7): 567–575

Kainuma Y, Ema S (1989) A study of the rice cooking (Part 7). Effect of the temperature during the murashi on rice cooking. J Home Econ Japan 40(10):897–904

Kainuma Y, Ema S (1990) A study of the rice cooking (Part 8). Effect of prolonging period of murashi on rice cooking. J Home Econ Japan 41(12):1151–1157

Kainuma Y, Ema S, Seki C (1988) A study of rice cooking (Part 6). The effect of murashi as a factor controlling rice cooking. J Home Econ Japan 39(9):969–977
Kainuma Y, Seki C (1986) A study of rice cooking (Part 5). Boiling time as a factor controlling rice cooking (2). J Home Econ Japan 37(12):1039–1047
Karizaki VM (2016) Ethnic and traditional Iranian rice-based foods. J Ethnic Foods 3(2):124–134
Kasai M, Lewis AR, Ayabe S, Hatae K, Fyfe CA (2007) Quantitative NMR imaging study of the cooking of Japonica and Indica rice. Food Res Int 40(8):1020–1029
Kashaninejad M, Maghsoudlou Y, Rafiee S, Khomeiri M (2007) Study of hydration kinetics and density changes of rice (Tarom Mahali) during hydrothermal processing. J Food Eng 79(4): 1383–1390
Kuwada H, Teramoto A, Jibu Y, Tabuchi M, Fuchigami M (2011) Texture and fine structure of cooked brown rice. J Cookery Sci Japan 44(2):137–144
Leelayuthsoontorn P, Thipayarat A (2006) Textural and morphological changes of Jasmine rice under various elevated cooking conditions. Food Chem 96(4):606–613
Li H, Prakash S, Nicholson TM, Fitzgerald MA, Gilbert RG (2016) Instrumental measurement of cooked rice texture by dynamic rheological testing and its relation to the fine structure of rice starch. Carbohydr Polym 146:253–263
Li H, Yang J, Gao M, Wang J, Sun B (2019) Washing rice before cooking has no large effect on the texture of cooked rice. Food Chem 271:388–392
Mabashi Y, Ookura T, Kasai M (2007) Effects of rate of temperature increase and holding temperature during cooking on the amounts of chemical components in rice grain. J Cookery Sci Japan 40(5):323–328
Machida Y, Tomita K, Shimizu T, Nakaoka F, Kobayashi A (2020) Preference survey for consumers in Tokyo and Fukui based on comparison of cooked rice. Hokuriku Crop Sci 55: 27–30
Maleki C, Oliver P, Lewin S, Liem G, Keast R (2020) Preference mapping of different water-to-rice ratios in cooked aromatic white jasmine rice. J Food Sci 85(5):1576–1585
Maruyama E, Sakamoto K (1992) A basic study on cooking of rice (part 1). *The influence of rice soaking by hot water*. J Home Econ Japan 43(2):97–103
Matsumoto Y (1982) Changes in automatic rice cookers. Sci Cookery 15(2):92–96
Meng L, Zhang W, Wu Z, Hui A, Gao H, Chen P, He Y (2018) Effect of pressure-soaking treatments on texture and retrogradation properties of black rice. LWT-Food Sci Technol 93: 485–490
Ministry of Education, Culture, Sports, Science and Technology. (2020) Standard tables of food composition in Japan - 2020 - (Eighth Revised Version)
Miwa S, Kuroda A, Oda H, Takaya T, Nishinari K (2002) The relationship between the sensory properties of cooked rice and the water uptake of raw rice. J Jpn Soc Food Sci Technol 49(1): 32–39
Morii S, Sakamoto K, Shirasugi N. (Kataoka)(2019) The reverse phenomenon in water absorption rate in Japonica rice grains caused by long-term soaking in water at different temperatures. J Cookery Sci Japan 52(3):159–168
Murakami T, Aida K (2010) Rice consumption and cooking conditions in the Tohoku and Hokkaido regions. J Cookery Sci Japan 43(5):319–321
Muramatsu Y, Tagawa A, Sakaguchi E, Kasai T (2006) Water absorption characteristics and volume changes of milled and brown rice during soaking. Cereal Chem 83(6):624–631
Nagai I (1959) History of rice. Shibundo Co., Ltd., Tokyo, Japan
Nagai T, Tamura M, Hidaka Y, Noda T, Yokoe M, Ogawa Y (2013) Relationships between physicochemical properties of rice grain and apparent viscosity of cooking water. J Jpn Soc Taste Technol 12(1):4–9
Nakamura T (2011) The trend of the IH rice cooker. J Inst Electr Install Eng Jpn 31(2):111–114
Ogawa Y, Yoneoka K, Tamura M, Yamamoto N (2012) Mechanical and histological properties of rice cooked with Varying amounts of water. J Jpn Soc Taste Technol 11(1):44–50

Ohisa N, Kodama I, Hoshino I, Turumaki H, Ohno T (2007) Quick cooking using low-amylose rice (*Awayuki-komachi*). J Jpn Soc Food Sci Technol 54(7):339–342

Patindol J, Gu X, Wang YJ (2010) Chemometric analysis of cooked rice texture in relation to starch fine structure and leaching characteristics. Starch-Stärke 62(3-4):188–197

Rousset S, Pons B, Martin J-F (1999) Identifying objective characteristics that predict clusters produced by sensory attributes in cooked rice. J Texture Stud 30:509–532

Sagum R, Arcot J (2000) Effect of domestic processing methods on the starch, non-starch polysaccharides and in vitro starch and protein digestibility of three varieties of rice with varying levels of amylose. Food Chem 70(1):107–111

Sakamoto K, Morii S, Ueda M (2015) Differences in water absorption of rice when soaked in warm and cold water. J Cookery Sci Japan 48(3):193–199

Sasagawa A, Naiki Y, Nagashima S, Yamakura M, Yamazaki A, Yamada A (2006) Process for producing brown rice with increased accumulation of GABA using high-pressure treatment and properties of GABA-increased brown rice. J Appl Glycosci 53(1):27–33

Sato Y (2003) The civilization of rice: When did humans acquire rice? PHP Institute, Inc., Kyoto, Japan

Sato Y (2020) Japanese History of Rice. Chuokoron-Shinsha Inc., Tokyo, Japan

Seki C, Kainuma Y (1982) A study of rice cooking (Part 2). Soaking time as a factor controlling rice cooking. J Home Econ Jpn 33(5):228–234

Seki C, Kainuma Y (1986) A Study of Rice Cooking (Part 4). Boiling time as a factor controlling rice cooking. J Home Econ Jpn 37(2):93–99

Siaw MO, Wang YJ, McClung AM, Mauromoustakos A (2021) Effect of protein denaturation and lipid removal on rice physicochemical properties. LWT-Food Sci Technol 150:112015

Son JS, Do VB, Kim KO, Cho MS, Suwonsichon T, Valentin D (2013) Consumers' attitude towards rice cooking processes in Korea, Japan, Thailand and France. Food Qual Prefer 29(1):65–75

Srisawas W, Jindal VK (2007) Sensory evaluation of cooked rice in relation to water-to-rice ratio and physicochemical properties. J Texture Stud 38(1):21–41

Suwannaporn P, Linnemann A (2008) Rice-eating quality among consumers in different rice grain preference countries. J Sens Stud 23(1):1–13

Suwannaporn P, Linnemann A, Chaveesuk R (2008) Consumer preference mapping for rice product concepts. Br Food J 10(6):595–606

Taira T, Shoji I (1999) Studies on simplified method to determine water uptake ratio and total solids of rice at 70 °C. J Jpn Soc Food Sci Technol 46(7):480–483

Takano K, Taniguchi A (eds) (2021) Science of Food and Health Series: Science of Rice. Asakura Publishing Co., Ltd., Tokyo, Japan

Tamura M, Kumagai C, Kaur L, Ogawa Y, Singh J (2021) Cooking of short, medium and long-grain rice in limited and excess water: Effects on microstructural characteristics and gastro-small intestinal starch digestion *in vitro*. LWT-Food Sci Technol 146:111379

Tamura M, Nagai T, Hidaka Y, Noda T, Yokoe M, Ogawa Y (2014a) Changes in histological tissue structure and textural characteristics of rice grain during cooking process. Food Struct 1(2): 164–170

Tamura M, Nagai T, Hidaka Y, Noda T, Yokoe M, Ogawa Y (2014b) Changes in Nonwaxy J aponica rice grain textural-Related properties during cooking. J Food Qual 37(3):177-184

Tamura M, & Ogawa Y (2010) A new approach of rice processing for control of compound contents and eating quality of cooked rice. Proceeding, XVIIth World Congress of the International Commission of Agricultural and biosystems Engineering (CIGR), CSBE100992

Tamura M, Ogawa Y (2012) Visualization of the coated layer at the surface of rice grain cooked with varying amounts of cooking water. J Cereal Sci 56(2):404–409

Tamura M, Singh J, Kaur L, Ogawa Y (2016) Impact of the degree of cooking on starch digestibility of rice—An *in vitro* study. Food Chem 191:98–104

Tani T, Yoshikawa S, Chikubu S, Horiuchi H, Endo I, Yanase H (1969) Physicochemical properties related to palatability evaluations of cooked rice (1). J Jpn Soc Food Nutr 22:452–461

Tomita H, Sakamoto K, John H, Takemori T (2015) Soaking time-related changes in microstructures and texture of cooked rice. J Cookery Sci Jpn 48:18–25
Tsujii Y, Kiyose N, Tatsuda N, Yaguchi Y, Uchino M, Takano K (2009) Change in the endospermous cell wall of cooked rice and its effects on the palatability of cooked rice. Food Preserv Sci 35(3):127–134
Tsuyukubo M, Ookura T, Kasai M (2013) Distribution of starch-degrading enzymes in rice grains of different cultivars and elution behavior during cooking. Food Sci Technol Res 19(2):303–311
Tsuyukubo M, Ookura T, Mabashi Y, Kasai M (2010) Different distributions of α-glucosidases and amylases in milling fractions of rice grains. Food Sci Technol Res 16(6):523–530
Tsuyukubo M, Ookura T, Tsukui S, Mitsui T, Kasai M (2012) Elution behavior analysis of starch degrading enzymes during rice cooking with specific antibodies. Food Sci Technol Res 18(5): 659–666
Wada T, Umemoto T, Aoki N, Tsubone M, Ogata T, Kondo M (2011) Starch eluted from polished rice during soaking in hot water is related to the eating quality of cooked rice. J Appl Glycosci 58(1):13–18
Wang L, Zhao S, Kong J, Li N, Qiao D, Zhang B et al (2020) Changing cooking mode can slow the starch digestion of colored brown rice: a view of starch structural changes during cooking. Int J Biol Macromol 155:226–232
Wang S, Li P, Zhang T, Yu J, Wang S, Copeland L (2017) *In vitro* starch digestibility of rice flour is not affected by method of cooking. LWT-Food Sci Technol 84:536–543
Wolever TM, Jenkins DJ, Kalmusky J, Jenkins A, Giordano C, Giudici S et al (1986) Comparison of regular and parboiled rices: explanation of discrepancies between reported glycemic responses to rice. Nutr Res 6(4):349–357
Wu J, Chen J, Liu W, Liu C, Zhong Y, Luo D et al (2016) Effects of aleurone layer on rice cooking: a histological investigation. Food Chem 191:28–35
Xu D, Hong Y, Gu Z, Cheng L, Li Z, Li C (2019) Effect of high pressure steam on the eating quality of cooked rice. LWT-Food Sci Technol 104:100–108
Yadav BK, Jindal VK (2007a) Dimensional changes in milled rice (Oryza sativa L.) kernel during cooking in relation to its physicochemical properties by image analysis. J Food Eng 81(4): 710–720
Yadav BK, Jindal VK (2007b) Water uptake and solid loss during cooking of milled rice (Oryza sativa L.) in relation to its physicochemical properties. J Food Eng 80(1):46–54
Yamakura M, Haraguchi K, Okadome H, Suzuki K, Tran UT, Horigane AK et al (2005a) Effects of soaking and high-pressure treatment on the qualities of cooked rice. J Appl Glycosci 52(2): 85–93
Yamakura M, Okadome H, Suzuki K, Tran UT, Homma S, Sasagawa A et al (2005b) Effects of high-pressure treatment and soaking on the cooked rice. J Jpn Soc Food Sci Technol 52(2): 60–67
Yamamoto A, Nakajima H, Tanida N (1990) Occurrence of abnormally cooked rice and water uptake by polished rice at high temperature. J Jpn Soc Nutr Food Sci 43(3):215–220
Yamazaki T, Fukuba H (1982) Influence of water addition ratio on quality of cooked rice of processed brown rice. Sci Cookery 15(3):184–188
Yokoo M (1995) The story of rice. Gihodo Shuppan Cp. Ltd., Tokyo, Japan
Yu Y, Ge L, Zhu S, Zhan Y, Zhang Q (2015) Effect of presoaking high hydrostatic pressure on the cooking properties of brown rice. J Food Sci Technol 52(12):7904–7913
Yu Y, Pan F, Ramaswamy HS, Zhu S, Yu L, Zhang Q (2017) Effect of soaking and single/two cycle high pressure treatment on water absorption, color, morphology and cooked texture of brown rice. J Food Sci Technol 54(6):1655–1664

Zhou Z, Robards K, Helliwell S, Blanchard C (2002) Composition and functional properties of rice. Int J Food Sci Technol 37(8):849–868

Zhu L, Bi S, Wu G, Zhang H, Wang L, Qian H et al (2020) Comparative analysis of the texture and physicochemical properties of cooked rice based on adjustable rice cooker. LWT-Food Sci Technol 130:109650

Zhu L, Cheng L, Zhang H, Wang L, Qian H, Qi X, Wu G (2019) Research on migration path and structuring role of water in rice grain during soaking. Food Hydrocoll 92:41–50

Zhu L, Zhang H, Wu G, Qi X, Wang L, Qian H (2021) Effect of structure evolution of starch in rice on the textural formation of cooked rice. Food Chem 342:128205

Chapter 9
Storage of Cooked Rice

Masatsugu Tamura

Abstract When rice is left unattended after cooking, its taste deteriorates markedly due to retrogradation, and its high starch and moisture content causes it to be spoiled. Thus, the storage of cooked rice is an important issue; in this part, the history of manufacturing processed rice is introduced in detail (including canned rice, retort-packed rice, dried rice, etc.). Then manufacturing methods for processed rice are demonstrated and the changes in the properties of rice (including moisture, physical and thermal properties, texture, and starch digestibility) during the production and storage processes are discussed. In particular, the development of processed rice and related technologies expected in the future are also described.

Keywords Cooked rice · Storage · Aseptically packaged · Retort-packed · Digestibility

Abbreviations

AP	Aseptically packaged
BQF	Block quick freezing
eGI	Estimated glycemic index
EVOH	Ethylene vinyl alcohol copolymer
GI	Glycemic index
GML	Glycerol monolaurate
IQF	Individual quick freezing
OTRs	Oxygen transmission rates
PVDC	Polyvinylidene chloride
RDS	Rapidly digestible starch
RP	Retort-packed
RS	Resistant starch

M. Tamura (✉)
Utsunomiya University, Utsunomiya, Japan
e-mail: m-tamura@cc.utsunomiya-u.ac.jp

J. Tian et al. (eds.), *Science of Rice Chemistry and Nutrition*,
https://doi.org/10.1007/978-981-99-3224-5_9

RSM	Response surface method
SDS	Slowly digestible starch
TS	Total starch
W/R	Water-to-rice
β-CD	β-Cyclodextrin

9.1 Introduction

Rice usually needs to be washed, soaked, and cooked for at least 1 h. When rice is left unattended after cooking, its taste deteriorates markedly due to retrogradation, and its high starch and moisture content causes it to be spoiled by heat-resistant bacteria and airborne organisms that are sources of secondary contamination. Usually, when rice is stored at 20–30 °C, it becomes foul-smelling, sticky, low in pH, and inedible. Since rice is heated to more than 100 °C during cooking, most of the microorganisms involved in deterioration are *Bacillus*, an aerobic heat-resistant spore bacterium; however *Lactobacillus* and yeast also affect secondary contamination that falls from the air after cooking. Therefore, to distribute and store high-quality rice, it is essential to prevent retrogradation and microbial contamination. In other words, processed rice can be stored for a long time by heating, freezing, or sterilization. Moreover, these processed rice products can be easily consumed by heating them in a microwave oven or by using boiling water for a few minutes.

A wide variety of processed rice products with different production methods, such as canned rice, retort-packed (RP) rice, dried rice, frozen rice, and aseptically packaged (AP) rice, have been developed and marketed. Figure 9.1 shows the trends in the production of processed rice in Japan. Among the processed rice products, the variety of products such as RP rice porridge and sets of AP rice and ingredients has increased, and health-conscious products containing germinated brown rice, millet,

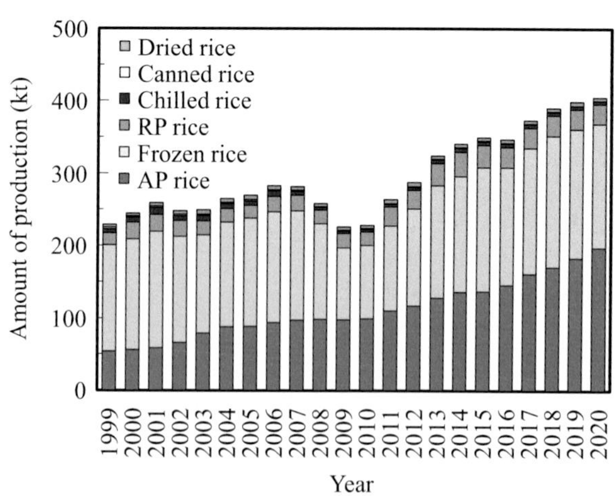

Fig. 9.1 Production trends of processed rice. (Reprinted with permission from the Ministry of Agriculture, Forestry and Fisheries of Japan 2021a)

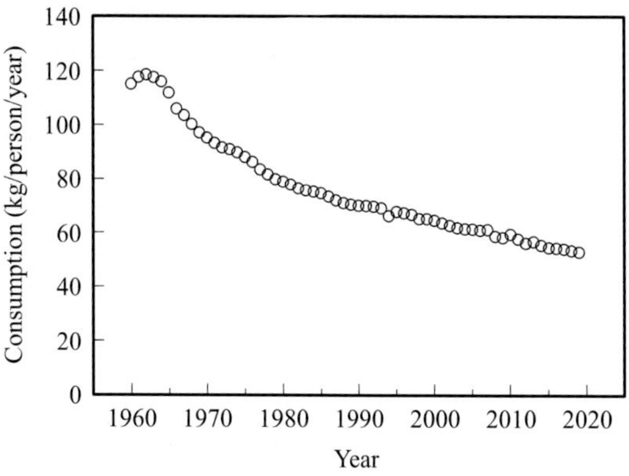

Fig. 9.2 Japanese rice consumption. (Reprinted with permission from the Ministry of Agriculture, Forestry and Fisheries of Japan 2021b)

and glutinous barley have increased. In addition, sales channels have expanded to meet the demand for rice in restaurants, home-delivered meals, and overseas markets where rice cookers are not available. In particular, the production of AP rice increases remarkably, exceeding the production of frozen rice in 2019. Figure 9.2 shows the per capita rice consumption in Japan. Consumption has steadily decreased each year. The contradictory relationship between the decrease in rice consumption and the increase in the consumption of processed rice is considered to be influenced by the convenience of processed rice, which does not require complicated cooking operations, labor, or time; can be stored for a long period of time; and is convenient to carry and transport. These characteristics are well suited to the growing need for simplified cooking due to the increase in single-person households and the social advancement of women, as well as the need for emergency food for stockpiling and safe, cheap, and tasty food. The processed rice market is expected to reach 100 billion yen in the near future. In this chapter, the history of food processing technology and processed rice in the world and Japan, the production method of processed rice, and the scientific changes in rice during storage are introduced.

9.2 History of Manufacturing Processed Rice

In this section, a review of the history of canned, RP, dried, frozen, and AP rice is provided, along with the history of food processing technology.

9.2.1 Canned Rice

Canning is considered the first technology in the history of food processing to be developed at the industrial level (Ishikawa 2001; Japan Canners Association, Retort Foods Section 1996). Emperor Napoleon Bonaparte was deeply involved in the birth of canned food. Napoleon was troubled by the problem of food supply every time he went on a European tour. At that time, long-term preserved foods were mainly salted, smoked, or pickled in vinegar, which tasted badly and often spoiled. In 1804, Nicolas Appert, a French, developed a new method of food storage in which food was placed in glass containers, sealed with cork stoppers, heatsterilized, degassed, and then completely sealed. This was approxomately 60 years before Louis Pasteur's academic discovery in 1864 that microorganisms are deeply involved in food spoilage. This technology was popularized by Peter Durand in England in 1810 as a method for storing food in tin cans. In 1821, the process was transferred to the United States, and canning began in earnest.

The detailed start of canned rice production in Japan is not clear, but it appeared as a military food at the start of the Sino-Japanese War (1937–1945) (Hieda 1971). However, its consumption has been decreasing since 1997 owing to the spread of other processed rice products (Naito 2020).

9.2.2 RP Rice

The first rotary sterilizer for retort sterilization was developed in France in 1917, and a fully automated version was developed by Stock in Germany in 1947. On the other hand, the retort pouch, which is a container for food, was first studied by Nelson and Seinberg at the University of Illinois around 1955. In 1959, the U.S. Army's Natick Development Center launched a research project to develop retort pouches as alternatives to canned foods for military use. RP foods were loaded onto the Apollo 11 lunar exploration ship in 1969 and eaten in space, which is one of the reasons why people began to pay attention to them. RP foods researched and developed in the United States were for special use as military and space food, so the product price was extremely high, and there were many problems in mass production. Furthermore, in the US market in the 1960s, the cold chain was well established and frozen foods were widely sold; therefore, the demand for retort foods that could be distributed and stored at room temperature was not necessarily high. In addition, owing to safety issues with the adhesives used in retort pouches, the Food and Drug Administration and the United States Department of Agriculture did not approve their manufacture until 1977 (Japan Canners Association, Retort Foods Section 1996; Yokoyama 1990).

In Japan, frozen food could not be distributed because refrigerators and cold chains were not sufficiently widespread. In 1969, Toyo Seikan Co., Ltd. and Otsuka Chemical Co., Ltd. jointly produced the world's first commercially available curry in a pouch. In 1970, RP rice was commercialized, and, tori kamameshi (cooked rice and chicken) and other products were launched in 1972. Currently, Japan is the world's largest producer of RP foods.

9.2.3 Dried Rice

In 1901, botanist Alexander Pierce Anderson accidentally discovered that rice swelled during experiment. He then marketed rice as puffed rice (Horigome 1990). At the Louisiana Purchase Expo in St. Louis in 1904, he shot puffed grain from a cannon and caught public attention (Miller 2019). It was studied in the United States before World War II and has been patented under the name of quick-cooking rice or precooked rice.

Dried rice in Japan originated in the fifth century as portable food for travel, hunting, and warfare. The development of pregelatinized rice among dried rice originated from an urgent request from the Japanese Navy in 1944 to develop rice that could be eaten without cooking. Based on the principle of α-starch, Haruyasu Onishi of the Onishi Food Research Institute and Jiro Nikuni of Osaka University developed pregelatinized rice as alpha rice, which can be eaten by pouring cold water over it. Pregelatinized rice was produced for military purposes around 1944, and postwar production was controlled by the government and municipalities. However, pregelatinized rice has attracted attention as a portable food for mountaineering since 1950, because of its ability to produce rice without heating. Pregelatinized rice began to be marketed as a general food starting in the 1960, and from the mid-1970s, pregelatinized rice was developed for use in rice cookers for home cooking, such as okowa (rice with red beans), and for commercial sake brewing. In recent years, pregelatinized rice has attracted attention as an emergency or stockpile food during disasters.

9.2.4 Frozen Rice

In the days before the invention of the refrigerator, natural cold was used to freeze food. Refrigeration machines were invented and put to practical use in the nineteenth century. Charles Tellier, a French, was the first person in the world to research a method for freezing and storing food. He began research on beef refrigeration in 1868 and successfully transported frozen beef and mutton by sea between France and Argentina twice in 1877 and 1878. In 1911, A.J.A. Ottesen in Denmark developed a method for freezing fish by soaking them in a brine solution that was cooled in a refrigerator. This was the forerunner of brine freezing and quick-freezing methods. Food freezing played a major role in the storage and transportation of food for military use during the First World War (1914–1918) in Europe, and military demand stimulated the progress of freezing technology. In the United States, quick-frozen foods were introduced to the market in 1930, whereas the history of quick-frozen foods in other countries began around 1940–1950 (Nippon Reizo Inc. Institute 1979; Ishitani et al. 1995).

In Japan, surrounded by the sea on all sides and with a thriving fishing industry, frozen foods began in 1899 with the freezing of fish by Kota Nakahara, who returned from the United States. Frozen foods were offered at the 1964 Tokyo Olympic Games and the 1970 Osaka World Fair and became popular in the early 1960s with the rise of supermarkets. The origin of frozen rice products was the use of old rice. Frozen pilafs for commercial use were developed around 1973 in anticipation of the increasing demand as the food service industry developed. Rice for pilaf should be large, not dissolve, and not stick together after cooking. However, there was a surplus of old rice at that time, and its properties were ideal for the production of frozen pilafs. Furthermore, the birth of individual quick freezing (IQF) technology in the 1980s, which could freeze food in pieces, accelerated the production of frozen pilaf and frozen fried rice for commercial use.

9.2.5 AP Rice

When canned and chilled rice were marketed in Japan as conveniently preserved foods, frozen rice was developed in 1972, RP red rice (rice mixed with red beans) in 1973, and RP white rice in 1975. However, the taste of these rice products was not as good as that of freshly cooked rice, and the market gradually shrank. In 1988, Sato Foods Industries Co., Ltd. (presently, Sato Foods Co., Ltd.) released AP rice, and since then, many companies have been developing and selling AP rice cooked in large pots with an oxygen scavenger. Since 1992, AP rice technologies have been developed, including containerized rice cooking using retort containers and sealed transportation in clean rooms, containerized rice cooking using micro-pressure, steam cooking in containers using pressurized heat, sterilization and cooking by pressurized steam in small retort containers, products that do not require oxygen absorbers. Because rice is cooked in the container through automation, these methods reduce the need for packaging materials, and environmental sanitation, leading to increased productivity and reduced costs (Masuda 2011).

9.3 Manufacturing Methods for Processed Rice

This section introduces production methods for canned, chilled, RP, dried, frozen, and AP rice. Table 9.1 shows an overview of processed rice.

9.3.1 Canned Rice

Canned rice is manufactured by measuring and mixing raw rice, ingredients, and seasonings; filling the can; heating; degassing; sealing; heat sterilization; and cooling (Fig. 9.3) (Naito 2020). In the United States, canned rice products are usually processed by the wet-pack procedure, where rice is subjected to severe processing

Table 9.1 Overview of processed rice (drawn by the author)

	Moisture content (%)	Storage temperature and duration	Adding water during cooking	Cooking method
Canned rice	60	Room temperature for 3–5 years	Unnecessary	Boiling for 5–15 min
Chilled rice	60	4–20 °C for 3–6 days	Unnecessary	Microwave heating for 2 min
RP rice	60	Room temperature for 6–19 months	Unnecessary	Microwave heating for 2 min or boiling for 5–10 min
Dried rice	5–10	Room temperature for 5 years	Necessary	Boiling for 15–30 min or pouring boiling water and holding for 5–20 min
Frozen rice	60	≤ −18 °C for 1 year	Unnecessary	Microwave heating for 5 min or frying for 3–6 min
AP rice	60	Room temperature for 6–12 months	Unnecessary	Microwave heating for 2 min or boiling for 10–15 min

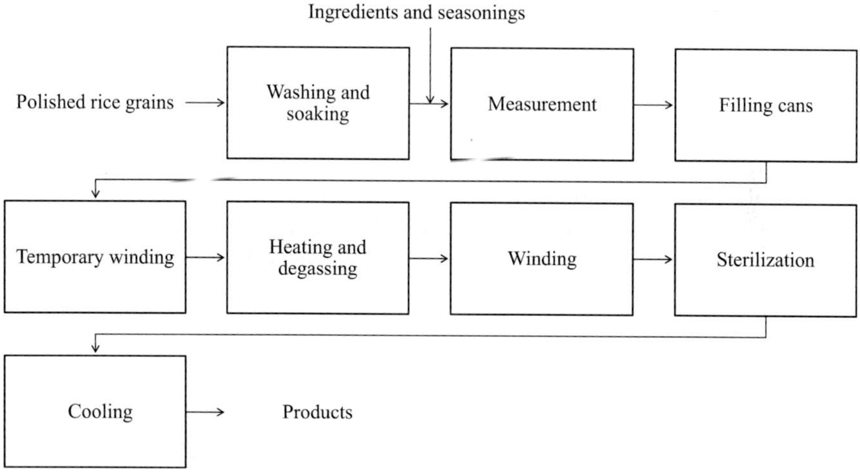

Fig. 9.3 Manufacture process of canned rice (drawn by the author)

steps (Patindol et al. 2007). Rice is cooked by heating before and after sealing, and the heating time and temperature determine the quality of rice. Sterilization is usually performed at 113 °C for 80 min. Because such high-temperature sterilization degrades the taste of the rice, some products are half cooked during manufacturing and the remaining cooking is completed when the rice is eaten. Most canned rice in Japan is rice porridge or rice mixed with beans or chicken. Because canned rice is stored at room temperature after production, it is always subject to retrograde. Therefore, retrograded canned rice must be reheated to serve as α-starch.

9.3.2 Chilled Rice

In Japan, most chilled rice products are sold for commercial use at temperatures ranging from −5 to 5 °C. The upper limit of 5 °C is effective in preventing harmful bacteria, whereas the lower limit of −5 °C is the lower limit of the maximum temperature range for the formation of surface crystals.

The upper limit of this temperature, 5 °C, is effective in preventing harmful bacteria, and the lower limit, −5 °C, is the lower limit of the zone of maximum ice crystal formation. Because the freezing rate is approximately 80% at −5 °C, rice with a freezing point (the temperature at which it begins to freeze) of −1 °C is physically frozen. On the other hand, retrogradation of α-starch in rice progresses quickly in the chilled temperature range. If the humidity in the air is low in this temperature range, the drying process is accelerated, and the rice loses its commercial value owing to drying before retrogradation. Therefore, appropriate humidification of chilled rice is necessary. Figure 9.4 shows the production process for chilled rice. Organic acids such as citric acid, malic acid, and lactic acid are added to polished rice before or after cooking to adjust the pH to 4–5. The rice is then sterilized at high temperatures for a short time, and inert gases such as carbon dioxide or nitrogen and oxygen scavengers are added when the rice is sealed. Thus, the initial microbial contamination of the chilled rice is minimized. Traditional cooling methods include blast, plate, and cold room cooling. Air blast (fast cooling speed) has less effect on cooked rice firmness, while cold room cooling (slow cooling speed) has some negative effects on cooked rice texture (Ma and Sun 2009). Vacuum cooling can cool cooked rice to chilling temperatures in a shorter time than the traditional cooling methods. However, since vacuum cooling loses the moisture content of the cooked rice, less weight of cooked rice should be vacuum cooled or cooled with a lower final pressure or with water spraying (Zhang and Sun 2006).

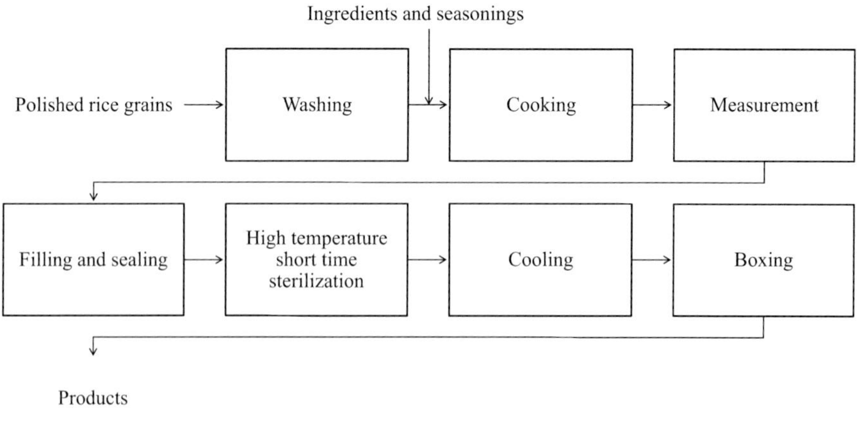

Fig. 9.4 Manufacture process of chilled rice (drawn by the author)

9.3.3 RP Rice

Manufacturing Methods

The term retort refers to a sterilization oven. Retortable foods are generally packaged in pouches or molded containers made of synthetic resin film or other light-impermeable materials laminated with aluminum foil, filled with the contents, completely sealed, and then pressure and heat sterilized (retort sterilized). The key point here is that the microorganisms that cause spoilage are sterilized by heat in the sealed package, enabling long-term storage. Therefore, the methodology of retort-packed foods is the same as that of rigid metal cans and glass bottles; however, the packaging used is different (Horigome 1990; Ishitani et al. 1995; Naito 2020; Tabuchi 1993; Yokoyama 1990).

Broadly, retort pouch foods are divided into retort pouch foods, retort container foods, and retort pack foods. Retort pouches are sealed on all sides and are available in two types: transparent pouches and aluminum foil pouches. Retort pouches have become mainstream food products with a large variety of products and production volumes. A retort container is a circular or rectangular transparent or aluminum foil container. Retort packs are made by clipping one end with aluminum wire, filling the packaging material with food, clipping the other end with the same aluminum wire, and retort sterilization. RP rice is primarily produced using retort pouches (Fig. 9.5a) or retort containers (Fig. 9.5b).

Figure 9.6 shows the manufacturing process of the RP rice. Rice washing is responsible for the removal of dust, bran, and other adhering materials from rice. Inadequate washing increases the amount of sugar and amino acids remaining in the rice, which affects its aroma after heating or reheating. There are three types of RP rice production methods: (1) steamed waxy rice or cooked rice is packed in a bag and

(a) (b)

Fig. 9.5 Product of (**a**) cooked rice porridge in retort pouch (Yawaraka-gohan), and (**b**) rice porridge and side dishes in retort container (Nico Nico Box). The photos are provided with permission from Kewpie Corp

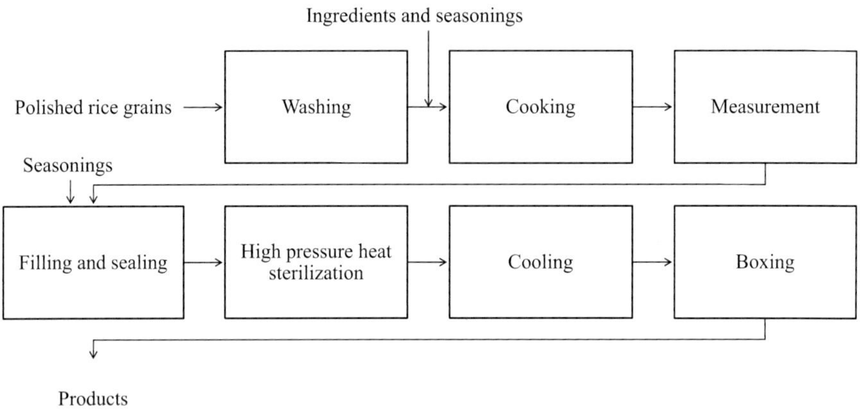

Fig. 9.6 Manufacture process of RP rice (drawn by the author)

then sterilized; (2) raw rice and water are packed in a bag and then processed for both cooking and sterilization; and (3) semi-steamed rice is packed in a bag and then sterilized under pressure. The method of packing rice in bags described in (1) has problems with the accuracy of weighing a certain amount of rice and the workability of packing it in bags because of the tendency of the rice to stick to each other and form aggregates, in addition to the occurrence of heating odor and coloring. Therefore, this method can be applied to pilafs, which can loosen cooked rice by adding large amounts of oil and fat. On the other hand, the method of packing raw rice in a pouch, as described in (2), is more accurate in weighing and easier to work with, but when the rice is left to sit and heat sterilized, the moisture content becomes uneven between the upper and lower layers of the pouch, and the texture also tends to be uneven. For this reason, the pouches were heat sterilized while rotating in a rotary retort. Method (3) is the most popular retort sterilization method and is an intermediate method between methods (1) and (2). RP rice with a high air content may deteriorate in quality due to oxidation during retort sterilization and storage; therefore, the air in the container is replaced with inert gases such as nitrogen gas. Regardless of the packaging type, retort sterilization involves pressurized heat treatment at 115 °C for 50–60 min or 120 °C for 20–45 min. During retort sterilization, the water absorbed by the rice accumulates in the packaging container as excess water. When the temperature decreases, the steam absorbs water from the rice surface. During this process, starch granules with little stickiness were disrupted. Waxy rice tends to absorb heating odors and packaging container odors during the high-temperature and high-pressure treatment of retort sterilization and releases these strange odors when reheated. On the other hand, waxy rice is suitable for retorting because it requires less water to be added, the starch granules do not break down even after high-pressure processing, and it has a unique strong aroma. Products such as sekihan (red rice) and okowa (rice with red beans) have been developed using the characteristics of waxy rice.

Packaging Materials

Packaging materials for retort-packed foods must have functions such as long-term blocking of light and oxygen, which adversely affects foods, moderate solubility for sealing by heat welding, good printing, and resistance to puncture and impact. Of course, because there is no plastic film that satisfies all of these characteristics, laminate structures or coextruded plastic films and material foils are used. Nylon/polypropylene and polyester/polypropylene are used as transparent pouches, and polyester/nylon/polypropylene is used as a packaging material to improve the mechanical properties. For opaque pouches, polyester/aluminum foil/polyolefin is used to block oxygen during retort sterilization (Ishitani et al. 1995; Japan Canners Association, Retort Foods Section 1996; Naito 2020). Polypropylene alone or aluminum foil (outer protective film/aluminum foil/polypropylene) is used for the tray-like containers. Aluminum foil containers have advantages such as almost no oxygen permeation and good aroma retention. Polyethylene and polypropylene are mainly used for the inner layer of retort containers, which are required to be heat resistant for heat sealing and safe for direct contact with food. Aluminum foil is used as the middle layer, and it must be impermeable to oxygen and light, as well as a barrier to prevent the transfer of ink and adhesives from the outer layer to the inner layer. For the outer layer, polyester or nylon is used, and it is important that the package is compatible with printing and that the metal foil is corrosion resistant and strong. In recent years, special film materials equipped with a mechanism to ventilate the water vapor generated inside the pouch have been developed and utilized to make the contents of the pouch visible and to meet the increasing demand for microwave oven compatibility.

9.3.4 Dried Rice

Dried rice is commonly referred to as instant rice; however, but there is no clear definition. In this section, the differences in the presence or absence of the gelatinization process (cooking) and drying method are categorized and explained as follows: (1) pregelatinized rice, in which the rice is cooked, dried to adjust the moisture content, and then swollen with hot air; (2) freeze-dried rice, in which the rice is cooked, pre-frozen, and then freeze-dried; and (3) puffed rice, in which the rice is cooked or uncooked, compressed under high-temperature and high-pressure conditions, and then released to evaporate the moisture content.

Pregelatinized Rice

Pregelatinized rice can be recovered as cooked rice by heating it with hot water for a short time or by adding hot water and holding it for 10–15 min. Pregelatinized rice is dried at high temperature for a short time in a gelatinized state where the double helix of the starch molecule is loosened, and its principle is the same as that of instant noodles and rice crackers. Properties such as shape, color, gloss, density, and texture of pregelatinized rice vary depending on the gelatinization and drying process (Horigome 1990; Ishitani et al. 1995; Tabuchi 1993).

Figure 9.7 shows the manufacturing process of alpha rice, which is pregelatinized rice. The washed rice is soaked until the water content saturated at a temperature below the gelatinization temperature to ensure that the rice is fully gelatinized. The soaked rice is gelatinized by steaming or cooking. In other words, fats and emulsifiers are added to the soaked rice grains before or after steaming/cooking, and the rice grains are soaked for a short time to allow the water on the surface of the grains to be absorbed by the inside of the grains, and then steamed and refrigerated for retrogradation. However, because soaked rice has a low moisture content of approximately 35%, steamed and dried rice has a dense structure, poor rehydration, and poor texture. On the other hand, alpha rice cooked after soaking in water has a relatively good texture after rehydration, but it is very sticky and tends to form rice clumps, making it difficult to single-grain, resulting in a large amount of loss during the grain preparation process. Table 9.2 shows the change in the incidence of rice clumps with the addition of emulsifiers (Horigome 1990). The addition of a small amount of emulsifier has little effect on the prevention of rice clumping, whereas the addition of a large amount of emulsifier affects the taste of the rice. The gelatinized rice is then dried in band or rotary dryers. The drying temperature is set from 100 to 120 °C in the early stage of drying when the moisture content is high and is gradually lowered to 80 °C and 60 °C as the moisture content decreases to reduce the color tone of the rice grains and the variation in moisture content among the grains. Eventually, the moisture content of the rice grains is dried to about 5–10%. Because of this low moisture content, alpha rice is not subject to much retrogradation. Normally, the presence of rice clumps prevents uniform drying and rehydration of the rice grains; therefore, they are separated into single grains during and after drying. The broken grains and rice clumps that occurred during processing are passed through a sieve after drying to prepare the whole grain.

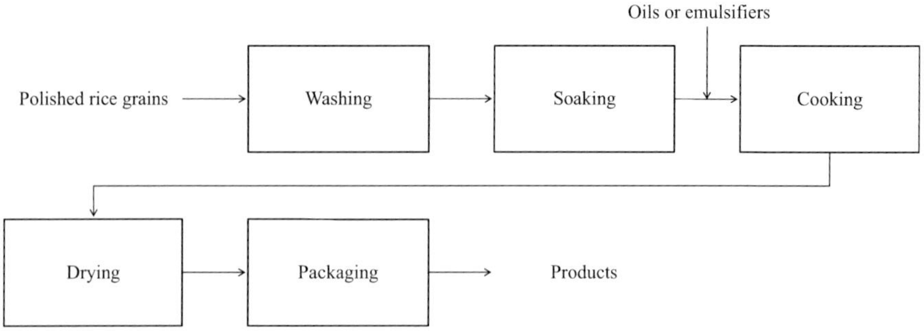

Fig. 9.7 Manufacture process of alpha rice (drawn by the author)

Table 9.2 Change in the incidence of rice clumps due to the addition of emulsifiers with slight modification (Horigome 1990). (Reprinted with permission from Kogyogijutsukai)

Emulsifiers	Incidence of rice clumps (%)
Fatty acid monoglycerides	5
Sorbitan fatty acid esters	6
Propylene glycol fatty acid esters	9
Sucrose stearate esters	5
Additive-free	43

Although the lipid content of pregelatinized rice is low (approximately 1%), the oxidation of lipids due to contact with air and drying of the processed rice increases the number of carbonyl compounds. Carbonyl compounds are the cause of the off-flavor and tend to diffuse inside hermetically sealed container with high gas barrier properties and remain until consumption. The following points are implemented to prevent the generation of off-flavors. The bran on the surface of the raw material, brown rice, or polished rice is rich in lipids that are easily oxidized; therefore, new rice is used and processed as soon as possible after polishing. The rice is then washed to sufficiently remove the bran. Because iron accelerates the oxidation of fats and oils, the water used in the soaking process should be low in terms of free metal ions. Because residual oxygen of more than 1% causes foul odors within 1 year, high-barrier films with high light shielding, water vapor barrier properties, and oxygen scavengers are used to suppress the effects of light and water activity, which affect oxidation, and to maintain residual oxygen below 0.1%.

Freeze-Dried Rice

The application of freeze-drying to rice has been limited to a few applications such as instant rice porridge for infants and the elderly because of the longer drying time and higher running costs compared to hot air-drying. Freeze-dried rice is obtained by freezing cooked rice at −30 to −40 °C and then sublimating it to 2–5% moisture content in a freeze-drying machine under vacuum conditions. Although freeze-dried rice is porous, its water rehydration capacity is poor. This is probably due to the fact that when boiling water is poured over the surface of the rice grains during rehydration, the micropores are quickly blocked by gelatinized starch, preventing the penetration of boiling water into the center of the rice grains. Furthermore, owing to its porous structure, freeze-dried rice oxidizes easily and develops an unpleasant odor after 2–4 weeks of storage at room temperature. Therefore, it is essential that freeze-dried rice is packaged with nitrogen gas replacement or oxygen scavengers for distribution and storage. Freeze-dried rice can be eaten within 1–2 min after being added to hot water, but it is softer than rice that has just been cooked.

Puffed Rice

Popping and puffing are the simplest, cheapest, and fastest conventional dry heat application methods for preparing instant rice, ready-to-eat snacks, and baby foods. Puffing can be carried out using various processes and is commonly classified as oil puffing, gun puffing, hot sand-bed puffing, microwave puffing, and fluidized bed puffing (Saha and Roy 2020). Explosion puffing technology is widely used in the world, especially in East Asia (Huang et al. 2018). Explosion puffing is the oldest method using a pressure cooker. When rice is sealed in a pressure-resistant container and heated, the moisture contained in the rice is in the form of compressed water, which is far above the boiling point. When the lid is opened, the water contained in the rice is suddenly released into atmospheric pressure, and with a sudden drop in pressure, it becomes water vapor with a volume 1200–2000 times larger and is vigorously released outside the rice grains. The rice grains are damaged by pressure during expansion and puffed up. Unmerous holes created on the surface of the puffed rice by the steam plume on the surface of the starch. This porosity is a characteristic of puffed rice and facilitates the penetration of water into rice, improving its immediacy. Therefore, water rehydration of puffed rice in boiling water is generally better than that of pregelatinized and freeze-dried rice. Microwave puffing is rarely used in actual production because of its high running cost and difficulty in uniform puffing, while heating oil and fat puffing are also uncommon because of the high oil and fat content of approximately 20% after puffing and the deterioration of the oil and fat due to the high temperature of approximately 170–200 °C during puffing (Tabuchi 1993).

9.3.5 *Frozen Fried Rice*

Manufacturing Methods

Frozen rice has been recognized as superior among rice products because its quality changes little over time, and microbial growth is inhibited. Many types of frozen rice products are available on the market, such as pilaf, fried rice, and rice balls, which are made by freezing and packaging cooked rice (Horigome 1990; Ishitani et al. 1995; Tabuchi 1993). Most frozen rice is prepared by seasonings and reheating at high temperatures, and the taste and texture are limited to a certain range, so that they can be patterned uniformly. Frozen rice is classified into two types according to its shape: individual pieces of frozen rice, such as pilaf, butter rice, cooked rice, and white rice, and blocks of frozen rice such as baked rice balls and doria. Production facilities are divided into two main types: IQF and block quick freezing (BQF). Examples of the production processes for frozen fried rice (Figs. 9.8 and 9.9) and frozen rice balls (Figs. 9.10 and 9.11) are shown. At the factory scale, rice is washed for a few minutes at a water temperature of 5–35 °C and then soaked in water at 5–35 °C for 30–120 min. The ratio of water to rice (W/R) is adjusted in the range of

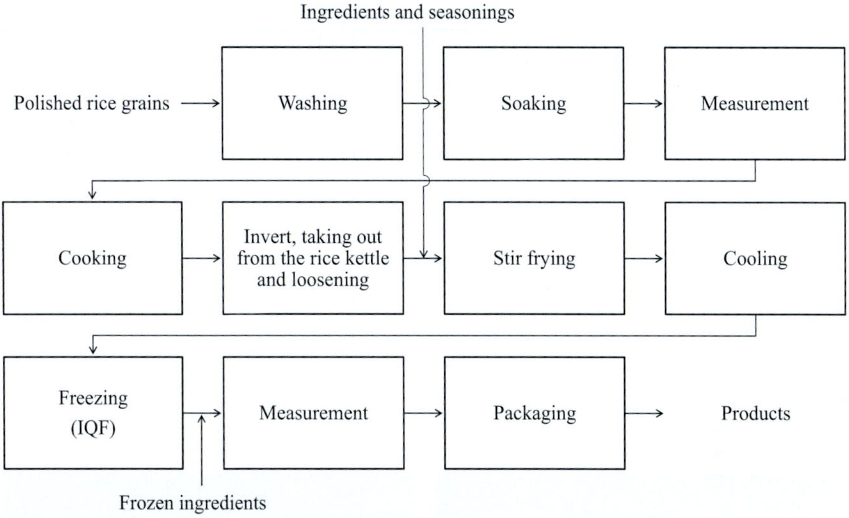

Fig. 9.8 Manufacture process of frozen fried rice (drawn by the author)

Fig. 9.9 Manufacturing process of frozen fried rice: (**a**) Washed rice; (**b**) steamed rice; (**c**) fried rice before freezing; (**d**) frozen fried rice; (**e**) measurement before packaging; (**f**) product (Honkaku-Itame Cha-Han). The photos are provided with permission from Nichirei Foods Inc.

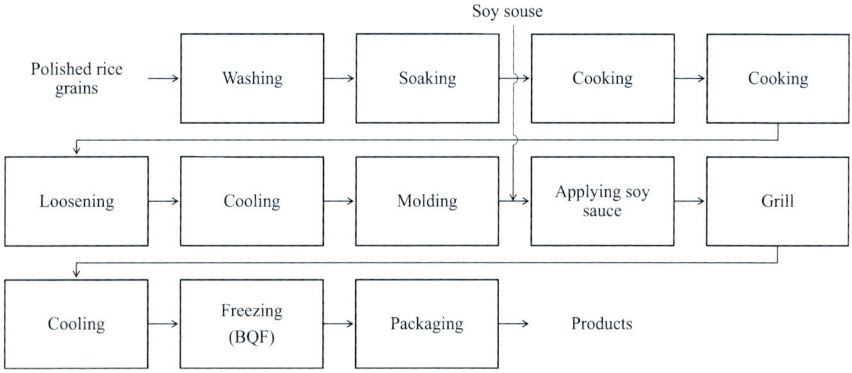

Fig. 9.10 Manufacture process of frozen rice ball (drawn by the author)

Fig. 9.11 Manufacturing process of frozen rice ball: (**a**) Cooked rice; (**b**) Cooked rice molded into rice ball; (**c**) Cooked rice ball coated with soy sauce before grill; (**d**) Product (Yaki-onigiri). The photos are provided by Nichirei Foods Inc.

1.4–1.5, depending on the variety of raw rice and the product. Rice is cooked either in a kettle or on a net conveyor. The rice is heated and steamed sufficiently so that the degree of starch gelatinization is evenly distributed.

Unlike other individual frozen foods, rice not only is sticky, but also has a small volume and weight per grain, is susceptible to impacts such as compression shock, and is mixed with ingredients of different bulk densities and sizes, all of which make it difficult to freeze individually. Because rice grains cannot be heated evenly when they are aggregated into a mass, IQF applies a method of freezing rice by loosening it in an air blast or agitating it in liquefied gas. Although IQF can be frozen in a very short time, it has a low tolerance to temperature changes and is susceptible to drying during freezing and storage. Once thawed and refrozen, the rice in bulk forms a block, which causes variations in heating during cooking and degrades the eating quality. On the other hand, block-shaped rice grains are assembled with weak adhesion to each other and to air, so they easily lose their shape and heat transfer is poor. For the BQF of rice, air blast-type quick-freezing equipment, which is also used for frozen foods such as croquettes, hamburgers, and fried foods, is used. In general, the temperature in the freezing machine is −45 to −40 °C, the freezing time is 3–5 min, and the temperature of the frozen rice is −20 °C. Because the retrogradation of cooked rice progresses rapidly at low temperatures (0–5 °C), the effect of retrogradation can be minimized if freezing and thawing can be performed in such a way that the rice passes through this temperature range rapidly. The higher the thermal insulation of the packaging, the longer the changes in moisture content, firmness, and adhesiveness of the thawed frozen cooked rice were delayed (Lee and Yoo 2020).

When the frozen rice balls were first developed, a binding agent was used to prevent cracking during the distribution process and reheating. Currently, frozen rice balls are made without the use of binding agents, using stickier rice varieties and a new manufacturing method. In some cases, surfactants and water-retaining materials are used for white rice and sushi because of problems in maintaining the texture, taste, and aroma after thawing.

Thawing and Reheating Methods

Thawing is an operation that melts ice by increasing the temperature of the food product. To maintain food quality, it is necessary to melt the ice quickly, while the product temperature does not rise above a certain level (Horigome 1990; Ishitani et al. 1995). The microwave oven is the most widely used heating apparatus and uses microwaves (915 or 2450 MHz). The microwave oven utilizes dielectric heating, which converts insulating (dielectric) food placed in a high-frequency electric field into heat energy by causing the constituent molecules to vibrate violently by changing the direction of the electric field. The amount of heat generated is proportional to the value of the dielectric loss factor inherent in the material. Microwaves hitting food generate heat according to the dielectric loss coefficient and penetrate the food while gradually weakening it. However, the microwave penetration distance and dielectric loss coefficient are inversely proportional. In other words, if the dielectric loss coefficient of the food is high, such as high-moisture food containing salt, the microwave penetration is shallow, and as a result, the surface layer of the food heats up and the internal temperature increases slowly. In the case of frozen foods, the dielectric loss coefficients of the frozen and thawed portions are very different; therefore, the thawed portion may be very hot, while the rest of the food is frozen. Furthermore, if the container has sharp corners, the microwaves are reflected by the corners and do not pass through the container, leaving some parts unheated. Therefore, the corners must be rounded. At least, microwave heating has an advantage of air thawing at room temperature in recovering the deteriorated quality of frozen cooked rice after storage at various temperature conditions (Kono et al. 2015). High-frequency heaters use high-frequency waves (10–300 MHz lower in frequency than microwaves) to heat the center of food, however, as the temperature of the surface of the food increases faster than that of the inside, the air inside the machine is cooled by a refrigerator, reducing the difference between the surface temperature and the center temperature. As with microwaves, the degree of absorption at high frequencies varies depending on the food type.

An air blast is a system that heats food by circulating air at 20–80 °C, sometimes in combination with condensation heat transfer and radiation. Air blasting is more suitable for mass heating than for small-volume heating, and because the atmosphere is air, there is less sudden application of heat energy and less damage to food due to overheating.

Steam convection can be used to heat and thaw food in a short time using steam at approximately 120 °C. Because of the high temperature, food may be overcooked, depending on its size and preparation before freezing.

During the soaking thawing, the product is immersed in a temperature-controlled liquid. Water is typically used as the medium. Small to large quantities can be thawed using simple equipment. Processed rice must be vacuum degassed to prevent external contamination and internal contamination when opened.

Steamers can be used as heating and thawing machines with small to large capacities. Usually, food is heated and defrosted in its packaging, but if there is air in the packaging, heat transfer is poor.

9.3.6 AP Rice

Manufacturing Methods

AP rice is a product that retains the physical and sensory characteristics of cooked rice, and its demand has increased annually. This compensates for the disadvantages of retort odor, wrapper odor, and discoloration due to overheating of RP rice, starch retrogradation of canned and frozen rice, and quality deterioration during drying of dried rice. The most important factor in the production of AP rice is to prevent contamination and inhibit the growth of microorganisms, since AP rice is generally not subjected to heat sterilization after cooking (Ishitani et al. 1995; Japan Canners Association, Retort Foods Section 1996; Naito 2020).

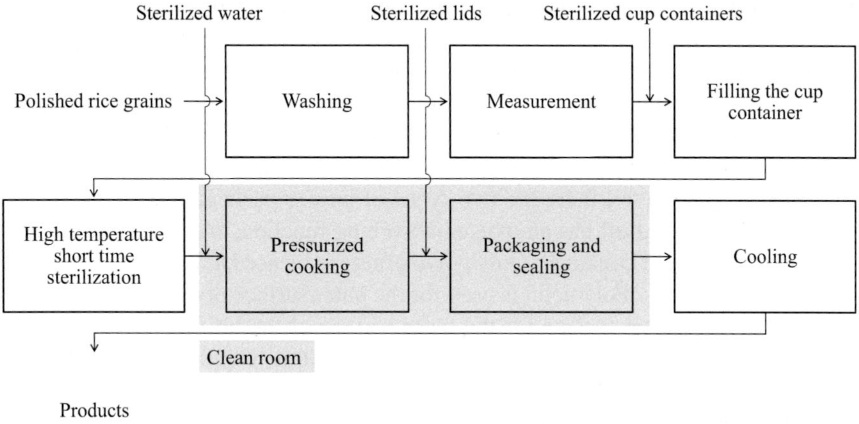

Fig. 9.12 Manufacture process of AP rice (drawn by the author)

The cooking methods for AP rice can be divided into steam and gas. In addition, gas cooking involves individual cooking and cauldron cooking. In individual cooking, rice is cooked in individual containers, which reduces the need for hygiene control of packaging materials and the environment and increases productivity through automation. Figure 9.12 shows the manufacturing process of AP rice. Specific example of the production of AP rice (Fukaya et al. 2010) is shown below. Polished rice, the raw material for AP rice, is washed, filled into cup containers, and sterilized in a steam-pressurized sterilizer at 130 °C or higher for approximately 20 s. This high-temperature and short-sterilization process minimizes heat damage to rice. The sterilized soaked rice and the cup containers are then transported to a clean room controlled to NASA Class 100 standards, where the pH is adjusted and the sterilized water is filled into each container in fixed quantities, and the rice is pressure cooked at approximately 105 °C. After cooking, the rice is transported again to the standard clean room, where it is covered with a lid that is treated in a steam-pressurized sterilizer and continuously sealed. After sealing, the rice is cooled to eliminate unevenness in the moisture content, and the product is finally manufactured. The key point of this aseptic packaging process is to keep the temperature as close as possible to the temperature history of the rice cooker to produce a superior product. Therefore, AP rice has better color, gloss, taste, and aroma than RP rice and less loss of nutrients due to heating.

Packaging Materials

Packaging materials for AP rice can be roughly classified into tray containers, tray lid films, and oxygen scavengers. The tray container may be sterilized separately from the contents, or it may be used as a steam sterilization container during aseptic packaging of rice, which has different requirements (Masuda 2011; Ishitani et al.

1995). On the other hand, the packaging container for AP rice must be made of microwave-permeable materials for use in microwave ovens. Furthermore, packaging materials are required to maintain a low oxygen concentration inside the container. To maintain a low oxygen concentration, there are two types of barrier packaging systems: one that inhibits the entry of oxygen from the outside during the storage period, and the other that actively absorbs the oxygen that enters through the container. For the latter, there are two types of barrier packaging systems: one in which the container itself has an oxygen-absorbing function, and the other which is attached to the barrier container. Only polyolefins can be used as barrier materials for food contact surfaces. Polyolefin is used for the outer surface of the retort container, and nylon or polyester is used for the outer surface of the lid film. Between these inner and outer layers, an ethylene vinyl alcohol copolymer (EVOH) or polyvinylidene chloride (PVDC) is usually placed. However, even if EVOH, which has the highest barrier property, is used as a packaging material for AP rice, it is not possible to keep the oxygen concentration below 1% for a storage period of several months. Therefore, for containers that rely solely on the inherent barrier properties of the resin, an oxygen scavenger that can be used in a microwave oven is sealed inside the container. The oxygen-absorbing container has polyolefin mixed inside the container, which eliminates direct contact between the iron powder and the food and simplifies the composition of the packaging material compared to the direct insertion of an oxygen scavenger.

The Oxyguard® cup, developed by Toyo Seikan Co., Ltd. in 1994, is the first tray with oxygen-absorbing capability to be commercialized by kneading iron powder into resin. Table 9.3 shows the changes in oxygen concentrations in the cup under different storage conditions (Suekane et al. 2006) with slight modifications. In the dark storage at 23 °C, the oxygen concentration in the containers of Oxyguard® cup and Lamicon Cup® with an oxygen scavenger decreases below 0.1% after 1 month

Table 9.3 Changes in oxygen concentrations in the cup under different storage conditions with slight modification (Suekane et al. 2006) (Reprinted with permission from Japan Canners Association)

	Concentration of oxygen (%)							
		In the dark storage at 23 °C				Under light storage at 23 °C		In the dark storage at 35 °C
Storage time (month)	0	1	2	3	6	1	3	1
Oxyguard® cup	0.21	tr.	tr.	tr.	tr.	tr.	tr.	tr.
Lamicon Cup® (O_2 1%)	0.96	1.07	1.14	1.95	2.58	tr.	tr.	1.53
Lamicon Cup® (O_2 3%)	3.51	2.98	4.05	4.49	4.76	1.58	tr.	3.92
Lamicon Cup® (O_2 10%)	8.95	9.66	8.56	9.00	9.75	6.55	2.51	8.97
Lamicon Cup® with an oxygen scavenger	0.15	tr.	tr.	tr.	tr.	tr.	tr.	tr.
PP single-layered cup	1.76	2.21	2.73	2.83	4.85	0.13	tr.	2.85

Oxyguard®: oxygen scavenger, PP and EVOH laminated plastic, Lamicon: PP and EVOH laminated plastic, PP: polypropylene, tr.: O_2 0.1% or less of storage, while the oxygen in the other containers remains unchanged or slightly increases. Under light storage at 23 °C, Oxyguard® cup and Lamicon Cup® with an oxygen scavenger, which have low initial oxygen concentrations, maintain oxygen concentrations below 0.1%, while the oxygen concentrations in the other containers decrease rapidly. On the other hand, among the volatile components in AP rice immediately after production, pentanal and hexanal carbonyl compounds are decomposed from free fatty acids under the influence of oxygen and increase with storage at initial oxygen concentrations of 1% or higher, leading to deterioration of the rice as an old rice odor. Therefore, it is suggested that the increase in carbonyl compounds is related to the decrease in oxygen concentration in the container under light storage at 23 °C. Thus, the initial oxygen concentration and the oxygen barrier of the container are factors that influence the lighting and storage temperature on the flavor of AP rice at room temperature. Therefore, to maintain the flavor of rice, it is necessary to have a high oxygen barrier and to reduce the initial oxygen concentration during production to at least 1% or less.

Oxygen Scavenger

According to Ishitani et al. (1995) and Masuda (2011), in the 1920s, various oxygen scavengers were developed in Europe and the United States using iron powder or iron sulfate, but none of them had been put to practical use because of problems in safety, oxygen absorption performance, cost, and handling. Ageless®, developed by Mitsubishi Gas Chemical Company Inc. in Japan in 1977, is the world's first oxygen absorber using iron powder to overcome these problems and to be used in food products. The Multisorb Filtration Group in the United States has also developed and marketed FreshPax® and FreshMax®, oxygen scavengers of the same type as Ageless®. In recent years, oxygen absorber materials have been replaced by low-odor types and sheet types that are resistant to acids and microwave heating.

In the manufacturing process of AP rice, the oxygen scavenger is sterilized beforehand by γ-rays and UV sterilization at the time of loading and then attached to the lid tray with hot melt adhesive. Since oxygen scavengers are designed to remove oxygen, the ability to remove oxygen is the most important factor. Vacuum packaging or inert gas replacement packaging such as nitrogen is not always sufficient to maintain freshness because of the low oxygen removal rate and the difficulty in removing oxygen that permeates through the film after packaging. Oxygen scavengers can continuously remove oxygen permeating through the film after packaging if the oxygen scavenging capacity is set to be large enough to maintain an oxygen-free state at all times. The most promising use of oxygen absorbers is to prevent microbial deterioration of foods. The removal of oxygen in the packaging bag is effective in preventing the growth of aerobic microorganisms, such as molds. However, the anaerobic condition in the bag may allow the growth of bacteria such as *Clostridium botulinus*, a food-poisoning organism that cannot grow under aerobic conditions.

9.4 Changes in the Properties of Rice during the Production and Storage Processes

In this section, the effects of processing and storage on the properties of rice are presented. Specifically, the moisture content, physical properties of starch, texture and composition, and starch digestibility, all of which are important to the quality of processed rice, are discussed.

9.4.1 Moisture

The moisture content of the rice that is sufficiently gelatinized and consumed is approximately 60%. For dried rice, for example, puffed rice, moisture in the rice is released into the atmosphere as water vapor due to high pressure and temperature during explosion. As a result, the moisture content is lower than that of untreated rice and is maintained at 5–10% to prolong shelf life. The moisture content of the other processed rice is the same as that of the fresh cooked rice. During storage, the moisture content of cooked rice decreases with an increase in temperature due to evaporation, while even with low-temperature storage at 4 °C, syneresis associated with starch retrogradation also decreases the moisture content in cooked rice (Tamura et al. 2019).

One effective method for extending the shelf life of rice while maintaining its quality is to cool it quickly to the optimum storage temperature. The cooling rate immediately after cooking has no significant effect on the moisture content of rice that has been refrigerated at 4 °C and reheated (Yu et al. 2010a). However, the moisture content of cooked rice frozen at −18 °C and reheated is different from the initial value as the cooling rate is slower (Yu et al. 2010b). An increase in the number of freeze-thaw cycles also decreases the moisture content of cooked rice (Lee and Yoo 2020). Moisture loss is related to the leaching of water during thawing from damaged cells and tissue structures caused by slow freezing. Based on the principle of minimum free energy, the nonuniform and fine ice crystals formed during the quick freezing of rice grow into large ice crystals during freezing and storage. At low freezing rates, fewer and larger ice crystals are formed inside the cell. On the contrary, at higher freezing rates, a large number of smaller ice crystals are formed both outside and within the cell (Kono et al. 2017). In general, the longer the storage period, the larger the diameter of the ice crystals. Kono et al. (2015) revealed that the value of the equivalent diameter does not increase more than about 120 μm at −5 °C after 30-day storage, and no growth in the size of ice crystals was observed during storage at −45 °C. The increase in the size of ice crystals leads to the compression and destruction of cellular tissues and ultimately to the deterioration of rice quality. Artificial neural network model analysis revealed that the maximum limits of the storage temperature and the mean equivalent diameter of ice crystals above the center score of palatability for more than 40 days are −25 °C and 20 μm in thawing by air at room temperature, and −15 °C and 40 μm by microwave heating, respectively (Kono et al. 2015). For frozen rice packages with different oxygen transmission rates (OTRs), Cho et al. (2019) measured the moisture content of rice stored at −20 °C and −80 °C and found that the moisture content of the rice significantly increases as increased from 3000 to 5000 mL/m^2/day, while it decreases when high-level OTR packaging is used. On the other hand, freezing rates in packaging with lower OTRs (0, 3000, and 5000 mL/m^2/day) were higher than those in packaging with higher OTRs (7000 and 20,000 mL/m^2/day). It is considered that changes in moisture content during freezing are influenced by the OTR of the packaging, likely due to differences in freezing rates that depend on OTR.

The rehydration of dried rice and the texture of rehydrated rice change significantly depending on the rice variety and processing conditions. Because of the relatively long shelf life of dried rice, more research has been conducted in recent years on the effects of rice variety and processing methods on water rehydration, expansion ratio, and texture, and on the optimization of the effects, rather than on changes associated with storage. Lee and Yang (2020) adjusted puffed rice with different degrees of gelatinization and amylose content and found that following the puffing process, waxy rice has a higher water-holding capacity than non-waxy rice, and the water-holding capacity is not significantly affected by partial gelatinization treatment. Puffed rice shows 2–3 times higher moisture absorption and moisture solubility than untreated grain (Huang et al. 2018). The pregelatinized rice cooked with the pressure cooker shows a higher dehydration rate and faster rehydration time than instant rice cooked with boiling and steaming because of the porous structures on the surface (Ritudomphol and Luangsakul 2019).

9.4.2 Physical and Thermal Properties of Starch

When starch is added to water and heated, the crystalline part of amylopectin melts, hydrates, and swells. This phenomenon is referred to as gelatinization. The rapid aggregation of amylose in the initial stage and the recrystallization of amylopectin in the long term are related to each other and change the degree of gelatinization of rice after cooking and storage below the gelatinization temperature. This phenomenon is known as retrogradation. Retrograded cooked rice during post-cooking storage causes higher firmness, enthalpy, and whiteness, and lower stickiness (Ikawa and Kouno 2005; Otahara et al. 1995; Tamura et al. 2019; Yu et al. 2010b), leading to lower eating quality than fresh cooked rice (Ikawa and Kouno 2005; Otahara et al. 1995). The retrogradation of cooked rice is mainly affected by temperature, moisture, pH, and coexisting substances.

With regard to storage temperature, cooked rice retrogradation progresses in the following order: refrigeration temperature (5 °C) > room temperature (20 °C) > freezing temperature (−18 °C) > warming temperature of rice cooker (70 °C) (Matsunaga and Kainuma 1981). The increase in retrogradation enthalpy of cooked rice resulting from the reassociation of amylopectin is observed rapidly in the first 3 days and gradually in the subsequent 14 days during storage at 4 °C (Yu et al. 2010b). Cooked rice with a high amount of amylose retrogrades quickly, while cooked rice with a low amount of amylose retrogrades slowly (Yu et al. 2010b). The retrogradation enthalpy of rice amylopectin increases with decreasing freezing temperature within 14 days of storage at 4 °C and varies due to the freezing rate, and the molecule and structure of amylopectin based on the variety (Yu et al. 2012). The microstructure of freshly cooked rice appeared to have a porous structure, while the microstructure of rice after quick freezing and 48-h storage shows that the starch granules shrank, gradually ruptured, and formed crystalline particles, possessing clearly angular, compact, continuous, and less spongy structure due to starch retrogradation and water loss (Huang et al. 2014). On the other hand, there is no amylopectin retrogradation during frozen storage at −18 °C for 5 months, regardless of the freezing rate (Yu et al. 2012). The retrogradation enthalpy of frozen cooked rice is influenced more by freezing storage time than by the degree of temperature fluctuation (Kumar et al. 2020).

Adjustment of pH during the production of processed rice not only inhibits the growth of microorganisms, but also changes the mechanical properties of the rice as it retrogrades. While the rice cooked with strong acid and weak alkali remains soft with strong adhesiveness both on the grain surface and whole, rice cooked with weak acid shows an acceleration tendency of retrogradation after refrigeration at 4 °C for 14 h in terms of physical properties, sensory evaluation, and brightness (Oishi et al. 2020). Among seasonings, vinegar, compared to sugar and salt, has the ability to lower pH and inhibit changes in firmness, stickiness, and L* of cooked rice under storage at 4 °C (Odahara et al. 2004; Otahara et al. 2020). In addition, vinegar has a strong inhibitory effect on rice retrogradation, even in the presence of sugar and salt. Although pregelatinized rice shows less reduction in the degree of gelatinization than cooked rice when stored at 6 and 20 °C, pregelatinized rice with salt, soy sauce, sugar, sake, and butter tends to retrograde more rapidly when stored at 4 °C for 24 h (Arai et al. 1982). Therefore, it is necessary to pay attention to the effects of seasonings on retrogradation as well as on eating quality when they are added to the processed rice production process. Therefore, when seasonings are added to processed rice during the manufacturing process, it is necessary to pay attention not only to the eating quality but also to the effect on retrogradation.

The decrease in the degree of gelatinization of cooked rice stored at 5 °C was observed after 24 h and was slower with higher moisture content (Maruyama et al. 1980). It is important to note that the changes in physical properties of cooked rice with retrogradation after about 24 h of refrigeration are evaluated differently depending on the analytical method. Rice refrigerated at 4–5 °C generally shows a gradual change in hardness and stickiness, an increase in L* and enthalpy after 6 h, and a peak in X-ray diffraction after 24–48 h (Ikawa and Kouno 2005; Otahara et al. 1995; Otahara et al. 2018a). Retrogradation of cooked rice as perceived by sensory testing is after 4–5 h of storage at 4 °C (Ikawa and Kouno 2005; Otahara et al. 1995). These results indicate that human sensory perception is more sensitive than instrumental measurement or chemical analysis to detect changes in rice due to retrogradation and that DSC and X-ray diffraction, which detect changes in crystalline properties, cannot detect changes in the amorphous region, such as rapid aggregation of amylose molecules in the early stages of retrogradation. Therefore, the evaluation of retrogradation of processed rice within 24 h of refrigeration requires attention, and an objective evaluation method that corresponds to sensory evaluation needs to be established.

9.4.3 *Texture and Chemical Components*

Table 9.4 provides the physicochemical properties and components of the processed rice. The eating quality of fresh cooked rice is generally evaluated based on its textural properties, such as firmness and stickiness (Rousset et al. 1999; Sesmat and Meullenet 2001). L* is also an important factor in the overall acceptability of cooked rice (Konishi et al. 1996). Texture and color are also considered to play a significant role in reheated rice after storage.

A study of japonica cooked rice stored in airtight containers at 4, 30, and 90 °C for 24 h revealed that the surface firmness of rice grain stored at 30 and 90 °C significantly decreases during storage, while it increases at 4 °C (Tamura et al. 2019). The overall firmness at 4 and 90 °C increases with an increase in the storage duration, while only a small change is observed at 30 °C. A greater increase in firmness during storage at ≤3 °C than storage at 20 and 36 °C was also shown (Perdon et al. 1999). On the other hand, the surface and overall adhesiveness of rice grains stored at 90 °C shows no significant difference during storage, although surface adhesiveness of rice stored at 4 °C and 30 °C and the overall adhesiveness of rice stored at 4 and 30 °C show a significant decrease during 24 h of storage (Tamura et al. 2019). Generally, when stored at 4 °C or higher, the lower the temperature and the longer the storage period, the higher the firmness and the lower the adhesiveness of cooked rice. In addition, when stored at 0 °C or lower, the higher the temperature and the longer the storage period, the greater the difference in the texture of cooked rice after reheating from that of fresh cooked rice. The firmness of cooked polished rice positively correlates with the amylose content and the enthalpy of cooked rice during cooling storage. Yu et al. (2009) reported that the firmness of fresh cooked rice is mainly caused by amylose retrogradation in a short time, and amylopectin retrogradation contributes to an increase in firmness during storage. On the other hand, adhesiveness is negatively correlated with amylose content and enthalpy of cooked rice during storage at 4 °C. The adhesiveness decreased rapidly on the first day, which may have been caused by amylose retrogradation, and then decreased slowly, possibly caused by amylopectin retrogradation. Furthermore, changes in the texture of cooked rice depend on both the cooling rate and rice variety, that is, amylose content during storage (Ma and Sun 2009). In other words, rapid cooling and low amylose content are important factors in suppressing changes in the texture of cooked rice during cold storage. Otahara et al. (2018b) applied the first-order rate equation to analyze the change in texture of rice during the storage process at 4 °C using kinetics and found the following: The

Table 9.4 Production method, storage method, physicochemical properties, and components of processed rice (drawn by the author)

	Preparation and storage conditions of rice samples	Sensory and instrumental physicochemical evaluation	References
Cooked rice stored under temperature control	Polished rice of 11 varieties with different amylose content was cooked, stored at 4 °C for 14 days, and incubated at 22 °C for 1.5–2.0 h	• Fresh cooked rice has the lowest firmness values, and these values increased significantly after 7 days of storage and slightly increased from 7 to 14 days. • Most adhesiveness decrease occurred during the first 3 days' storage but was then slower from 7 to 14 days.	Yu et al. (2009)
	Polished japonica, non-waxy rice (cv. Nipponbare) soaked at 25 and 60 °C, cooked, stored at 8 °C for 24 h	• Compared with soaking at 25 °C, soaking at 60 °C suppressed the decrease in weight and the change in surface firmness and stickiness of chilled rice.	Kainuma (2000)
	Japonica, non-waxy rice (cv. Nipponbare) cooked, stored at −7– −37 °C for 7 days, and heated by microwave at 500 W for 2 min	• Freezing temperature of −13.0 to −31.0 °C, storage temperature of −31.0 °C, and storage period within 7 days prevented increase in white cloudiness and yellowness and maintained the breaking stress of the rice balls compared with the values of freshly cooked rice ball.	Yanai et al. (2001)
	Japonica, non-waxy rice (cv. Koshihikari and Nipponbare) cooked, stored at −30– −25 °C for 2–10 days, and heated by a microwave	• The changes in firmness and adhesiveness of frozen rice tended to decrease with decreasing storage temperature. These changes were almost completely eliminated by thawing at room temperature or in a microwave oven. However, the degree of elimination for the samples stored at −5 °C was smaller than that for the other samples. • The storage temperature should be 15 °C or higher to maintain the sensory firmness for 1 day and −20 °C or lower to maintain the firmness for a long time. Similarly, the storage	Sagara (2009)

(continued)

Table 9.4 (continued)

	Preparation and storage conditions of rice samples	Sensory and instrumental physicochemical evaluation	References
		temperature should be 10 °C or higher to maintain the adhesiveness for 1 day and −25 °C or lower to maintain it for a long time. • The degradation of the eating quality of rice after heating and thawing was influenced more by the variety of rice than by the storage temperature conditions.	
	Glutinous rice (Koganemochi) was cooked; cooled at −23 °C; stored at −20 °C for 1, 5, 10, 30, and 60 days; and thawed by a household steamer or incubation at room temperature (25 °C)	• Frozen stored glutinous rice after 60 days shows darker and more unpleasant color than freshly cooked one by both thawing methods. • Natural thawed glutinous rice tends to show dark and unpleasant color by a relatively short time storage.	Araki et al. (2018)
	Japonica milled rice (cv. Koshihikari) was cooked, frozen, and stored at −5, −15, and −45 °C and air thawed at room temperature (25 °C) and microwave heated	• While the surface firmness of the sample frozen at −5 °C and thawed by air tended to increase more quickly than those stored at −15 °C and −45 °C, the same behavior was not observed for the samples thawed by microwave heating. • Surface and overall adhesiveness for the samples thawed by air sharply decreased by freezing at −5 °C and gradually decreased during the storage. • Sample stored at −5 °C showed difficulty maintaining the viscoelasticity, and the samples stored at above −10 °C showed difficulty improving the palatability score for longer term storage regardless of the thawing methods.	Kono et al. (2016)

(continued)

Table 9.4 (continued)

	Preparation and storage conditions of rice samples	Sensory and instrumental physicochemical evaluation	References
	Three varieties of rice grain (low-, middle-, and high-amylose content) cooked, cooled by air blast (fast cooling speed) and cold room cooling (slow cooling speed), stored at 4 °C for 12 days, and reheated by microwave oven (750 W) for 40 s	• Firmness of all samples cooled by cold room cooling increased significantly during storage. • While low-amylose rice cooled rapidly could keep the firmness during whole storage, middle-amylose rice could maintain firmness for the first 4–5 days and high-amylose rice could only maintain firmness for 2–3 days.	Ma and Sun (2009)
	Japonica milled rice grain was cooked, pre-cooled (0.40, 1.11, 1.42, 1.64, 2.47, and 3.36 °C/min^{-1}), stored at 4 °C for 0–14 days, and incubated at 22 °C	• Firmness of cooked rice increased rapidly during the first 7 days and changed slowly from 7 to 14 days. • The slower the cooling rate, the more rapidly the firmness increased and adhesiveness decreased within 1 day. • Retrogradation enthalpy of cooked rice chilled at different cooling rates increased significantly from 0 to 7 days, whereas no such difference was observed between 7 and 14 days. • There was an inverse proportional relationship between the cooling rate and the retrogradation enthalpy of cooked rice during storage.	Yu et al. (2010a)
	Japonica milled rice grain was cooked, pre-cooled (0.09, 0.26, 0.33, 0.53, and 1.45 °C/min^{-1}), stored at −18 °C for 0–7 months, and incubated at 22 °C	• Rapid freezing can effectively maintain the firmness and adhesiveness of cooked rice for at least 7 months. • Retrogradation enthalpy of cooked rice increased gradually from 0 to 7 months of frozen storage. • There was an inverse proportional relationship between the freezing rate and the retrogradation enthalpy of cooked rice during storage.	Yu et al. (2010b)

(continued)

Table 9.4 (continued)

	Preparation and storage conditions of rice samples	Sensory and instrumental physicochemical evaluation	References
	Japonica milled rice was cooked, cooled at −20 and −80 °C in a freezer for 7 and 14 days in the pouch with OTR (0, 3000, 5000, 7000, and 20,000 mL/m^2/day), reheated for 1 min, and cooled for 10 min	• L* and total color difference (ΔE) values were the highest in OTR 20,000 mL/m^2/day packaging, whereas ΔE values were the lowest in OTR 5000 mL/m^2/day packaging. • Firmness and cohesiveness of frozen cooked rice gradually increased from OTR 0 to 5000 mL/m^2/day but decreased from OTR 5000 to 20,000 mL/m^2/day.	Cho et al. (2019)
Retort-packed (RP) rice	Rice and water filled in retort pouch, placed in the retort, and processed at different retort temperatures and process times (112 °C, 30 min; 115 °C, 20 min; 118 °C, 8 min; 118 °C, 12 min; and 121 °C, 8 min) to achieve a minimum F_0 of 3 min and cooled	• RP rice was not significantly different from control (cooked in household pressure cooker) with respect to sweetness, aroma, chewiness, plumpness, and whiteness. • Retorted rice samples tended to show lower firmness and higher stickiness than control. • The instrumental firmness parameter correlated well with the sensory firmness and overall quality for the optimum processing condition (118 °C, 8 min).	Prakash et al. (2005)
Dried rice	Cooking non-waxy pregelatinized rice and polished rice after 16 months of storage at room temperature and 35 °C	• Pregelatinized rice stored at room temperature was less affected by storage period in terms of appearance, taste, firmness, stickiness, and overall evaluation and was rated higher than polished rice.	Kawabata et al. (1979)
	Jasmine rice (cv. Khao Dawk Mali 105) was cooked by four methods (boiling in a pressure cooker (BPC), steaming in a pressure cooker, boiling water method, and steaming method) and dried by microwave oven to the moisture content to lower than 14%	• Firmness, stickiness, volume expansion, and whiteness of rehydrated rice cooked by different methods had no significant difference with freshly cooked rice. • Instant rice grain cooked by BPC after hydration appeared to resemble freshly cooked rice and be individually separated.	Phukasmas and Songsermpong (2019)

(continued)

Table 9.4 (continued)

	Preparation and storage conditions of rice samples	Sensory and instrumental physicochemical evaluation	References
	Thai jasmine rice (cv. Khao Dawk Mali 105) was soaked, cooked, and dried at 80 °C before being puffed at different moisture contents (15–20%), temperatures (200–220 °C), and time intervals (20–30 s)	• Whiteness index tended to increase with increased moisture content, while lower index was observed when puffing temperature and time increased. • Volume expansion increased significantly with the increase in moisture content of the pre-puffed rice and also the puffing temperature and time. • The products with less expansion ratio indicated higher bulk density and longer rehydration time. • Hardness of the rehydrated rice tended to decrease with increase in moisture content, puffing temperature, and time. • The stickiness of the rehydrated rice was significantly lower than that of the normal cooked rice using electric cooker.	Wongsa et al. (2016)
	Jasmine rice (cv. Khao Dawk Mali 105) was soaked to obtain moisture contents of 35–60%, cooked under pressures of 8.00×10^4 to 19.58×10^4 Pa for 5 min, and dried at 166.4–233.6 °C to obtain moisture content of less than 12%	• The firmness and chewiness of rice decreased as moisture content and pressure increased. • Higher drying temperatures caused increases in firmness and chewiness. • Only pressure and moisture content affected density, rehydration ratio, and increase in the volume of instant rice. • Factors of moisture content, drying temperature, and especially pressure influenced the whiteness index.	Prasert and Suwannaporn (2009)
	To make freeze-dried rice porridge, fresh milled japonica rice was cooked; frozen for 2 days at −5, −20, −40, and −70 °C; and freeze-dried	• Slow freezing produced porous freeze-dried rice porridges with large pores and more brittle structure than rapid freezing. • Initial rehydration ratio and	Rhim et al. (2011)

(continued)

Table 9.4 (continued)

	Preparation and storage conditions of rice samples	Sensory and instrumental physicochemical evaluation	References
		dissolution time of the freeze-dried rice porridge decreased with increase in water temperature. • Effect of freezing temperature on rehydration ratio and dissolution time indicated that slow freezing is more preferable for the freeze-dried rice porridge.	

firmness and stickiness of varieties with different amounts of amylose change the slope of the straight line in the process of storage and can be divided into the first and second halves with different rate constants. In other words, the rate constant for the increase in firmness during the first half of the storage period is greater for rice with higher amylose content, while the rate constant for the second half of the storage period tends to be greater for rice with higher amylopectin content. On the other hand, the rate constant for decreasing stickiness is larger in the second half than in the first half, and both the first and second halves tend to be larger for rice with higher amylose content.

Research on the quality and taste of frozen rice has been systematically elucidated compared to research on the effects of milling rice varieties, rice cooking methods, and storage on physicochemical properties and taste (Sagara 2009). Artificial neural network model analysis constituted from the dataset of storage conditions and palatability scores revealed that to maintain the palatability score more than the central value for more than 40 days of storage, storage temperatures should be lower than −25 °C for air-thawed samples and −15 °C for microwave-thawed samples (Kono et al. 2016). Lee and Yoo (2020) evaluated the impact of repeated freeze-thaw and insulation levels of packaging on the physical properties of frozen cooked rice. As the number of freeze-thaw cycles increases, the firmness of the frozen rice initially decreases and then increases. Furthermore, less insulation packages reach the starting point of a firmness increase more quickly. It has been considered that this is due to both a decrease in firmness caused by moisture loss and destruction of tissue structures and an increase in firmness caused by the rapid retrogradation of frozen rice in low-insulation packaging. On the other hand, the adhesiveness of frozen rice increases significantly with increasing freeze-thaw cycles, while the adhesiveness of frozen rice in low-insulated packaging shows a tendency to increase, probably related to the starch leached by the freeze-thaw cycle attaching to the surface of the frozen rice. L* decreases significantly with increasing number of freeze-thaw cycles because of the recombination of starch molecules by retrogradation. Different additives, including glycerol monolaurate (GML), β-cyclodextrin (β-CD), and alcohol, have different effects on the textural and physicochemical properties of frozen rice (Huang et al. 2014). Optimization of additives using the response surface method (RSM) revealed that the addition of GML and alcohol has a significant influence on the comprehensive quality, while β-CD has no significant influence. In addition, this model has 97.72% of the capacity to analyze and predict the quality of frozen rice, which obtains the optimum conditions: 0.025% GML, 0.58% β-CD, and 0.41% alcohol. The optimized additives can effectively reduce the crystallization of amylose and delay the short-term reversal of gelatinized starch.

The texture of dried rice varies greatly depending on the rice variety and processing conditions. For example, puffed waxy rice showed lower firmness than puffed non-waxy rice (Lee and Yang 2020). Lee and Yang (2020) reported that the firmness of the puffed samples using partially gelatinized rice was lower than that of the non-partially gelatinized samples, regardless of the rice type. Mridula et al. (2017) investigated pregelatinized rice with different salt types (calcium chloride, sodium chloride, acetic acid, and sodium citrate) and their concentrations (0, 0.5, 1.0, and 1.5%) during soaking. The optimal cooking time is the minimum value for rice with sodium chloride and decreases with decreasing concentration of sodium chloride. On the other hand, the optimum cooking time was greatest for rice with acetic acid and sodium citrate. This was probably due to the low pH affecting moisture diffusion and the degree of gelatinization. Peak viscosity and final viscosity measured for rice with calcium chloride by rapid visco-analyzer are minimum followed by sodium chloride. L* is higher in rice prepared with acetic acid than in rice prepared with salt, while b* is higher in rice with acetic acid and sodium chloride. Rice with the addition of sodium chloride and calcium chloride shows the highest values for texture, flavor, taste, and overall acceptability in the sensory test without significant differences from the control (pregelatinized rice without salt). Mir et al. (2016b) constructed a mathematical model of the maximum popping expansion of brown rice using the RSM. The model indicated that the maximum expansion ratio (6.85) can be obtained at a salt content of 1.75 g/100 g raw material, moisture content of 15 g/100 g raw material, and a popping temperature of 225 °C. This is consistent with the experimental results for the verified expansion ratio (6.79). Wongsa et al. (2016) also reported that the best puffed rice product, which is the maximum whiteness index (higher than 73), minimum rehydration time of 5 min (without case hardening), and optimum range for hardness (31.38–50.99 N) and stickiness (0.54–1.08 N·s), can be produced under the condition of final moisture 16.83–18.83% wb and puffing temperature of 209.4–211.5 °C for 22.92–25.11 s. The structure of the freeze-dried rice porridge is influenced by the freezing rate; the freeze-dried rice porridge produced through rapid freezing has a denser structure with smaller pore sizes. As a result, freeze-dried rice porridge frozen rapidly has higher firmness than those frozen slowly (Rhim et al. 2011). In general, large, nonuniform ice crystals produced by slow freezing are not preferred because they damage the tissue cells of food products, such as vegetables, fruits, and meat, resulting in quality deterioration (Zhu et al. 2019). However, the damage to the tissue structure caused by the large ice crystals formed by slow freezing is not as important as it facilitates the rehydration and restoration of the freeze-dried porridge (Rhim et al. 2011).

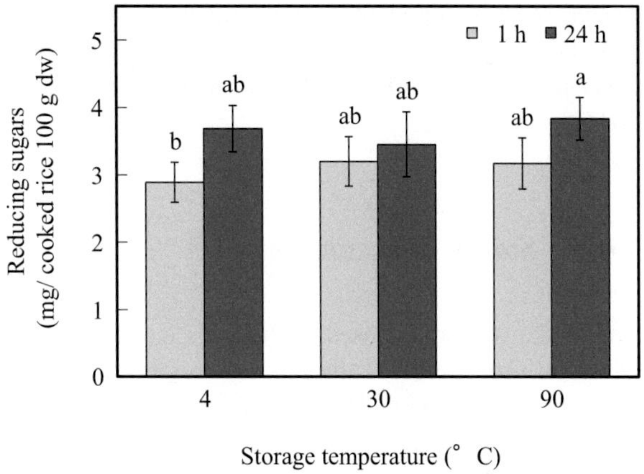

Fig. 9.13 Changes in the amount of reducing sugars in cooked grain samples during storage (drawn by the author) (The histogram with bars indicates the mean ± standard deviation ($n = 3$–4). Different letters indicate significant differences ($P < 0.05$). Cooked grain samples were prepared as described by Tamura et al. (2019). Reducing sugars content was measured according to Oshima et al. (2012) and Kitamura and Nakaya (2012)

The amount of starch, the main component of rice, changes with sugar content due to leaching and degradation into small molecules depending on the cooking process and methods such as boiling and puffing (Huang et al. 2018; Tamura et al. 2021; Zhu et al. 2020). During storage at 4, 30, and 90 °C for 24 h, the total starch (TS) content of cooked rice does not change (Tamura et al. 2019). Changes in the amount of reducing sugars in the cooked grain samples during storage at different temperatures are shown in Fig. 9.13. Note that reducing sugars are generic term for monosaccharides and oligosaccharides that impart a sweet taste of cooked rice (Kasai et al. 2000). Although no significant differences were observed during storage, the average amount of reducing sugars in the grains stored for 24 h was slightly higher than that stored for 1 h. Reducing sugars in polished rice, particularly glucose, is mainly generated by enzymatic degradation of starch during cooking (Mabashi et al. 2010). The main endogenous enzymes in rice grain are amylase and α-glucosidase, which are rapidly deactivated at 70–80 °C (Awazuhara et al. 2000). Thus, the enzymes that could generate reducing sugar in the rice grain were deactivated during continuous boiling (100 °C), which explains the lack of change in reducing sugar content over the storage period, irrespective of the storage temperature. Extractable phenols, flavonoids, anthocyanins, and antioxidant activity of cooked black rice decrease with storage period at 4 °C (Meng et al. 2019).

Huang et al. (2018) reported that explosion puffing (heating in an expansion chamber under a pressure of 0.8–1.5 MPa for 70–80 s) of rice results in the nutritional composition with higher total sugar content and lower starch content. Although explosion puffing has no significant effect on the total protein content, there is a significant decrease in the solubility of the protein. This is presumably due to the breakdown of starch into small molecules, resulting in an increase in sugar content and protein aggregation in the puffing process, which is a more severe condition than cooking rice under atmospheric pressure. In a series of puffing processes of brown rice, total phenolic content, total flavonoid content, and reducing power decrease upon puffing treatments, whereas the DPPH radical scavenging activity and amount of minerals such as calcium, potassium, magnesium, manganese, phosphorus, sulfur, and zinc increase with parboiling treatment (Mir et al. 2016a).

9.4.4 Starch Digestibility

The intake of starch causes fluctuations in blood glucose levels after consumption, which in turn increases the risk of obesity, type 2 diabetes, hyperlipidemia, and cardiovascular disease. This glycemic response is quantified by not only the glycemic index (GI) in an in vivo test, but also classification into three types of starch based on its digestion rate: rapidly digestible starch (RDS), slowly digestible starch (SDS), and resistant starch (RS) contents (Englyst et al. 1992) and kinetic analysis of estimated glycemic index (eGI) (Goñi et al. 1997). GI of rice and processed rice products is 19–116, the second highest value after potatoes (Atkinson et al. 2021). Therefore, in recent years, intensive research has been conducted to reduce starch digestibility in rice by processing. Table 9.5 presents the production method, storage method, and starch digestibility of the processed rice. The glycemic response of processed rice is affected by gelatinization, recrystallization, and associated microstructural changes during processing and storage.

Starch granules heated to gelatinization temperature in water lose crystal order and swell due to water absorption, while at the same time leaching amylose and small amylopectin molecules in the granules occur during rice cooking and processing. Amorphous starch and eluted starch increase the reactivity with amylolytic enzymes, resulting in higher starch digestibility. In other words, rice samples that are highly gelatinized by cooking increase starch digestibility (Lee and Yang 2020; Tamura et al. 2016). In contrary, retrograded starch is resistant to amylolytic enzymes and is classified as RS3, a type of the resistant starch (Raigond et al. 2015). RS3 is formed in gelatinized starch in moist-heated foods such as cooked rice during cooling storage and is thermally stable. Refrigerated storage of cooked rice at 4 °C gradually decreases RDS content, increases SDS and RS content, and significantly decreases rice starch digestibility when stored for more than 10 h (Lu et al. 2016). In addition, retrograded rice reduces not only starch digestibility, but also plasma cholesterol, liver cholesterol, and triacylglycerol contents in adipose tissue when compared to those in fresh rice (Ha et al. 2012). In the case of frozen rice, there are both reports that starch digestibility is reduced compared to fresh rice and that it is unchanged (Huang et al. 2014; Shibata-Ishiwatari et al. 2017). These differences are considered to be related to the freezing rate at approximately 4 °C, which causes retrogradation. In other words, when the sample temperature quickly passes around 4 °C, retrogradation is less likely to occur, and the starch digestibility of rice is less different from that of rice immediately after cooking. It has been reported that some resistant starches are not able to recover to the state of freshly cooked rice upon reheating (Huang et al. 2014). Since the preparation method of the samples also

Table 9.5 Production method, storage method, and starch digestibility of processed rice (drawn by the author)

	Preparation and storage methods	Preparation of digestion samples	Digestive process	Starch digestibility	References
Cooked rice stored under temperature control	Cooked waxy and non-waxy rice grains with different W/R ratios (1.0–1.8) were stored at 2 °C for 96 h	Cooked polished waxy and non-waxy rice grain is used immediately after storage	Diastase treatment	• The starch digestibility of waxy and non-waxy rice grain decreased with decreasing W/R ratio. • Starch digestibility of the waxy rice was significantly higher than that of the non-waxy rice at 0 h of storage.	Ozaki (1960)
	Cooked brown rice stored at 4 °C for 0 and 24 h	Cooked brown rice slurry of six varieties with different amylose content	Simulated gastric digestion with pepsin from porcine gastric mucosa followed by simulated small intestinal digestion with α-amylase from porcine pancreas	• Storing the cooked samples at 4 °C for 24 h led to a reduction of HI and eGI. • eGI ranged between 68 and 109 for cooked rice and between 64 and 87 for stored rice containing retrograded starch.	Frei et al. (2003)
	Cooked rice stored at 4 °C for 0 and 24 h	Cooked polished rice slurry of 12 varieties with different amylose contents	Simulated gastric digestion with pepsin from porcine gastric mucosa followed by simulated small intestinal digestion with porcine pancreatic α-amylase	• Retrogradation in cooked samples led to a reduction of HI and eGI for all varieties.	Hu et al. (2004)
	Cooked rice stored at 4, 30, and 90 °C for 1 and 24 h and then kept at 30 °C for 5 min	Cooked polished rice grain (non-waxy, japonica, cv. Koshihikari)	Simulated gastric digestion with pepsin followed by simulated small intestinal digestion with pancreatin,	• TS and RS contents were not significantly different between samples. • The cooked rice stored at 4 and 90 °C for 24 h showed	Tamura et al. (2019)

(continued)

Table 9.5 (continued)

	Preparation and storage methods	Preparation of digestion samples	Digestive process	Starch digestibility	References
			invertase, and amyloglucosidase	significantly lower starch hydrolysis (%) and eGI than those stored for 1 h at similar temperatures. • Little difference was found in the starch hydrolysis curve between sample grains stored at 30 °C for 1 and 24 h.	
	Cooked rice stored in household freezer (at <−18 °C) for 1 day and 1 month and heated by microwave at 600 W for 2 min	Freeze-dried rice powder (non-waxy, japonica, cv. Koshihikari)	α-Amylase treatment	• There were no significant differences in eGI values between freshly cooked rice and frozen cooked rice samples.	Shibata-Ishiwatari et al. (2017)
	Rice grain cooked, frozen, and kept at −18 °C for 48 h prior to microwave reheating	Freshly cooked rice and reheated rice (cv. Longjingxiang)	Simulated gastric digestion with pepsin followed by simulated small intestinal digestion with porcine pancreatic α-amylase	• Frozen rice had lower starch digestion rate and final digestibility than fresh rice.	Huang et al. (2014)
Dried rice	To make pregelatinized rice, three Thai rice varieties were cooked at 82, 86, and 90 °C with 100, 145, and 190 mL of water content (W/R of 1.1, 1.45, and 1.91) and dried to moisture content of less than 12%	Rehydrated ground pregelatinized rice	Porcine pancreatic α-amylase and amyloglucosidase treatment	• Cooking at 86–90 °C resulted in higher RDS content than cooking at 82 °C. • SDS content was the highest when rice was cooked at the lower three temperatures (82 °C) with the higher W/R ratio (1.9) in all three varieties. • With increasing cooking	Ritudomphol and Luangsakul (2019)

				temperature and decreasing ratio of W/R, there was a high level of disintegration of the starch granules resulting in increased RDS content and decreased RS content in all three cultivars. • The instant rice, which was produced at 82 °C with W/R of 1.9, exhibited a lower eGI as compared to the other cooking conditions.	
	To make puffed rice, three varieties of rice grains (rice, black rice, and glutinous rice) were heated in an expansion chamber under a pressure of 0.8–1.5 MPa for 70–80 s	Defatted rice powder samples with/without puffing	Trypsin and glucoamylase treatment	Starch hydrolysis rate of three varieties of puffed rice increased during in vitro digestion.	Huang et al. (2018)
	To make instant rice, rice grain was soaked in water with various W/R ratios (1.0–1.4), cooked in water bath for varying amounts of time (22–33 min) at several temperatures (83–89 °C), stored at 4 °C for 0–3 days, and dried at 50 °C for 3 h and then at 95 °C for 1 h	Rehydrated ground instant rice (cv. Taikeng 9, Taichung Sen 10, and Taichung Sen 17)	According to Englyst et al. (1992), amyloglucosidase, pancreatin, and invertase treatment	• Cooking at 86 °C for 28 min in a 1.0 times volume of water significantly reduced RDS content and increased SDS content compared to rice cooked on a traditional electronic cooker in all three varieties of rice. • As a result of comparing RDS content of regular rice (not instant rice), instant rice without retrogradation, and instant rice with 1-day retrogradation treatment, the	Hsu et al. (2015)

(continued)

Table 9.5 (continued)

	Preparation and storage methods	Preparation of digestion samples	Digestive process	Starch digestibility	References
				combination of retrogradation treatment and drying was necessary to minimize the RDS content. • No significant changes were observed when retrogradation was extended from 1 day to 3 days, except for an increase in RS in Taichung Sen 17 after 2 days of retrogradation. There were significant interactions between variety and retrogradation duration for RDS, SDS, and RS.	

affects starch digestibility, it is difficult to make a simple comparison between samples prepared by different methods, and further studies are necessary.

In frozen rice, the higher the freezing temperature and the longer the storage period, the more ice crystals are characterized by large pores and less compact structures on the surface of the grain (Cho et al. 2019). On the other hand, puffing hollows out the interior of the rice, making the rice a loose and porous structure (Huang et al. 2018). This increase in the surface area of rice increases the accessibility of digestive enzymes, leading to an increase in starch digestibility. Ritudomphol and Luangsakul (2019) optimized cooking condition for producing pregelatinized rice containing maximum SDS and RS and minimum RDS and GI using a RSM. They concluded that pregelatinized rice grain cooked at 82 °C with W/R ratio of 1.9 shows the lowest RDS content and eGI remains intact with a more compact periphery. Hsu et al. (2015) also reported that cooking of pregelatinized rice with a W/R ratio of 1.0 at 86 °C for 28 min significantly decreased the RDS content and increased the SDS content compared to rice cooked in a conventional electronic rice cooker. This is presumably due to the low cooking temperature, short cooking time, and limited amount of water added, which inhibit the destruction of the cells. High-temperature drying after low-temperature drying would have contributed to the decrease in starch digestibility due to the crystallization of starch molecules.

9.5 Summary and Future Direction

The revolution of processed rice is deeply related not only to the technological development of rice processors, but also to the convenience and preference of the users. Processed rice started with canned rice in the postwar period when microwave ovens were not widely available. The cooking time for canned rice was approximately 20 min, while it was reduced to approximately 10 min for RP rice. This simplicity led to the widespread use of RP rice as a substitute for canned rice. However, with the widespread use of microwave ovens, the mainstream of processed rice has completely shifted from RP rice to frozen rice, which has a better taste and can be prepared in a shorter time. Furthermore, the production quantity of AP rice surpassed that of frozen rice in 2019, largely due to the fact that AP rice is not only microwavable but also can be stored at room temperature with the convenience of the container doubling as a tableware. On the other hand, when looking at the history of processed rice from the standpoint of containers, it has changed from cans and retort pouches to plastic molded containers. While the production of plastic, which accounts for the majority of food containers, has been increasing, the amount of waste has also been increasing. Stakeholders are more concerned about disposal, biomass, and recyclability, and as a result, the processed rice industry will be required to take more appropriate actions for processed rice and its production system.

To maintain the quality of the processed rice, all processing steps need to be properly controlled. Previous studies on rice storage and quality evaluation technologies have investigated the effects of individual processing operations on rice quality, and few studies have examined the effects of multiple processes. The optimization of the quality of processed rice by combining each process will be increasingly studied in the future. On the other hand, the demand for low-GI staple foods has greatly increased with the rise of lifestyles and health consciousness in the modern world. Since low GI in processed rice, especially in AP rice, can be an important quality factor as well as storability, palatability, and convenience, further accumulation of knowledge is necessary. The development of processed rice and related technologies will be expected in the future.

Acknowledgements This chapter was written with the assistance of Yoshimasa Sasahara (Nichirei Foods Inc.), Hideaki Kobayashi (Kewpie Corp.), and Akifumi Mikami (Kewpie Jyozo Co., Ltd.). Photographs were provided by Nichirei Foods Inc. and Kewpie Corp. Kogyogijutsukai, Japan Canners Association and Elsevier gave permission to reprint. Nwaeche Chiemerie Famous (Tokyo University of Agriculture and Technology) took care of the grammar proofreading. Author would like to express gratitude to them.

References

Arai T, Sawayama S, Kawabata A (1982) Effect of seasonings on retrogradation of gelatinized cooked rice. J Home Econ Japan 33(10):559–562

Araki T, Du K, Ueno S, Kono S, Do G, Maeda T (2018) Effects of frozen storage and thawing conditions on physical properties of glutinous rice: (part 1) ice crystal and color measurement. Transactions of the Japan Society of Refrigerating and Air Conditioning Engineers:231–236

Atkinson FS, Brand-Miller JC, Foster-Powell K, Buyken AE, Goletzke J (2021) International tables of glycemic index and glycemic load values 2021: a systematic review. Am J Clin Nutr

Awazuhara M, Nakagawa A, Yamaguchi J, Fujiwara T, Hayashi H, Hatae K, Chino M, Shimada A (2000) Distribution and characterization of enzymes causing starch degradation in rice (*Oryza sativa* cv. Koshihikari). J Agric Food Chem 48(2):245–252

Cho CH, Lee HG, Jeong S, Yoo S (2019) Influence of packaging oxygen transmission rate on physical characteristics of frozen cooked rice under various freezing conditions. J Food Sci 84(12):3483–3493

Englyst HN, Kingman SM, Cummings JH (1992) Classification and measurement of nutritionally important starch fractions. Eur J Clin Nutr 46:S33-50

Frei M, Siddhuraju P, Becker K (2003) Studies on the *in vitro* starch digestibility and the glycemic index of six different indigenous rice cultivars from The Philippines. Food Chem 83(3):395–402

Fukaya T, Sakuma K, Tsutsumi R (2010) Development the ready-to-heat foods with cooked rice aseptically packed (Kagome deli). Japan J Food Eng 11(1):13–18

Goñi I, Garcia-Alonso A, Saura-Calixto F (1997) A starch hydrolysis procedure to estimate glycemic index. Nutr Res 17(3):427–437

Ha AW, Han GJ, Kim WK (2012) Effect of retrograded rice on weight control, gut function, and lipid concentrations in rats. Nutr Res Pract 6(1):16–20

Hieda F (1971) Use of rice for preservation food -canned rice and frozen rice. Food Manag Monthly Rep 23(8):37–44. (in Japanese)

Horigome H (ed) (1990) Rice technology and its applications. Kogyogijutsukai, Tokyo. (in Japanese)

Hsu RJC, Chen HJ, Lu S, Chiang W (2015) Effects of cooking, retrogradation and drying on starch digestibility in instant rice making. J Cereal Sci 65:154–161

Hu P, Zhao H, Duan Z, Linlin Z, Wu D (2004) Starch digestibility and the estimated glycemic score of different types of rice differing in amylose contents. J Cereal Sci 40(3):231–237
Huang M, He G, Chen S, Cui M, Ma L, Liu Y (2014) Optimisation of a quality improver for instant rice and its quality properties. Int J Food Sci Technol 49(2):606–615
Huang R, Pan X, Lv J, Zhong W, Yan F, Duan F, Jia L (2018) Effects of explosion puffing on the nutritional composition and digestibility of grains. Int J Food Prop 21(1):2193–2204
Ikawa Y, Kouno C (2005) Multiple evaluation of cooked rice retrogradation in early stage of storage. J Appl Glycosci 52(3):293–298
Ishikawa Y (2001) Canned and bottled goods (in Japanese). Food Encyclopedia, National Food Research Institute Ed., pp.318–324. Asakura Publishing Co., Ltd, Tokyo
Ishitani T, Egawa K, Yokoyama M (eds) (1995) Manufacturing system and quality assurance measures for processed rice under HACCP control (in Japanese). Science Forum Inc., Tokyo
Japan Canners Association, Retort Foods Section Ed. (1996). Understanding retort foods. Maruzen Co. Ltd., Tokyo (in Japanese)
Kainuma Y (2000) Effect of soaking rice in hot water on the textural properties of cooked rice stored at a low temperature. J Home Econ Japan 51(2):129–135
Kasai M, Ishiguro K, Kyouda H, Hamazono T, Hatae K, Shimada A (2000) Change in the amounts of reducing sugars and free amino acids in rice during the cooking processes. J Home Econ Japan 51(7):579–585
Kawabata A, Sawayama S, Kamoi I, Shinozaki T, Tanimura W, Obara T (1979) Effect of storage temperature and storage period on the sensory attributes of gelatinized rice. J Jpn Soc Food Sci Technol 26(6):233–238
Kitamura S, Nakaya M (2012) Determination method of sugar. J Biosci Bioeng 90:790–793
Konishi M, Ide K, Hatae K, Yoshimura R, Hatae K, Shimada A (1996) Eating quality evaluation for cooked rice (2). Objective measurement including taste-active-components. J Cookery Sci Japan 29(4):264–274
Kono S, Kawamura I, Araki T, Sagara Y (2016) ANN modeling for optimum storage condition based on viscoelastic characteristics and sensory evaluation of frozen cooked rice. Int J Refrig 65:218–227
Kono S, Kawamura I, Yamagami S, Araki T, Sagara Y (2015) Optimum storage temperature of frozen cooked rice predicted by ice crystal measurement, sensory evaluation and artificial neural network. Int J Refrig 56:165–172
Kono S, Tobari Y, Araki T, Sagara Y (2017) Investigating the ice crystal morphology in frozen cooked rice based on size, fractal dimension and ANN modeling. Int J Refrig 84:210–219
Kumar PK, Joyner HS, Tang J, Rasco BA, Sablani SS (2020) Kinetics of starch retrogradation in rice (*Oryza sativa*) subjected to state/phase transitions. Food Bioprocess Technol 13(9): 1491–1504
Lee HG, Yoo S (2020) Changes in the physical properties of frozen cooked rice depending on thermal insulation levels of packaging during freeze-thaw. J Food Sci 85(12):4342–4350
Lee NY, Yang JH (2020) Influence of partial gelatinization treatment on the quality changes of puffed rice. Prev Nutr Food Sci 25(2):225–231
Lu LW, Monro J, Rush E (2016) Effect of cold storage, reheating, and particle sizes on in vitro glucose release and starch digestibility among five rice products in Auckland, New Zealand. J Rice Res 4(3):1000171
Ma Y, Sun DW (2009) Hardness of cooked rice as affected by varieties, cooling methods and chill storage. J Food Process Eng 32(2):161–176
Mabashi Y, Miwa Y, Ookura T, Kasai M (2010) Role of endogenous enzymes in milled rice of various cultivars in the accumulation of chemical components in rice grains during cooking. J Cookery Sci Japan 43:228–236
Maruyama E, Higuchi H, Terada K, Kajita T (1980) Studies on the cooking of rice (part 3). Effect of seasonings on retrogradation of cooked rice. J Home Econ Japan 31(8):568–573
Masuda T (2011) Aseptically packaged rice. J Packaging Sci Technol Japan 20(5):447–458. (in Japanese)

Matsunaga A, Kainuma K (1981) Studies on the retrogradation of starchy foods (part 1) retrogradation of cooked rice. J Home Econ Japan 32(9):653–659
Meng L, Zhang W, Zhou X, Wu Z, Hui A, He Y et al (2019) Effect of high hydrostatic pressure on the bioactive compounds, antioxidant activity and *in vitro* digestibility of cooked black rice during refrigerated storage. J Cereal Sci 86:54–59
Miller BM (2019) Race and region: tracing the cultural pathways of rice consumption in the United States, 1680-1960. Glob Food Hist 5(3):183–203
Ministry of Agriculture, Forestry and Fisheries. (2021a). Food industry survey. https://www.maff.go.jp/j/zyukyu/jki/j_doutai/doutai_top.html Access 8th Jun. 2021 (in Japanese)
Ministry of Agriculture, Forestry and Fisheries. (2021b). Food supply and demand table. https://www.maff.go.jp/j/tokei/kouhyou/zyukyu/index.html Access 8th Jun. 2021 (in Japanese)
Mir SA, Bosco SJD, Shah MA, Mir MM (2016a) Effect of puffing on physical and antioxidant properties of brown rice. Food Chem 191:139–146
Mir SA, Bosco SJD, Shah MA, Mir MM, Sunooj KV (2016b) Process optimization and characterization of popped brown rice. Int J Food Prop 19(9):2102–2112
Mridula D, Kaur H, Goswami D, Gupta RK, Gurumayum S, Tyagi SK (2017) Development of quick cooking rice: application of salts and enzyme pretreatments. Proc Natl Acad Sci, India Section B: Biol Sci 87(3):637–646
Naito S (2020) Science of food deterioration—causative microorganism and management. Saiwai Shobo Co. Ltd., Tokyo
Nippon Reizo Inc. Institute (1979) Summary of frozen foods (in Japanese). Kenpakusha, Tokyo
Odahara M, Sokooshi H, Takahashi T, Okadome H, Ohtsubo K (2004) The effect of sushi vinegar on texture of sushi rice before and after storage under low temperature. J Jpn Soc Food Sci Technol
Oishi K, Kanari H, Otahara M, Kasai M (2020) Correlation of physical properties and initial staling of cooked rice with the pH of cooking water. J Cookery Sci Japan 53(2):98–106
Oshima T, Tamura M, Isaka A, Ogawa Y (2012) Changes in reducing sugar, amylose and protein content of rice grain during model cooking. J Japanese Soc Taste Technol 11(2):22–28
Otahara M, Hatae K, Shimada A (1995) Objective evaluation of staling of cooked rice. J Home Econ Japan 46(9):841–848
Otahara M, Kitahara M, Ohishi K, Kasai M (2018a) Evaluation of staling of cooked rice by spectrophotometric and image analyses of squashed cooked rice grains. J Jpn Soc Food Sci Technol 65(4):170–182
Otahara M, Kitahara M, Ohnishi K, Kasai M (2020) Evaluation of early staling of seasoned cooked rice using squashed cooked rice grains. J Cookery Sci Japan 53(3):187–196
Otahara M, Sato Y, Kasai M (2018b) Evaluation of staling of cooked rice by kinetic analysis of texture changes. Food Sci Technol Res 24(3):427–434
Ozaki N (1960) Studies on retrogradation of boiled rice. J Agric Chem Soc Japan 34(12): 1054–2057
Patindol JA, Gonzalez BC, Wang YJ, McClung AM (2007) Starch fine structure and physicochemical properties of specialty rice for canning. J Cereal Sci 45(2):209–218
Perdon AA, Siebenmorgen TJ, Buescher RW, Gbur EE (1999) Starch retrogradation and texture of cooked milled rice during storage. J Food Sci 64(5):828–832
Phukasmas P, Songsermpong S (2019) Instant rice process development: effect of rice cooking methods on the quality of jasmine instant rice dried by industrial microwave oven. J Microbiol Biotechnol Food Sci 9(2):330–334
Prakash M, Ravi R, Sathish HS, Shyamala JC, Shwetha MA, Rangarao GCP (2005) Sensory and instrumental texture measurement of thermally processed rice. J Sens Stud 20(5):410–420
Prasert W, Suwannaporn P (2009) Optimization of instant jasmine rice process and its physicochemical properties. J Food Eng 95(1):54–61
Raigond P, Ezekiel R, Raigond B (2015) Resistant starch in food: a review. J Sci Food Agric 95(10):1968–1978

Rhim JW, Koh S, Kim JM (2011) Effect of freezing temperature on rehydration and water vapor adsorption characteristics of freeze-dried rice porridge. J Food Eng 104(4):484–491

Ritudomphol O, Luangsakul N (2019) Optimization of processing condition of instant rice to lower the glycemic index. J Food Sci 84(1):101–110

Rousset S, Pons B, Martin JF (1999) Identifying objective characteristics that predict clusters produced by sensory attributes in cooked rice. J Texture Stud 30(5):509–532

Sagara Y (2009) Optimum condition of preservation temperature for frozen cooked rice based on sensory evaluation of food preference —correlation analysis method between food preference of the aged and viscoelastic properties. J Jpn Soc Food Sci Technol 56(11):558–572

Saha S, Roy A (2020) Puffed rice: a materialistic understanding of rice puffing and its associated changes in physicochemical and nutritional characteristics. J Food Process Eng 43(9):e13479

Sesmat A, Meullenet JF (2001) Prediction of rice sensory texture attributes from a single compression test, multivariate regression, and a stepwise model optimization method. J Food Sci 66(1): 124–131

Shibata-Ishiwatari N, Hirose J, Udagawa E, Nakazawa K, Matsuda H (2017) Effects of post-cooking cooling and freezing conditions on the quality of frozen cooked rice. J Cookery Sci Japan 50(6):264–271

Suekane S, Sumitani H, Okiura A, Naka T (2006) Evaluation of aroma compounds in cooked rice aseptically packaged by using large volume static headspace sampling-GC/MS. Canners J 85(1):143–153

Tabuchi M (1993) Technology of convenience rice products. J Jpn Soc Starch Sci 40(2):169–175

Tamura M, Kumagai C, Kaur L, Ogawa Y, Singh J (2021) Cooking of short, medium and long-grain rice in limited and excess water: effects on microstructural characteristics and gastro-small intestinal starch digestion *in vitro*. LWT 146:111379

Tamura M, Singh J, Kaur L, Ogawa Y (2016) Impact of the degree of cooking on starch digestibility of rice–an *in vitro* study. Food Chem 191:98–104

Tamura M, Singh J, Kaur L, Ogawa Y (2019) Effect of post-cooking storage on texture and *in vitro* starch digestion of Japonica rice. J Food Process Eng 42(2):e12985

Wongsa J, Uttapap D, Lamsal BP, Rungsardthong V (2016) Effect of puffing conditions on physical properties and rehydration characteristic of instant rice product. Int J Food Sci Technol 51(3): 672–680

Yanai K, Miura M, Nakamura R, Nishinomiya T, Harada T, Kobayashi S (2001) Effect of freezing, storage and thawing conditions on quality of frozen cooked rice. J Jpn Soc Food Sci Technol 48(10):777–786

Yokoyama M (1990) Retort-packed food. J Home Econ Japan 41(8):791–796

Yu S, Ma Y, Liu T, Menager L, Sun DW (2010a) Impact of cooling rates on the staling behavior of cooked rice during storage. J Food Eng 96(3):416–420

Yu S, Ma Y, Sun DW (2009) Impact of amylose content on starch retrogradation and texture of cooked milled rice during storage. J Cereal Sci 50(2):139–144

Yu S, Ma Y, Sun DW (2010b) Effects of freezing rates on starch retrogradation and textural properties of cooked rice during storage. LWT-Food Sci Technol 43(7):1138–1143

Yu S, Ma Y, Zheng X, Liu X, Sun DW (2012) Impacts of low and ultra-low temperature freezing on retrogradation properties of rice amylopectin during storage. Food Bioprocess Technol 5(1): 391–400

Zhang Z, Sun DW (2006) Effects of cooling methods on the cooling efficiency and quality of cooked rice. J Food Eng 77(2):269–274

Zhu L, Bi S, Wu G, Gong B, Zhang H, Wang L et al (2020) Study of the migration and molecular structure of starch and protein in rice kernel during heating. Int J Biol Macromol 147:1116–1124

Zhu Z, Zhou Q, Sun DW (2019) Measuring and controlling ice crystallization in frozen foods: a review of recent developments. Trends Food Sci Technol 90:13–25

Chapter 10
Commercial Processing of Rice

Yukiharu Ogawa

Abstract In this part, different form of commercial rice food is introduced in detail, including the grain form (freshly cooked rice, frozen cooked rice, and puffed rice), the flour form (bread, noodle, cake cracker), the liquid form (alcoholic beverages, alcoholic beverages for cooking use, rice vinegar, rice milk), and the bran by-products (rice bran oil and rice bran-related products).

Keywords Rice · Flour · Liquid · Grain · Commercial processing · By-product

10.1 Introduction

Rice, a staple food mainly in Asian countries, is usually consumed in grain form by simple boiling, often called freshly cooked rice. The demand for ready-to-eat products according to changes in social circumstances and lifestyles promotes novel technologies for food manufacturing. These technological innovations have also been applied to commercial cooked rice products. Although products wrapped in bamboo leaves have been traditionally or regionally produced, hermetically or semi-hermetically packed or film-wrapped cooked rice is now provided as a commercial product for the freshly cooked form. For this type of product, hygiene control and sensory properties are critical indices for commercial products. Meanwhile, many processed products made from rice flour are also recognized and consumed as flour-based commercial products, such as rice noodles and rice crackers. This type can easily reduce the water activity of products after processing and thus extend their shelf life if they are hermetically sealed. In other words, commercial rice products can be classified by their forms, including liquid-type derivatives, such as brewed beverages, and by-products, such as bran (Fig. 10.1). Rice varieties, i.e., glutinous (sticky or waxy) and non-glutinous (non-sticky or non-waxy), are also related to the classification of commercial products due to their physicochemical

Y. Ogawa (✉)
Graduate School of Horticulture, Chiba University, Chiba, Japan
e-mail: ogwy@faculty.chiba-u.jp

J. Tian et al. (eds.), *Science of Rice Chemistry and Nutrition*,
https://doi.org/10.1007/978-981-99-3224-5_10

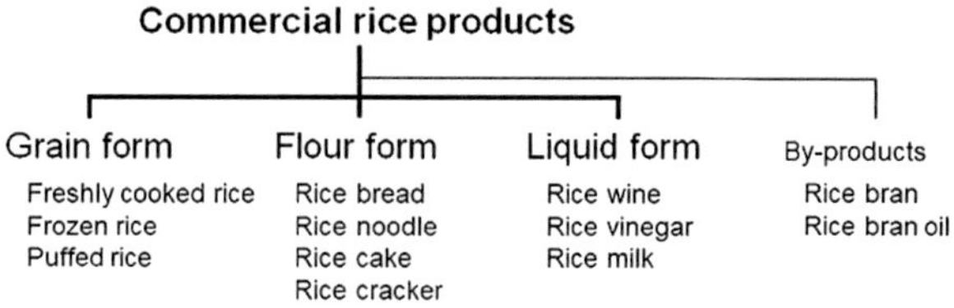

Fig. 10.1 Classification of commercial rice products (drawn by the author)

properties. In this chapter, commercially processed rice products are classified based on their forms and characteristics, such as physical properties. Standard and novel processing methods and technologies for commercial rice products have also been introduced.

10.2 Grain Form

Freshly cooked rice contains comparatively high moisture, which can easily result in deterioration by microorganisms. Although it has been commercially provided as a part of the lunch box-style packed meals, called "bento" in Japan, appropriate quality control is necessary to extend the shelf life as commercial products. For this purpose, chilling, freezing, frying, and coating with antibacterial ingredients such as vinegar and semi-aseptic packaging during manufacturing are applied to the commercial cooked rice for the grain form. This commercial product type is mainly categorized into three groups: (1) freshly cooked rice, which is packed by appropriate approaches and can be recognized as a ready-to-eat style, including packed sushi products; (2) frozen materials after processing, such as frozen fried rice; and (3) puffed rice as a type of rice snack. In general, frozen products tend to be produced in comparatively larger food factories. In contrast, fresh ready-to-eat products are typically made at a smaller business site, such as a kitchen area of a supermarket, except for semi-aseptic packed cooked rice and film-wrapped cooked rice balls.

10.2.1 Freshly Cooked Rice Products

Semi-aseptic packed cooked rice, which can be categorized as ready-to-eat type, is sometimes called precooked rice or packed rice in Japan because it can easily be consumed by microwave reheating. The manufacturing process is usually controlled under semi-aseptic conditions to maintain shelf life by reducing microbial contamination of products. Thus, the final products need to be hermetically sealed and preserved mostly for 1 year.

The production procedure for packed cooked rice follows the Japanese-style rice cooking method, which combines soaking with an optimum amount of water and

Fig. 10.2 Appearance of commercial rice ball, onigiri. (Photographed by the author)

heating until the water disappears (https://www.tablemark.co.jp/english/products/packed_cooked_rice.html). Because all cooking procedures are automatically batch-wise processing, the heat-tolerant packaging container or casting mold for the individual product is applied to replace the rice cooker inner pan. The soaked white rice is placed in the container and heated by pressurized heating for uniform heat treatment, thereby reducing the cooking time. The container is hermetically sealed using plastic film and steamed (aging or incubating cooked rice) by self-derived steam from the rice to ensure uniform moisture in the cooked rice grain.

The final packed products can be distributed and stored at normal temperatures and subjected to microwave heating before consumption. This manufacturing process can be applied to not only white rice but also brown rice and mixed rice with other cereals. The eating quality of packed cooked rice has improved continuously and is currently recognized as the same as freshly cooked rice. The digestive properties of packed cooked rice have also been studied, suggesting that the reheating conditions of microwave heating can depend on the digestibility (Li et al. 2014; Li et al. 2019).

Film-packed rice ball, called "onigiri" in Japan, was commercialized as a direct film-packed cooked rice at the end of 1970 and has been consumed as a ready-to-eat product (Fig. 10.2). Historically, rice balls have been made as a carriable ration. "Onigiri" in Japanese means packing and forming the cooked rice in triangular shape by the palm and fingers, although there are various shapes like spherical and cylindrical by the cultural background at the local area in Japan (https://www.kikkoman.com/en/foodforum/close-up-japan/30-3.html). Because cooked white rice is sticky by gelatinization, a diluted salt solution is generally used as a lubricant agent while shaping onigiri at the household level. A dried laver seaweed called "nori" is traditionally used as an enclosing material for onigiri to deter the sticking between rice and fingers when the rice ball is held to eat.

For plant-scale manufacturing of onigiri, a small amount of oily additive is sprayed into the cooked rice to reduce its surface stickiness when forming the shape by the manufacturing apparatus. This treatment could also reduce the deterioration caused by the aging of starch, which occurs because of moisture evaporation, and the possibility of microbial contamination. Traditionally, a salted Japanese

Fig. 10.3 Photograph of packed nigiri-zushi, hand-pressed sushi. (Photographed by the author)

apricot pickle called “umeboshi” is filled into a rice ball to prevent microbial activity during short-term portability. Today, various fillings, including umeboshi, are usually placed into commercialized rice balls to produce higher and tasty value-added products as ready-to-eat fast food. Glutinous rice stuffed with various fillings and wrapped in flat leaves (like bamboo leaves) called “zongzi” in China, and “ba chang” in Thailand, etc., can also be considered as a kind of commercially packed rice balls. However, it tends to be regarded as a homemade product based on its cultural background.

Packed sour rice, called “sushi” in Japan, is one of the well-known Japanese cuisines (Feng 2012). “Sushi” in Japanese means vinegared rice; thus, the taste is sour. This is the origin of the name “sour rice,” similar to vinegared rice. Sour rice is prepared by sprinkling sugared vinegar into freshly cooked rice and mixing it. Dried kelp called “kombu” in Japan is usually added to the cooking water to generate a taste of “umami” to prepare sour rice, particularly for the “nigiri-sushi (−zushi)” use. Sour taste historically resulted from fermentation with fish meat; however, rice vinegar is used to make sour rice nowadays.

Such sour rice forms several types of sushi meals: the hand-pressed sushi called “nigiri-zushi” (the pronunciation of “su” in Japanese sometimes changed to “zu,” like “-zushi” depending on the previous words) (Fig. 10.3), the wrapped and rolled sushi called “maki-zushi” (Fig. 10.4), and the scattered sushi called “chirashi-zushi” or “bara-zushi” (Fig. 10.5) (https://www.maff.go.jp/e/policies/market/k_ryouri/search_menu/menu/bara_zushi_kyoto.html). Nigiri-zushi is made by placing sliced raw fish on a small chunk of sour rice and shaping it manually. Maki-zushi is a type of sushi rolled with a sheet of dried laver seaweed, “nori,” in the shape of a tube. Ingredients of maki-zushi, such as sliced raw fish and other items, are rolled in vinegared rice, with nori as the outer layer. Chirashi-zushi is sour rice topped and mixed with various colorful ingredients, such as seafood, eggs, and vegetables. Currently, the robotic actuator can imitate the hand-pressing movement to form a small chunk of sour rice and can be applied to commercialized sushi manufacturing. Automatic manufacturing systems to make sushi chunks are used at the plant and sushi bars like kaiten-zushi (https://www.japan-guide.com/e/e2036_kaitenzushi.html).

Fig. 10.4 Photograph of packed maki-zushi, wrapped and rolled sushi. (Photographed by the author)

Fig. 10.5 Photograph of packed bara-zushi, scattered sushi. (Photographed by the author)

Freshly cooked rice is typically provided as a hot meal. However, it must be cooled if mixed with raw fish (sashimi) to reduce microbial activity (Kulawik and Dordević 2022). Thus, sushi made with raw fish meat can generally be regarded as chilled-cooked rice. This attribute influences the physical properties of rice. The acidic conditions of rice can also prevent microbial activity and affect the physical and digestive properties of cooked rice. The Japanese horseradish, "wasabi," is often used in sushi products because of its antibacterial properties (Lu et al. 2016). These characteristics are appropriate for short-term logistics and storage, although they must be controlled in a cold chain. Thus, simple packed products have been widely commercialized, despite mixing raw fish meats with cooked rice (Figs. 10.3, 10.4 and 10.5).

10.2.2 Frozen Cooked Rice Products

Although cold chain logistics and storage systems are required, frozen foods quickly cooked by microwave heating have been widely consumed as ready-to-eat foods because of their convenience. Cooked rice products, mainly frozen fried rice, are commercialized as frozen ready-to-eat meals.

Fried rice is typically consumed as a freshly cooked product. It can mainly be classified as fried "cooked rice," like stir-fried rice in East and Southeast Asia, and boiled "fried precooked rice," like pilaf in India, the Middle East, and Europe. Stir-fried rice is panfried, adding a small amount of oil and usually seasoned using seasonings such as soy sauce and black pepper. Because it has an oily grain surface than boiled products, it inhibits moisture migration from the grain core or outer circumstances. Therefore, stir-fried rice could be more applicable to the freezing and thawing process, which influences changes in rice grain texture compared to freshly cooked rice products. It has been commercialized as frozen stir-fried rice called "frozen yaki-meshi" in Japan (https://www.youtube.com/watch?v=Tcmi48KSvpI). A grilled rice ball, called "frozen yaki-onigiri" in Japan, has also been manufactured as a commercial frozen product (https://www.youtube.com/watch?v=ICbOR8f3xyg).

Recently, a novel technology to prevent texture degradation of frozen fried rice has been developed and applied to manufacture packaged frozen fried rice. A higher heating temperature during panfrying is applied as a manufacturing condition that could mostly maintain the physical properties of the original cooked rice after thawing.

This technology contributes to the better texture of frozen fried rice than the previous method. Stir-fried rice showed a comparatively lower enzymatic starch hydrolysis rate and higher resistant starch content than steamed rice and pilaf; however, the digestive properties of thawed fried rice require further investigation (Reed et al. 2013).

10.2.3 Puffed Rice (Popped Rice)

Puffed rice, also called popped rice, is consumed as a breakfast cereal and snack (Saha and Roy 2020). As a snack food, like popcorn, it is sometimes made with sugar or by spraying sugar syrup after puffing (Fig. 10.6). Although puffed rice was traditionally made by oil frying, it is currently produced using a heat-pressure chamber combined with pressure heating and quick pressure release mechanisms.

Fig. 10.6 Photograph of puffed brown rice, freshly puffed and syrup-coated type. (Photographed by the author)

The pressure chamber could rotate for uniform heating of the rice grain material. The moisture in the grain expands in a single moment when pressure is quickly released, resulting in puffed grains. Because puffed rice is directly processed from uncooked white rice, it maintains the grain shape, but the grain size is expanded. The texture is changed to crispy, which is the specific attribute of puffed rice (Wongsa et al. 2016). Puffing alters the nutritional composition of white rice. Puffed rice shows comparatively higher total sugar content and lower moisture, starch, and fat contents than white rice. In contrast, the protein content barely changes, but its solubility declines significantly. Puffing also changes the physical properties of white rice, including the water absorption index, water solubility index, and degree of gelatinization. It has also been reported that the digestibility of starch and protein in grains indicated by the starch hydrolysis rate and free amino acid content is significantly increased (Huang et al. 2018). Therefore, puffing can improve the nutritional value of white rice. Yukwa, a traditional Korean snack supplied for anniversary days, is made by puffing glutinous rice; thus, it has a unique texture. Yukwa is usually coated with honey or sweet syrups and is a commercial rice product produced and consumed in Korea (Ryu 2011).

10.3 Flour Form

Rice is generally consumed in grain form, but flour (or powder) form as rice flour, made from finely ground white rice, is also used to produce several types of rice-related processed foods such as rice crackers. Rice flour made from brown rice, called "brown rice flour," is specifically produced, as well. Rice flour can be classified according to starch type. Glutinous rice flour, called "shiratama-ko (or mochi-ko)" in Japan and "nuòmǐ fěn" in China, is made from glutinous rice and consists of amylopectin only or very low amylose content. This flour type is frequently used to produce sweets such as sweet-filled dumplings. In contrast with glutinous rice flour, non-glutinous rice flour, called "joshin-ko" in Japan and "zhānmǐ-fěn" in China, is made from non-glutinous rice and consists of amylose and amylopectin. This flour type is used in rice noodles, bakery products, etc. Rice flour can also be classified by the milling method, such as dry milling or wet milling. The dry milling method is usually applied to prepare rice flour, but the wet milling method is also used for soaked rice grains because of accessible grain disruptions. In Japan, glutinous rice flour milled using the wet milling method is called "shiratama-ko." In contrast, flour milled by the dry milling method is called "mochi-ko."

As rice flour consists of almost all rice starch, it can be regarded as a gluten-free starch base. Therefore, rice flour is currently used as raw materials to produce not only traditional commercial products like rice cakes and noodles but also bread, etc., as an alternative to wheat flour. Such flour-based commercial rice products are mainly categorized into three groups: (1) rice bread, which is made from fine rice flour; (2) rice noodles, whose product type varies by country and region; and (3) rice cake, which is made by mashing and pounding steamed glutinous rice, dumpling as a type of processed sweets, and rice crackers.

10.3.1 Rice Bread

Rice bread made from rice flour has received attention as a gluten-free bakery product. Unlike wheat flour, which contains gluten as a thickener of dough, rice flour is difficult to retain the gas produced during the dough-rising process and form dough. Therefore, as an additive, wheat flour or gluten powder needs to be mixed with rice flour to promote the dough-rising process. However, it is hard to be regarded as gluten-free. A short baguette used for "banh mi" in Vietnam and many Chinese bakery products can be regarded as mixed rice bread (Yamauchi et al. 2004). To avoid such circumstances, several methods that apply gelatinized rice flour, polysaccharides, or joshin-ko instead of gluten powder have been developed to improve the dough-rising process. For these methods, it was reported that the rice variety, flour particle size, dough water content, and microstructure of the dough could influence the quality of gluten-free rice bread (Hera et al. 2013). The particle size and proportion of damaged starch granules, which vary according to the milling principles and methods, have been studied and regarded as indices of rice flour characteristics (Wu et al. 2019; Qin et al. 2021a). The surface of flour particles, related to the water absorption rate, influences the flour characteristics (Qin et al. 2021b). The amylose content of rice flour and the milling method also influence dough quality (Han et al. 2012). Finely milled rice flour, which can be obtained using advanced flour milling technology such as a jet mill to reduce damaged starch particles, has been used commercially to produce rice bread. The particle size of the fine powder is finer than an average of several tens of micrometers or less, and it is better to make quality rice bread than the conventional rice flour, joshin-ko. Magnetic resonance imaging (MRI) can be applied to measure the water content dynamics in rice flour dough and thus evaluate the characteristics of rice bread (Bosmans et al. 2012).

10.3.2 Rice Noodle

Rice noodles made from rice flour are a popular cooking ingredient consumed mainly in East, Southeast, and South Asian countries. Rice noodles can be classified by noodle shape and thickness, which vary according to the processing method and local culture (Li et al. 2015). Thus, the name of the rice noodles is altered by the region. Because rice flour has no gluten, no salt is needed to add when kneading the dough. This indicates that rice noodles are comparatively healthy food. However, there are no network structures, such as the gluten network in the dough, and the noodle texture after cooking tends to be crumbly. Therefore, tapioca starch or cornstarch is sometimes added to rice flour to improve noodle physical properties. Starch molecule properties depending on rice variety, such as the ratio of amylose and amylopectin content in the material, are related to noodle texture. The noodle shape also influences textural attributes (Low et al. 2020).

Rice noodles are commercially manufactured and provided as dried, frozen, or fresh products. Dried- and frozen-type products have longer shelf life. In contrast, fresh products are usually made at home and only have a short shelf life (e.g., a few days).

Rice vermicelli ("bi-fun" in Japan and Taiwan, "mi fen" in China, "mee hoon" in Thailand, etc.) is a needlelike thin noodle. The appearance of rice vermicelli is similar to that of a glass noodle made from mung bean flour. Rice vermicelli is made from wet milled rice flour in which boiling water is added to form dough by gelatinization of rice starch. The dough is extruded through a mold with numerous tiny holes (https://www.youtube.com/watch?v=6Q8GHfOsKVw). After the dough formed a needlelike shape, it is soaked in boiling water for 1 or 2 min and quickly placed into cold water to fix the noodle shape. After cooling, the noodles are washed to remove extra starch on the noodle surface, dried, and packaged as a commercial product. Dried rice vermicelli can be rehydrated by soaking in water before cooking (Yoenyongbuddhagal and Noomhorm 2002a). The particle size of rice flour is an important parameter for estimating the noodle quality. Starch granule swelling and paste or gel conditions of the dough are also important parameters (Yoenyongbuddhagal and Noomhorm 2002b). Rice noodle quality is evaluated by uniformity, cooking, and eating quality.

Flat rice noodle ("shahe (he)-fen" or "qie-fen" in China) is mainly used in stir-fried cuisines and soups. The preparation process of dough is mostly the same as that of rice vermicelli, but the noodle-shaping process is different. The rice flour dough is pressed into sheets by heating and cut into long rectangular strips as a flat noodle (https://www.youtube.com/watch?v=4pLceTpeUK8). Other starches are often added to form a flat shape. Steaming is usually applied to the sheet-press process to shape the noodles. After shaping, the noodles are dried and packaged as a commercial product. Dried flat rice noodles can be cooked after rehydration. Flat rice noodles are used in many ethnic cuisines such as chow fun (China), pad thai (Thailand), and pho (Vietnam).

10.3.3 Rice Cake

Rice cakes are commonly made from kneaded rice flour ("dango" in Japan, "jian dui" in China, "chaltteok" in Korea, etc.) or by mashing and pounding steamed glutinous rice ("mochi" in Japan). It has many varieties depending on the country, region, and culture, particularly in Asian countries, which include cake-like sweets and rice dumplings stuffed with sweet fillings. Such rice cakes have specific textures characterized by raw materials: glutinous or non-glutinous rice. The raw materials can be classified by powdered flour or mashed cooked grain. Grain or flour of glutinous rice contains a higher amylopectin content. Thus, it can form its shape by glutinous attributes, which gelatinization generates after heating, even though it has no gluten. The formation of the product shape must be completed during the gelatinization of starch. Gelatinized starch loses its glutinous property when it changes to beta starch after cooling, called retrogradation. However, the addition of sugar can retain its glutinous properties even though it is cooled. This characteristic is suitable for sweets. Therefore, processed rice cakes are typically categorized as confections. Because rice cake is processed from thermally treated raw materials, such as steamed rice grains, as mentioned above, appropriate heating treatments are usually applied during the manufacturing process. Therefore, processed rice cake can be a commercially applicable product because of its comparatively longer shelf life, even as a fresh and soft glutinous product.

Rice cake made from cooked glutinous rice, which is mashed and pounded, called "mochi" in Japan, is a significant commercial rice cake product (https://www.youtube.com/watch?v=4q-FzuKk4PU). Glutinous rice material contains almost 100% amylopectin. Therefore, it can be converted into a glutinous paste when cooked (or steam heated) and pounded, resulting in a glutinous texture (Chuang and Yeh 2006). Such pounded glutinous rice products can be formed by clipping or molding immediately after pounding. After forming the shape and cooling, it can be packaged as a commercial product. The cooled or packaged mochi is normally hardened by starch retrogradation, but reheating can produce a glutinous texture again. Mochi is generally consumed in the Japanese New Year because of its cultural and historical background. Mashed glutinous rice coated with red bean paste called "ohagi" or "bota-mochi" in Japan that can also be categorized into rice cakes is a traditional Japanese sweet and is currently a commercial product. A similar sweet coated with sweetened soybean powder called "abekawa-mochi" or "kinako-mochi" has also been made and consumed in Japan (Fig. 10.7).

Rice dumplings, made from glutinous cooked rice or kneaded glutinous rice flour with various fillings and wrapped using flat leaves, are commercial rice products, although they have traditionally been homemade products and consumed on

Fig. 10.7 Photograph of bota-mochi and kinako-mochi. (Photographed by the author)

Fig. 10.8 Photograph of Japanese dango covered with sweetened paste. (Photographed by the author)

anniversary or festival days in Asian countries. Chinese rice dumpling made from glutinous rice stuffed with various fillings and wrapped in flat leaf is called "zongzi." The wrapped material is directly cooked via steaming or boiling. A glutinous rice dumpling called "bachang" in Southeast Asian countries is also a rice dumpling, such as zongzi.

Cooked and mashed glutinous rice stuffed with sweet fillings wrapped in bamboo leaves called "chimaki" in Japan can also be regarded as a commercial rice cake product. It is usually consumed during a seasonal festival in May. Chimaki is similar to zongzi in China but is sometimes made from glutinous rice flour and filled with sweet red bean paste. Thus, it can be considered a different type of rice cake.

Rice flour is also used to produce rice cakes. Japanese "dango," which looks like mochi but is formed by kneaded rice flour, is one of the powder-based rice cakes. Dango is sometimes covered with a sweetened paste as a commercial product

(Fig. 10.8). Non-glutinous rice flour (joshin-ko) is typically used to make dango. However, a mixture of glutinous rice flour can be used to modify the textural properties of dango. “Shiratama-dango” in Japan, “tang-yuan” in China, made from glutinous rice flour, is one of the glutinous flour-based rice cakes. The glutinous dough is treated in water to reduce the sticky conditions on the dough surface during the product-shaping process. The texture of shiratama-dango and tang-yuan is chewy, but its surface is silky. “Jian-dui” or “ma-tuan,” made from glutinous rice flour, is a Chinese oil-fried rice cake filled with sweet fillings and coated with sesame seeds on the cake surface. Oil-fry conditions could influence starch digestibility due to structural damage to starch by high-temperature heating (Din et al. 2018).

10.3.4 Rice Cracker

Rice crackers, a commercial rice-related product, are also produced and consumed as baked products, mainly in Japan. Crackers, starch-based dried snacks, are normally made from wheat flour. In contrast, rice crackers are made from rice flour. Therefore, they can be categorized in a dried Japanese confectionery, called “beika,” but there are a lot of derivations. Because rice does not contain gluten, starch gelatinization is required to form dough during the making process of rice crackers, similar to other rice products. Thus, rice flour is mixed with hot water for kneading and steamed to shape the dough before baking. This is a processing step specific to rice crackers (Takei et al. 2019). Such rice crackers can be classified by their raw materials: glutinous (mochi-ko) or non-glutinous (joshin-ko) rice flours. The physical properties of rice crackers vary according to the raw materials used.

Rice cracker made from non-glutinous rice flour (joshin-ko) is called “senbei,” a flat and thin disklike shape. After steaming, the process of making senbei consists of pounding and rolling out the dough, cutting out the shape, and drying. The dried dough is then grilled. Soy sauce with a small amount of sweet sake is usually applied to the dried dough before and during baking as a seasoning (Fig. 10.9) (https://www.youtube.com/watch?v=0cNa9MOmj7E). Because soy sauce is salty, it tends to be seasoned rather than sweet. Senbei is sometimes coated with coarse sugar as a sweet or wrapped by a dried seaweed “nori” as a seasoned snack. The texture of senbei is comparatively crunchy because of the characteristics of non-glutinous material.

Rice cracker in a bite size made from glutinous rice flour (mochi-ko) is called “arare” or “okaki.” The making process after steaming involves pounding well and then cooling to harden the dough. The hardened dough is cut into small pieces of bite size and baked. Baked materials are usually flavored by applying soy sauce with sweet sake, called “mirin.” The difference between the arare and okaki is normally regarded as the difference in the size of one piece. Arare has bite-size dimensions. In contrast, the size of one piece of okaki, even depending on the cultural background, is larger than arare (Fig. 10.10) (https://www.youtube.com/watch?v=k26Eh-20NHY). The texture is crispy because of the presence of glutinous material. “Kaki-no-

Fig. 10.9 Photograph of rice cracker made from non-glutinous rice flour, senbei. (Photographed by the author)

Fig. 10.10 Photograph of rice cracker made from glutinous rice flour, okaki. (Photographed by the author)

tane" or "kaki-pi" (Fig. 10.11), which is mixed with peanuts, is one of the varieties of arare produced in Japan. A piece of kaki-no-tane has a crescent-like shape. Today, kaki-no-tane has become a primary commercial rice cracker product as a seasoned snack because of its wide variety of flavors and suitability for serving alcoholic beverages.

10.4 Liquid Form

Rice contains 70% starchy carbohydrates. Therefore, alcoholic beverages can be produced using an appropriate fermentation process. Rice wine, specifically called "sake" or "nihon-shu" in Japan, "huangjiu" or "mijiu" in China, "cheongju" in Korea, "sato" in Thailand, etc., is one of such alcoholic beverages made from rice.

Fig. 10.11 Photograph of kaki-pi. (Photographed by the author)

There are several variations in rice wine depending on its processing method, fermentation degree, and usage. Sweet sake and rice vinegar are derivatives of alcoholic beverages produced from rice. As a commercial product, alcoholic beverages are generally distinguished by their alcohol content according to tax regulations. Therefore, the classification of alcoholic beverages made from rice depends on the regulations of each country. In other words, they need to be classified according to such social regulations, even if they are made using a similar processing method. In this section, however, rice-related beverages are mainly classified by the processing method, such as (1) alcoholic beverages, that is, rice wine; (2) alcoholic beverages for cooking use (e.g., mirin); and (3) two-stage fermented acidic ingredients, i.e., rice vinegar. Nonalcoholic beverages, such as (4) rice milk, can also be categorized as commercially rice-related liquid products.

10.4.1 Alcoholic Beverages

Alcoholic beverages can be divided into brewed and distilled liquors. Because rice contains mainly starch and not sugar, the normal fermentation process using yeast cannot produce alcohol. Therefore, starch must be converted into sugar and then fermented into alcohol. These conversions, such as brewing, can be achieved by combining saccharification and alcoholic fermentation. Multiple parallel fermentations, which can simultaneously convert starch into sugar and sugar into alcohol, are a conversion process to produce brewed beverages for rice. The maximum alcohol concentration produced during brewing by this fermentation depends on the alcohol resistance of the yeast, that is, up to around 20%, depending on the strain. Therefore, to produce beverages with higher alcohol concentration, distillation must be applied to brewed materials. Distilled liquors have a much higher alcohol concentration than brewed liquors do. Alcoholic beverages made from rice can be distinguished by their appearance. One is a clear beverage obtained by filtration after fermentation. The other is a milky beverage with no filtration of the fermented material.

Rice wine, of which commercial products have been developed depending on historical and cultural backgrounds in each country, is the most popular brewed alcoholic beverage made from rice (Teramoto et al. 2000). They can be categorized according to the processing method and alcohol content.

Japanese rice wine, called "sake" or "nihon-shu" or "sei-shu," is one of such brewed clear beverages made from only non-glutinous rice (https://www.nrib.go.jp/English/sake/sakeinfo.html). For the production of sake, rice is usually polished to a high percentage to reduce the protein or lipid content, which could influence the sensory qualities. For the preparation of sake processing, "koji, *Aspergillus oryzae*," is inoculated on a small amount of steamed grain to cultivate spores of koji-fungi, called "tane-koji," as a fermentation starter for the primary saccharification stage. This product is called "koji rice" or "kome-koji," which can be more easily handled than fungi spore and is used to proceed with the saccharification of the steamed rice material. The main saccharification process involves mixing koji rice with a large amount of steamed non-glutinous white rice and water to convert gelatinized rice starch into glucose. The "yeast, *Saccharomyces cerevisiae*," is prepared as seed mash, called "shubo," to significantly increase the amount of yeast before the main fermentation process. Yeast has a lower competitive potential than the other microorganisms but adapts to higher acidic conditions. To brew sake, thus, seed mash is prepared under highly acidic conditions with lactic acid bacilli. For the main fermentation process, steamed rice is mixed with koji rice and water, and seed mash is added stepwise. The main mash, called "moromi," in which the enzymes of koji dissolve the steamed rice and the yeast ferments the liberated glucose by saccharification simultaneously, is formed and produces ethanol. The thermal conditions of this process are strictly controlled for product quality. Proteins in rice are also digested into oligopeptides and amino acids by the proteolytic enzymes secreted by koji. Minerals in rice are eluted and partially assimilated by yeast and converted into the flavor, aroma, and taste components of sake. After the fermentation process is completed, the main mash is initially filtered to separate the liquid and the undissolved rice called "sake-kasu," a filtered sake cake. Although sake-kasu is a by-product of the sake-making process, it can be used as an ingredient for cooking. Because some turbidity remains with the liquid after the initial filtration, it is filtered again to remove it, including sedimented dissolved proteins using persimmon tannin or colloidal silica. After sedimentation and filtration, the clear liquid is pasteurized at 60–65 °C prior to storage. This process can also terminate the enzyme activity. The pasteurized liquid is aged 6–12 months, adjusted to an alcohol content of approximately 15% by dilution, bottled, and shipped as commercial sake products. The quality of sake also depends on the state of raw rice materials; thus, a few cultivars, which tend to have a chalky white core in the center of the grain, are mainly used to produce current Japanese sake (Okuda 2019). Distilled liquor made of rice is commercially available in Japan. Japanese distilled liquor is called "shochu." Rice in Japanese is "kome"; thus, the distilled liquor made from rice is called "kome-shochu." For kome-shochu production, pot still is generally applied. The appearance

Fig. 10.12 Photograph of ama-zake. (Photographed by the author)

of kome-shochu is clear. Its alcoholic concentration is approximately 25%, but it allows up to 45% depending on the distillation.

Chinese rice wine, mainly called "huangjiu" or "yellow wine" in English, is also a brewed beverage produced from glutinous rice or other grain mixture. It applies koji from wheat to make saccharification and then alcoholic fermentation. Shaoxing wine is a major variety of huangjiu that is traditionally produced and commercially provided using an earthenware jar. "Mijiu" is a clear alcoholic beverage like sake. Chinese distilled liquor, called "baijiu," has several commercial brands made from rice and other grain mixture.

Korean rice wine, called "cheongju," is a brewed clear beverage made from rice. Steamed rice mixed with nuruk and water is fermented and filtered. Nuruk is a fermentation starter for Korean alcoholic drinks prepared from grains. It contains several microorganisms that contribute to its fermentation. Some variations of Korean distilled liquor, called "soju," were made from rice and other grains traditionally.

"Ama-zake," a Japanese traditional low-alcoholic sweet beverage made from fermented rice, is also a commercially available rice-related liquid product (Fig. 10.12). It can be produced as a mixture of sake-kasu, water, and sugar. Ama-zake can also be prepared by saccharification of steamed rice using koji; thus, it has a sweet taste and a milky appearance. The nutritional properties of ama-zake, including amino acids and vitamins, are much higher than those of unfermented cooked rice because of enzymatic reforming of rice ingredients. Therefore, ama-zake consumption is considered to have many health benefits (https://www.morinaga.co.jp/amazake/en/). "Doburoku" or "nigori-zake" or "moromi-zake" in Japan is a milky brewed alcoholic beverage, which is made by adding koji to steamed rice and water and then applying no filtration after fermentation. "Makgeolli," a raw Korean milky rice wine, is also a non-filtered milky alcoholic beverage fermented using nuruk. Makgeolli has a slightly sweet taste and somewhat carbonated drink-like characteristics. "Choujiu," a Chinese milky alcoholic beverage brewed from glutinous rice, is also fermented using "qu" as a fermentation starter. The taste of these milky brewed beverages tends to be sweet because of their unfermented glucose content.

10.4.2 Alcoholic Beverages for Cooking Use

Alcoholic beverages can be used as ingredients in cooking. Because the taste or smell of meals can be improved by adding alcohol during cooking, there are many uses for cooking. Alcohol also acts on the water retention of proteins and tenderizes the meat after cooking. Alcohol can also eliminate the unfavorable smell of foods. Furthermore, adding alcohol to foodstuffs can result in a burnished appearance and characteristic flavor after cooking (Inoue et al. 2016). Alcoholic beverages are generally categorized as luxury items and thus regarded as a comparatively expensive commodity owing to the taxation system in each country, although it depends on public regulations. Therefore, it is not suitable for daily use, such as cooking ingredients. To avoid such a taxation system problem, several constituents such as salt and sugar are often added to the alcoholic products for cooking use.

Japanese sweet rice wine for cooking use, called "mirin" or "sweet sake," is one of the commercial alcoholic products made from rice. Mirin is a common ingredient in Japanese cuisine named "washoku." The manufacturing process for mirin is similar to that for sake. However, it uses a glutinous rice cultivar as the raw material, and 40% ethanol or shochu is added before the moromi aging process to avoid starting alcoholic fermentation. The glutinous rice starch in moromi is decomposed into reducing sugars by koji enzymes. However, the alcohol content is comparatively lower owing to the avoidance of alcoholic fermentation by the addition of ethanol (approximately 14%). Instead, the sugar content is 40–50%. After aging, the processed mixture is filtered to separate the liquid phase. The resulting liquid product, mirin, has a clear light-brown appearance. There are three types of commercial mirin: (1) "hon-mirin," a traditionally manufactured product with an alcohol content of approximately 14%, as mentioned above; (2) "fermented seasonings" or "shio-mirin," a salt added-type cooking's sake of which salt content is around 2% and alcohol content is 10–14%; and (3) "mirin-style seasoning" or "nonalcoholic mirin-like seasonings" made from grain mixtures containing less than 1% alcohol and retaining a mirin flavor (https://www.kikkoman.com/en/cookbook/glossary/mirin.html).

Chinese cooking wine made from rice is categorized into regular rice wines, such as "shaoxing wine" in "huangjiu." "Liaojiu" is also one of the rice-related cooking wines and a type of "huangjiu," flavored and seasoned with various herbs. Lower grade production is often converted into commercial cooking wine, supplemented with salt, and manufactured. The lower grade of "mijiu," a clear brewed rice beverage, is also commercially used as cooking wine. It contains 1.5% salt and less than 20% alcohol.

10.4.3 Rice Vinegar

Vinegar is an aqueous ingredient mainly used as a culinary seasoning containing 5–8% acetic acid. It is a fermented product of various materials that contain starch or sugar. The vinegar production process, usually a two-stage fermentation, converts simple sugars of the raw material to ethanol using yeast. Acetic acid bacteria subsequently produce acetic acid by oxidizing ethanol during fermentation. Depending on the source materials, various vinegar types have been commercially produced and used as condiments or garnishes, including balsamic vinegar and malt vinegar.

There are several methods for vinegar production. The traditional production process requires a long fermentation period of around a month because natural vinegar is used as a starter culture. In contrast, industrial vinegar can be manufactured in approximately 1 day. Currently, two main methods are being commercially employed. One is a traditional method classified as a "surface method" in which the culture of acetic acid bacteria grows on the surface of wood shavings and provides oxygen at the surface. The other, classified as a "submerged culture," is a method in which oxygen is supplied in fermentation to accelerate industrial production (Budak et al. 2014). Vinegar made from rice is commercially manufactured and is widely consumed in East Asia and Southeast Asia. The first stage of rice vinegar production is similar to that of rice wine brewing. After alcohol fermentation by yeast, the natural vinegar is mixed with the main mash. The mixture is heated, and acetic acid bacteria are cultured to start acetic acid fermentation in the second stage. This process needs approximately 2 weeks. Fermented production needs to be aged for 1 or 2 months to reduce the irritating odor of acetic acid.

Rice vinegar has several variations owing to its raw materials: white (or light yellow), red, black, etc. Japanese rice vinegar is generally made from polished non-glutinous rice, similar to sake. Thus, it appears white or light yellow and has a mild acidity. Sushi rice is seasoned with Japanese rice vinegar by the addition of sugar and salt. Chinese rice vinegar, usually made from glutinous brown rice similar to huangjiu, has a more robust acidity than Japanese rice vinegar. The color range varies from clear to red, brown, and black, depending on the source rice materials. Similar to cheongju, Korean rice vinegar is made of either white or brown rice. Because rice bran contains many functional ingredients, colored vinegar made from brown rice tends to have superior functional properties.

10.4.4 Rice Milk

Rice milk, which is a mixture of pulverized rice flour and water, is a commercially processed beverage made from rice. It can be categorized as plant-based milk, which is necessary for consumers with lactose intolerance or an allergy to cow's milk (Aydar et al. 2020). In general, oilseed and nut milk are produced and consumed as the main varieties of plant-based milk. The consumption of milk substitutes from plants has positive effects because of their rich antioxidant activity and fatty acids; however, it lacks protein and calcium content compared with cow's milk. Meanwhile, rice milk, made from brown rice, contains slightly more proteins than other milk substitutes (Amagliani et al. 2017).

The typical processing procedure for commercial rice milk consists of several unit operations such as soaking in water, steam cooking, wet milling, filtration, sterilization, and homogenization, which are different from fermented alcoholic beverages (Aydar et al. 2020). The soaking process softens the rice kernel and reduces the cooking time. Steam cooking reduces the initial microbial load and enzyme activation. Steaming is also more effective than boiling in increasing total solids and protein yields. Wet milling is generally applied to the grinding of softened rice. The processing conditions of wet milling, such as the amount of water added, milling temperature, pH, milling type, and feed rate, are critical factors that affect the final product quality. Sterilization by heat treatment, similar to cow's milk processing, is applied to extend the shelf life of the final products. Finally, homogenization is applied to enhance the physical stability of the final product, and then the product is aseptic packed and stored at 4 °C. Xanthan gum is often used as an additive to rice milk to inhibit particle precipitation during milling. Alpha-amylase, which hydrolyzes starch granules, is also added to inhibit gelatinization during heat treatment. Fortification and enrichment of protein, vitamin, and mineral contents are usually applied to the final rice milk products.

10.5 Bran by-Products

Rice as a plant seed is a typical caryopsis species. Therefore, the morphological structure consists of pericarp, seed coat, germ, and endosperm. The pericarp typically has a light-brown color. Thus, seeds enclosed within the pericarp are called brown rice. In general, brown rice is polished to remove the pericarp, seed coat, germ, and part of the aleurone layer of the endosperm, because those are the main causes of the low sensory traits of cooked rice. Those removed are called "rice bran," which can be regarded as a by-product of rice processing. It is well known that rice bran contains many nutritional substances, such as proteins, oils, and vitamins, including substances with antioxidant potential. As one of such substances, rice bran oil is commercially produced from the bran by-product. The bran itself is also commercially applied to process Japanese pickles, called "nuka-zuke" ("nuka" means rice bran and "zuke" means pickling in Japanese). Rice bran is also used commercially as additional material to enhance the nutritional value of baked products such as cookies.

10.5.1 Rice Bran-Related Products

Rice bran is commercially used to enrich processed baked products such as bread, breakfast cereals, and cookies due to its nutritional ingredients, including dietary fiber. In Japan, rice bran is used as a base for traditional Japanese fermented pickles made of fresh vegetables, such as eggplant, Japanese radish, cabbage, cucumber, and Japanese ginger. The rice bran base (or rice bran bed) is usually mixed with salt, called "nuka-doko," and naturally fermented by bacterial flora, mainly lactic acid bacteria (Iuchi et al. 2012). The bacterial flora saccharifies and utilizes the starch of rice bran as a carbon source to grow themselves and then produces lactic acids by fermentation (Irisawa et al. 2014). Therefore, the resulting pickles, called "nuka-zuke" in Japan, taste sour and salty. The functional ingredients in rice bran, including vitamin E and oryzanol and perhaps transformed functional substances by fermentation, are absorbed in the pickles during the fermentation process. Thus, the "nuka-zuke" is regarded as healthy food, although it contains a lot of salt.

10.5.2 Rice Bran Oil

Rice bran contains approximately 10–20% crude bran oil. Crude rice bran oil contains 90–96% saponifiable lipids and approximately 4% unsaponifiable lipids. The principal component of the unsaponifiable fraction is gamma oryzanol, which is one of the primary antioxidants in rice bran oil (Sharif et al. 2014). Several methods are available for extracting oil substances from rice bran. Solvent extraction using hexane is the most common method used for the extraction of commercial rice bran oil (Ghosh 2007). After extraction, the solvent is recovered by distillation, and the rice bran is divided into crude bran oil and defatted residues. Crude bran oil contains triacylglycerols as its primary components and various impurities. Such impurities in crude bran oil can lead to high refining losses when a normal refining process is employed. Therefore, various refining processes must be incorporated into crude bran oil processing to remove impurities. The critical refining processes for producing commercially acceptable quality rice bran oil are filtration of fines, degumming, dewaxing (or winterization), deacidification (or refining), bleaching, and deodorization.

Filtration is preliminary processing prior to degumming to remove bran fines in crude bran oil. Solvent-extracted filtered bran oil contains phosphatides and other mucilaginous materials, which increases refining losses by emulsifying neutral oil. Degumming removes such phosphatide components and is thus needed as a refining process for crude bran oil. There are several methods for degumming crude bran oil, such as water degumming, acid degumming, super degumming, enzymatic degumming, simultaneous degumming/dewaxing, and membrane degumming. Water degumming is generally applied in the commercial refining of solvent-extracted bran oils. Before refining, crude bran oil becomes cloudy in cold areas because of its glutinous composition. Dewaxing separates such waxes at low temperatures. In the normal dewaxing process, oil is heated to approximately 90 °C to destroy any existing crystal nuclei. The oil is then cooled by stirring to appropriately 20 °C and allowed to mature. The resulting waxy fractions are then separated by filtration. Dewaxing can be applied to both oil and miscella phases. Degummed and dewaxed bran oil contains a comparatively large amount of free fatty acids that quickly deteriorate after production. To decrease the acidity of the bran oil, deacidification is applied to crude bran oil. There are a few applicable deacidification methods for rice bran oil, such as alkali and steam refining. Alkali refining is commonly employed as a deacidification process after degumming and dewaxing, depending on the oil usage. Bleaching aims to remove pigment content, oxidation products, and traces of metals in refined bran oil using adsorbent substances. Activated clay is employed as a common adsorbent. Finally, refined bran oil is subjected to deodorization. The treatment temperature of the usual deodorization process is between 200 and 220 °C, and the pressure is 6–10 mmHg, after which heat-labile pigments are destroyed. In steam refining, deodorization is performed simultaneously with deacidification. The treatment conditions used in steam refining are comparatively more intense than those used in the usual method. The deodorized oil is then cooled to approximately 50 °C. The cooled oil is passed through a polishing press to provide transparency. Resulting in these processes, commercially applicable purified rice bran oil is produced.

10.6 Summary and Future Direction

As a starchy carbohydrate source for the human diet, rice is one of the most important crops in the world. When polished, rice has a comparatively lower protein content than wheat, specifically no gluten; therefore, it can be considered a healthier food material against gluten-related disease. It means that there are a lot of commercial products using rice materials, although these have mainly been consumed in Asian countries. Generally, standard processing methods based on unit operations, such as heating, drying, and pulverization, are applied to raw food materials to enhance their shelf life and sensory quality. Such processing methods are also applied to rice to produce commercial products mentioned in this chapter.

Today, more healthy and functional attributes are required as a trend for the commercial food products. As a starch food, commercial rice products need to fit the requirement of social and medical backgrounds, such as modifying or adjusting starch digestibility, which relates to decreasing type 2 diabetes. Thus, the commercial processing method for rice materials has to focus on not only shelf life extension and sensory quality enhancement but also an increase in healthy and functional properties. Therefore, advanced processing technologies to enhance the healthy and functional properties of rice must be developed and applied to commercial rice products.

References

Amagliani L, O'Regan J, Kelly AL, O'Mahony JA (2017) The composition, extraction, functionality and applications of rice proteins: a review. Trends Food Sci Technol 64:1–12. https://doi.org/10.1016/j.tifs.2017.01.008. 673

Aydar EF, Tutuncu S, Ozcelik B (2020) Plant-based milk substitutes: bioactive compounds, conventional and novel processes, bioavailability studies, and health effects. J Funct Foods 70:103975. 676. https://doi.org/10.1016/j.jff.2020.103975

Bosmans GM, Lagrain B, Deleu LJ, Fierens E, Hills BP, Delcour JA (2012) Assignments of proton populations in dough and bread using NMR relaxometry of starch, gluten, and flour model systems. J Agric Food Chem 60:5461–5470. https://doi.org/10.1021/jf3008508

Budak NH, Aykin E, Seydim AC, Greene AK, Guzel-Seydim ZB (2014) Functional properties of vinegar. J Food Sci 79:R757–R764. https://doi.org/10.1111/1750-3841.12434

Chuang GC-C, Yeh A-I (2006) Rheological characteristics and texture attributes of glutinous rice cakes (mochi). J Food Eng 74:314–323. https://doi.org/10.1016/j.jfoodeng.2005.03.001

Din Y, Yang L, Xia Y, Wu Y, Zhou Y, Wang H (2018) Effects of frying on starch structure and digestibility of glutinous rice cakes. J Cereal Sci 83:196–203. https://doi.org/10.1016/j.jcs.2018.08.014

Feng CH-I (2012) The tale of sushi: history and regulations. Compr Rev Food Sci Food Saf 11:205–220. https://doi.org/10.1111/j.1541-4337.2011.00180.x

Ghosh M (2007) Review on recent trends in rice bran oil processing. J Am Oil Chem Soc 84:315–324. https://doi.org/10.1007/s11746-007-6951047-3

Han HM, Cho JH, Kang HW, Koh BK (2012) Rice varieties in relation to rice bread quality. J Sci Food Agric 92:1462–1467. https://doi.org/10.1002/jsfa.4727

Hera EDL, Martinez M, Gómez M (2013) Influence of flour particle size on quality of gluten-free rice bread. LWT Food Sci Technol 54:199–206. https://doi.org/10.1016/j.lwt.2013.04.019

Huang R, Pan X, Lv J, Zhong W, Yan F, Duan F, Jia L (2018) Effects of explosion puffing on the nutritional composition and digestibility of grains. Int J Food Prop 21:2193-2204. 705. https://doi.org/10.1080/10942912.2018.1514634

Inoue Y, Katsumata T, Watanabe H, Hayase F (2016) Mechanisms of d-amino acid formation during maturation of sweet rice wine (mirin). *Food Sci Technol Res* 22:679–686. https://doi.org/10.3136/fstr.22.679

Irisawa T, Tanaka N, Kitahara M, Sakamoto M, Ohkuma M, Okada S (2014) Lactobacillus furfuricola sp. nov., isolated from Nukadoko, rice bran paste for Japanese pickles. Int J Syst Evol Microbiol 64:2902–2906. https://doi.org/10.1099/ijs.0.063933-0

Iuchi A, Haruguchi S, Mongkolthanaruk W, Arima J, Nagase M, Khanh HQ, Ichiyanagi T, Yamaguchi T, Shimomura N, Aimi T (2012) Characterization of novel amylase from amylolytic lactic acid bacteria Pediococcus ethanolidurans isolated from Japanese pickles (nuka-zuke). Food Sci Technol Res 18:861–867. https://doi.org/10.3136/fstr.18.861

Kulawik P, Dordević D (2022) Sushi processing: microbiological hazards and the use of emerging technologies. Crit Rev Food Sci Nutr 62:1270–1283. https://doi.org/10.1080/10408398.2020.1840332

Li J, Han W, Xu J, Xiong S, Zhao S (2014) Comparison of morphological changes and in vitro starch digestibility of rice cooked by microwave and conductive heating. Starch/Stärke 66:549–557. https://doi.org/10.1002/star.201300208

Li N, Cai Z, Guo Y, Xu T, Qiao D, Zhang B, Zhao S, Huang Q, Niu M, Jia C, Lin L, Lin Q (2019) Hierarchical structure and slowly digestible features of rice starch following microwave cooking with storage. Food Chem 295:475–483. https://doi.org/10.1016/j.foodchem.2019.05.151

Li Y, Liang J-F, Yang M-Y, Chen J-Y, Han B-Z (2015) Traditional Chinese rice noodles: history, classification, and processing methods. Cereal Foods World 60:123–127. https://doi.org/10.1094/CFW-60-3-0515

Low YK, Effarizah ME, Cheng LH (2020) Factors influencing rice noodles qualities. Food Rev Intl 36:781–794. https://doi.org/10.1080/87559129.2019.1683747

Lu Z, Dockery CR, Crosby M, Chavarria K, Patterson B, Giedd M (2016) Antibacterial activities of wasabi against Escherichia coli O157:H7 and Staphylococcus aureus. Front Microbiol 7:1403. 739. https://doi.org/10.3389/fmicb.2016.01403

Okuda M (2019) Rice used for Japanese sake making. Biosci Biotechnol Biochem 83:1428–1441. 742. https://doi.org/10.1080/09168451.2019.1574552

Qin W, Lin Z, Wang A, Chen Z, He Y, Wang L, Liu L, Wang F, Tong LT (2021b) Influence of particle size on the properties of rice flour and quality of gluten-free rice bread. LWT 151: 112236. https://doi.org/10.1016/j.lwt.2021.112236

Qin W, Lin Z, Wang A, Xiao T, He Y, Chen Z, Wang L, Liu L, Wang F, Tong L-T (2021a) Influence of damaged starch on the properties of rice flour and quality attributes of gluten-free rice bread. J Cereal Sci 101:103296. https://doi.org/10.1016/j.jcs.2021.103296

Reed MO, Ai Y, Leutcher JL, Jane J-L (2013) Effects of cooking methods and starch structures on starch hydrolysis rates of rice. J Food Sci 78:H1076–H1081. https://doi.org/10.1111/1750-3841.12165

Ryu G-H (2011) Yukwa-A Traditional Korean puffed rice snack. Cereal Foods World 56:104–108. https://doi.org/10.1094/CFW-56-3-0104

Saha S, Roy A (2020) Puffed rice: a materialistic understanding of rice puffing and its associated changes in physicochemical and nutritional characteristics. J Food Process Eng 43:e13479. https://doi.org/10.1111/jfpe.13479

Sharif MK, Butt MS, Anjum FM, Khan SH (2014) Rice bran: a novel functional ingredient. Crit Rev Food Sci Nutr 54:807–816. https://doi.org/10.1080/10408398.2011.608586

Takei R, Maruyama K, Washio H, Watanabe T, Takahashi T (2019) Effects of 762 rheological properties of rice dough during manufacture of rice cracker on the quality of the end product. J Texture Stud 50:139–147. 764. https://doi.org/10.1111/jtxs.12380

Teramoto Y, Hano T, Ueda S (2000) Production and characteristics of an ancient form 766 of sake made with shitogi. J Inst Brew 106:95–100. https://doi.org/10.1002/j.2050-0416.2000.tb00044.x

Wongsa J, Uttapap D, Lamsal BP, Rungsardthong V (2016) Effect of puffing conditions on physical properties and rehydration characteristic of instant rice product. Int J Food Sci Technol 51:672–771. https://doi.org/10.1111/ijfs.13011

Wu T, Wang L, Li Y, Qian H, Liu L, Tong L, Zhou X, Wang L, Zhou S (2019) Effect of milling methods on the properties of rice flour and gluten-free rice bread. LWT 108:137–144. https://doi.org/10.1016/j.lwt.2019.03.050

Yamauchi H, Noda T, Matsuura-Endo C, Takigawa S, Saito K, Oda Y, Funatsuki W, Iriki N, Hashimoto N (2004) Bread-making quality of wheat/rice flour blends. Food Sci Technol Res 10: 247–253. https://doi.org/10.3136/fstr.10.247

Yoenyongbuddhagal S, Noomhorm A (2002a) Effect of raw material preparation on rice vermicelli quality. Starch/Stärke 54:534–539. https://doi.org/10.1002/1521-379X(200211)54:11%3C534:AID-782STAR534%3E3.0.CO;2-O

Yoenyongbuddhagal S, Noomhorm A (2002b) Effect of physicochemical properties of high-amylose Thai rice flours on vermicelli quality. Cereal Chem 79:481–485. https://doi.org/10.1094/CCHEM.2002.79.4.481

Chapter 11
Germinated Brown Rice

Sukanya Thuengtung

Abstract The germinated brown rice attracted more and more attention for its high GABA content, gamma-oryzanol, phenolic compounds, dietary fiber, free amino acids, vitamins B and E, and minerals. In this part, the germination process of brown rice is introduced, including the factors affecting the germination and the advanced technologies that aid germination. Then, the physicochemical properties of germinated rice (color, texture, functional, thermal, crystalline structure) are discussed. After that, the nutritional properties of germinated rice (macronutrients, micronutrients, GABA, phenolic compounds, etc.) are analyzed.

Keywords Rice · Germination · γ-aminobutyric acid · γ-oryzanol · Phenolic compounds

Abbreviations

DPPH	Hydroxyl radical scavenging activity
FRAP	Ferric reducing antioxidant power
GABA	Gamma-aminobutyric acid
GBR	Germinated brown rice
GI	Glycemic index
RC	Relative crystallinity
RDS	Digestible starch (RDS)
RS	Resistant starch
SDS	Slowly digestible starch
TPC	Total phenolic content
XRD	X-ray diffraction

S. Thuengtung (✉)
Graduate School of Horticulture, Chiba University, Chiba, Japan
e-mail: sukanya.thu@chiba-u.jp

J. Tian et al. (eds.), *Science of Rice Chemistry and Nutrition*,
https://doi.org/10.1007/978-981-99-3224-5_11

11.1 Introduction

Brown rice is known as unpolished whole-grain rice that merely removes hull or husk, and the outer bran layer is still retained. Several kinds of nutrients are mostly contained in the outer bran layer such as lipids, proteins, vitamins, and minerals as well as bioactive compounds (e.g., phenolic acids, flavonoids, and gamma-aminobutyric acid (GABA)) (Saleh et al. 2019). Thus, consumption of brown rice could serve potential health benefits such as lowering the risk of type 2 diabetes and antioxidant, neuroprotective, anticancer, and cardiovascular disease prevention (Babu et al. 2009). However, the low demand for brown rice consumption is due to its unpleasant taste and chewy texture as well as quick rancidity when stored for a long period. Consequently, the germination process received much attention for addressing this problem since germinated brown rice (GBR) can be cooked in an ordinary rice cooker, which has soft texture, is tastier, as well as requires less cooking time. Additionally, consumption of GBR is increasing in several Asian countries because it is much more nutritious and provides numerous beneficial health effects (Cho and Lim 2016; Patil and Khan 2011).

Brown rice seeds are subjected to soaking and subsequently germinated at a specified temperature and time to obtain the optimum length of sprout of 0.5–1 mm. After germination, the GBR is dried to reduce the moisture content and the appearance of the final product is similar to the ordinary brown rice (Patil and Khan 2011). The germination process can modify the physicochemical properties of GBR, including its digestibility. Numerous nutrients and bioactive substances, including their biological properties, will also be enhanced during germination. Specifically, GABA content, gamma-oryzanol, phenolic compounds, dietary fiber, free amino acids, vitamins B and E, and minerals are the major bioactive substances found in the GBR. The production and increment of these substances in GBR have been confirmed to exert several physiological effects on human health (Tables 11.1 and 11.2) (Cho and Lim 2016; Donkor et al. 2012). Studies on *an* in vivo model, feeding the pregnant mice with supplementation of a high-fat diet with GBR and GBR-derived gamma-oryzanol-rich extract for 8 weeks, showed the potential effects on lowering the glucose and insulin levels in their blood circulation (Adamu et al. 2017). Moreover, a high-fat diet fortified with GBR extracts could play important roles with anticancer and antioxidative properties and increasing the antioxidant enzyme activities in mice (Chung et al. 2018). Additionally, the GBR extract exhibited an anti-amyloidogenic effect via transcriptional regulation of amyloid precursor protein metabolism. Thus, GBR can serve as an effective product in preventing the risk of Alzheimer's disease and improving brain function (Azmi et al. 2015).

Table 11.1 Example of the in vivo physiological effects of germinated brown rice (GBR). Reprinted with permission from Cho and Lim (2016)

Physiological effects	Subjects	Dietary interventions	Main results
Anti-obesity	Obesity and hypercholesterolemia-induced male Sprague-Dawley rats	AIN-76 (lard 15%, Ch 0.5%); WR (50% diet), BR (50% diet), or GBR (50% diet) substitution in AIN-76	↓ Increased TG, TC, HDL-Ch, and total lipid (liver). ↑ Fecal excretion of Ch, lipid, and TG.
Hypotensive	Spontaneously hypertensive rates	AIN-76; BR (50% diet) or GBR (50% diet) substitution in AIN-76	↓ FER, SBP, TG.
Antidepression-like	Male ICR mice	AIN-93G; WR or GBR substitution for cornstarch in AIN-93G	↓ Immobility time (forced swimming test), escape failures (learned helplessness paradigm), 5-HIAA/5-HT (frontal cortex). ↑ 5-HT (frontal cortex).
Hypocholesterolemic	Hepatoma-bearing male Donryu rats	AIN-93G; WR, BR, or GBR substitution for cornstarch in AIN-93G	↓ TBARS, TC, (VLDL-Ch + LDL-Ch), AI. ↑ HDL-Ch, fecal steroid excretion, hepatic Ch 7-α-hydroxylase activity.
	Hypercholesterolemic male Sprague-Dawley rats with high-cholesterol diet	High-cholesterol diet; high-cholesterol+WR, BR, or GBR diet AIN-76; BR (81.1 g/100 diet) or GBR (83.2 g/100 diet) substitution for cornstarch in AIN-76	↓ TC, LDL-Ch. ↑ HDL-Ch. ↓ Liver, kidney, lung weights, plasma AI, TBARS, levels in plasma, liver, heart, kidney, lung, plasma GOT and GPT levels. ↑ Plasma HDL-Ch, fecal Ch excretion, bile acid excretion, hepatic antioxidant enzyme (SOD, CAT, GSH-Px) activities.

11.2 Germination Process of Brown Rice

GBR can be prepared by submerging brown rice in water at varying temperatures and time to induce slight germination at the embryo of seed grains. Water is drained after soaking, and rice is kept in moist conditions for further germination. The overall process of seed germination including biochemical process that occurs

Table 11.2 Example of in vivo physiological effects of germinated brown rice (GBR) (cont.). (drawn by the author)

Physiological effects	Subjects	Dietary interventions	Main results
Antidiabetic	Subject with type 2 diabetes	Cooked WR, GBR	↓ TC, TG, FBG, fructosamine level. ↑ HDL-Ch.
	Healthy subjects	Cooked WR, BR, GBR	↓ GI, IAUC-Glc120.
	Male OLETF rats	AIN-93G; WR or GBR substitution for cornstarch in AIN-93G	↓ Blood glucose levels, HbA1c, plasma PAL-1, TNF-α adipocytokine concentration. ↑ Adiponectin.
	Streptozotocin-induced diabetic male Wistar rats	AIN-93G; WR or GBR substitution for cornstarch in AIN-93G	↓ Blood glucose levels, plasma PAL-1, lipid peroxide concentration.
Learning and memory deficits	β-Amyloid peptide-treated male ICR mice	AIN-93G; WR or GBR substitution for cornstarch in AIN-93G	↓ Escape latencies (water maze test). ↑ Spontaneous alternation behavior (Y-maze test).
Anticancer	Azoxymethane-induced colon cancer male Sprague-Dawley rats	Commercial basal diet+2.5, 5.0, 10.0 g (rice)/kg (rat weight)	↓ Aberrant crypt foci (ACF), β-catenin expression, COX-2 expression which inhibit colon cancer.

AI atherogenic index ([total cholesterol]-[HDL cholesterol]), *AIN* American Institute of Nutrition, *BR* brown rice, *CAT* catalase, *Ch* cholesterol, *COX-2* cyclooxygenase 2, *FBG* fasting blood glucose, *FER* food efficiency ratio [bogy weight gain (g/week)/food intake (g/week)], *GBR* germinated brown rice, *GI* glycemic index, *GOT* glutamic oxaloacetic transaminase, *GPT* glutamic pyruvic transaminase, *GSH-Px* glutathione peroxidase, *HbA1c* hemoglobin A1c, *HDL-Ch* high-density lipoprotein cholesterol, *5-HIAA* 5-hydroxyindoleacetic acid, *5-HT* 5-hydroxytryptamine, *IAUC-Glc-120* incremental areas under the glucose curves over 120 min, *MDA* malondialdehyde, *OLETF* Otsuka Long-Evans Tokushima fatty rat, *PAL-1* type 1 plasminogen activator inhibitor, *SBP* systolic blood pressure, *SOD* superoxide dismutase, *TBARS* thiobarbituric acid-reactive substances, *TC* total cholesterol, *TG* triglyceride, *TNF-α* tumor necrosis factor-α, *VLDL* very-low-density lipoprotein, *WR* white rice

during germination phases is summarized in Fig. 11.1. Generally, the period of soaking time will be inversed with soaking temperature; a cooler temperature prolongs germination time. However, microorganisms can be produced and multiplied during soaking, affecting the fermentation and the undesirable odor of soaked water. Thus, water is recommended to change every 4–6 h to prevent fermentation and maintain consistent water temperature (Patil and Khan 2011). Additionally, the combination treatment of steaming for 20 min and then soaking in ethanol (70% v/v) for 3 min has been reported as an effective method to eliminate the microbial growth during soaking (Komatsuzaki et al. 2007).

The germination process enhances several nutrients in rice grains, especially GABA content and a much more beneficial health effect. However, the optimum condition to achieve the maximum GABA content and other nutrients is varied

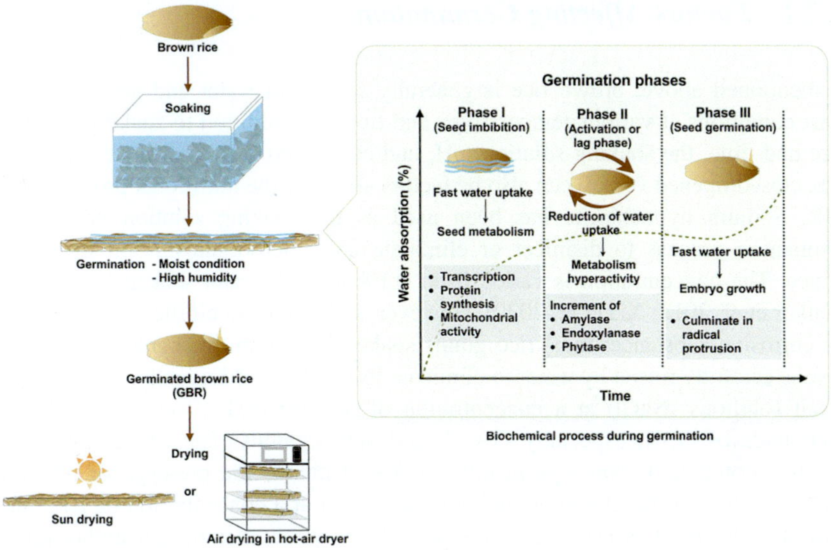

Fig. 11.1 The overall process of seed germination including biochemical process that occurred during germination phases. The scheme of process occurring during the germination phase was reproduced with an open-access license from do Espirito Santo Pereira et al. (2021)

depending on several factors. Soaking at 30–36 °C and germination at 35 °C or 36 °C for 21–96 h have been reported as the optimum conditions to promote the highest GABA content along with thiamine, total sugar content, free amino acid, and bioactive compounds in GBR (Chaijan and Panpipat 2020; Lin et al. 2015; Zhang et al. 2014). Moreover, some previous studies revealed that 24 h of germination at 35 °C could elevate the GABA content of GBR, approximately four times, compared to non-germinated rice (Jirapa et al. 2016). In addition to temperature and time, other conditions during the soaking and germination processes, for instance, closed vessel, neutral pH of the soaking solution, and darkness condition, can also be significant factors in promoting these bioactive compounds in the GBR (Chaiyasut et al. 2017; Lin et al. 2015).

Recently, several technologies have been applied to improve the germination of rice and aid the accumulation of nutrients in the GBR. For instance, the application of low-pressure plasma and nanopriming could increase the α-amylase activity, resulting in lifting of soluble sugar content for stimulation of germination and seedling vigor (Chen et al. 2016; Mahakham et al. 2017). Besides, techniques such as ultrasonic or ultrahigh pressure could also improve the texture and flavor of GBR (Ding et al. 2018a; Xia et al. 2017). Further details of these advanced technologies are mentioned below.

11.2.1 Factors Affecting Germination

As mentioned above, brown rice is generally soaked in water and germinated in moist conditions at various temperatures and times. In addition to soaking temperature and time, the soaking solution, pH, and conditions of treatments (e.g., light/dark, closed/opened vessel) can also be factors affecting the nutritional properties of GBR. Sodium hypochlorite has been used as the soaking solution before the germination process to disinfect or eliminate any microorganisms on the grain surface. The concentration is varied from 0.1% to 1.25% and soaked for 30 min (Donkor et al. 2012; Xia et al. 2020). However, sodium hypochlorite is a dangerous and corrosive substance; thus, rice grains soaked in sodium hypochlorite solution must be carefully rinsed by water to eliminate its residue. Additionally, rice soaking in salt solutions (NaCl) at a concentration of 50–200 mM affects the bioactive compounds of GBR, especially GABA and anthocyanins. The concentration of bioactive compounds was significantly increased due to the presence of 50 mM NaCl regardless of the duration, and applying NaCl at a concentration of 200 mM showed to be most effective to increase GABA, γ-tocotrienol, and anthocyanin content in GBR (Chaiyasut et al. 2017). The pH of the soaking solution is also a parameter affecting the bioactive compounds in the GBR. The previous study demonstrated that the highest level of GABA content was found in acidic pH conditions of the soaking solution, whereas the lowest level was found in the neutral and alkaline pH conditions of the soaking solution. They also found that the optimum germination time to maximize the GABA content was shortened by 6 h when rice was soaked in acidic soaking water (Zhang et al. 2014). These findings might justify that the type of soaking solution and different pH of soaking solution have the predominant effect on bioactive compounds in the GBR.

The surrounding conditions of the germination processes such as darkness and closed-vessel conditions could enhance the bioactive compounds and antioxidant activity of GBR, compared to the lightness condition. The GABA and tocotrienol content ranged from 12.37 to 44.11 mg/100-g and 0.42 to 0.73 mg/100 g, respectively, wherein the highest level of these compounds was found in brown rice germinated at 36 °C for 72 h in the darkness and closed-vessel conditions (Lin et al. 2015). Besides, brown rice germinated in the darkness with conditions of high humidity has increased starch and protein contents as well as arabinoxylan and GABA content (Donkor et al. 2012).

11.2.2 Advanced Technologies Aid Germination

Recently, several technologies have been applied to alter the germination process and improve the nutritional properties of the GBR. They can be applied both before and after the germination process.

Plasma Treatment

The plasma process, one of the physical type advanced technologies, applies an ionized gas that is flammable at low or atmospheric pressure to form a complex mixture of ions, molecules, atom, and UV photons. This process can be used in numerous plant seeds such as rice, wheat, legume, and vegetable seed and is suggested as a cost-effective and environmentally friendly method in terms of improving the germination rate, seed's vigor, and nutritional quality (Waskow et al. 2021; Zargarchi and Saremnezhad 2019). Plasma treatment greatly induces the water imbibition of brown rice during the germination period, which results from etching of the rice surface after plasma exposure. Consequently, the germination percentage and seedling length of brown rice are significantly elevated, wherein the exposure of low-pressure plasma at 3 kV for 10 min has been reported as the most effective. The percentage of germination and seedling length could be increased up to 62% and 37%, respectively, after 24 h of germination compared with the unexposed brown rice (Chen et al. 2016). Moreover, plasma application has been indicated to alter the internal cells of seeds by penetration of plasma energy through porous regions on the seed coat. This results in improved enzymatic activities, modified protein structure, and enhanced physiological properties such as α-amylase activity, fatty acids, GABA accumulation, and phenolic compounds and its antioxidant capacity, which the efficiency depends upon the intensity of plasma conditions cooperating with the germination period (Chen et al. 2016; Sookwong et al. 2014). Plasma processing has been indicated to impact the activity of saccharolytic enzymes of GBR as well that causes improved production of the free phenolic compounds (Sookwong et al. 2014). Nevertheless, an excessive period of germination for brown rice pre-treated by plasma could generate a reverse effect on phenolic compounds and antioxidant capacities. For instance, plasma-treated brown rice that was germinated for over 2 days showed a decreasing trend of total phenolic compounds and anthocyanin content, and thereby reduction in antioxidant activities (Sookwong et al. 2014; Yodpitak et al. 2019). Additionally, a key enzyme in phenolic biosynthesis such as phenylalanine ammonia lyase also has the possibility to get inactivated during plasma processing (Zargarchi and Saremnezhad 2019).

High-Pressure Technology

High pressure is an advanced nonthermal and additive-free food preservation technology that gained much attention for application in various fields related to biology. This technology is generally performed at room temperature as an environmentally friendly process, with low energy consumption and low contamination risk. Additionally, it can promote the potential effects on the functionality, extractability, and bioavailability of micronutrients and components in several types of foods (Barba et al. 2015; Huang et al. 2020; Huang et al. 2017). In particular, high hydrostatic pressure and ultrahigh pressure are considered the effective processes to enhance the functional components and micronutrients used in germinated rice products (M. Y. Kim et al. 2015; Xia et al. 2017). For instance, the GABA content, total arabinoxylan, γ-oryzanol content, and tricin 4′-*O*-(threo-β-guaiacylglyceryl) ether content in germinated grains can be improved by extending the duration of high hydrostatic pressure (M. Y. Kim et al. 2015). Moreover, the application of mild high hydrostatic pressure before the germination process improves the physicochemical properties of the GBR, such as maintaining the acceptable sprouting rate, color stability, prolonging the shelf life, and retention of some minerals. Simultaneously, it can reduce the digestible and resistant starch content, which means enhancing the enzymatic reaction and subsequently increasing starch digestibility (Xia and Li 2018). Besides, high-pressure treatment using the ultrahigh pressure at 300 MPa in the GBR has been reported as the optimum condition for mineral bio-accessibility such as manganese, magnesium, and phosphorus due to cell structural disruption of the treated GBR, which leads to the release of the intercellular components and the occurring of the synergetic effect between these components. In addition to mineral bio-accessibility, use of ultrahigh pressure at 300 MPa or above induces starch gelatinization in GBR, and this can be responsible for decreasing hardness, cohesiveness, resilience, and gumminess of GBR (Xia et al. 2017).

Ultrasonic

Ultrasonic or ultrasound treatment is a nonthermal process widely applied in several kinds of food for modifying physicochemical properties, preservation, and maintaining the quality of food products. Classification of this technique can be categorized by the range of ultrasonic waves as low-power (high frequency) diagnostic ultrasound (1–10 MHz), extended-range ultrasound (100 kHz to 1 MHz), and power ultrasound (16–100 kHz) (Bonto et al. 2021; Patist and Bates 2008). An ultrasonic bath system is generally applied in rice grains, by placing the rice grains in a beaker containing water before immersing in the ultrasonic bath. A high water level inside the bath must stabilize a beaker. However, rice grains can also be immersed directly in an ultrasonic bath. The treatment temperature and time are generally monitored (Bonto et al. 2021). Ultrasonic treatment has been revealed to soften the texture (Park and Han 2016), enhance nutritional properties (Xia et al. 2020), and change physicochemical properties (Dang et al. 2019) in brown rice. Additionally, the application of ultrasonic treatment and its impacts on GBR have also been reported in several studies. Recent studies found that the application of power ultrasound treatment at 28 and 35 kHz for 5 and 15 min, respectively, could improve the germination rate of GBR, which the values varied during soaking and germination times (Chatchavanthatri et al. 2020; Xia et al. 2020). Similarly, use of power ultrasound treatment at 25–28 kHz could enhance the accumulation of GABA content, riboflavin (vitamin B2), nutritional index of free amino acids, and bio-accessibility of some minerals. Simultaneously, the starch granule size and total starch content of treated GBR were reduced, whereas the starch hydrolysis, reducing sugar, and glucose content were increased (Ding et al. 2018a, b; Xia et al.

2020). Furthermore, this technique affected the rheological behavior, reduced the viscosity and energy use during the germination process, as well as improved the texture and flavor of GBR (Ding et al. 2018a). In vice versa, applying the ultrasonic treatment along with prolonging the germination time could erode the microstructure of the pericarp layer and subsequently interfere with pericarp pigments of treated rice, which might impact the color of the GBR (Ding et al. 2018a; Xia et al. 2020).

Nanopriming Technology

Nanopriming is a nanotechnology that referred to the combination between seed priming and nanoparticle treatment, mainly applied in aged seeds or seeds that should germinate. This technique uses the suspensions or nano-formulations as the media, which the nanoparticles may be taken up by crop seeds (do Espirito Santo Pereira et al. 2021). Several nanoparticles such as silver (Ag), silicon oxide (SiO_2), copper (Cu), and zinc oxide (ZnO) have been used in this technique for various crop seeds including rice (Mahakham et al. 2017; Rai-Kalal et al. 2021; Raja et al. 2019). The application of nanopriming technology can promote seed germination, growth, and quality and enhance tolerance to the environmental stresses (Song and He 2021). Moreover, previous literature has applied the nanopriming technique using phytosynthesized silver nanoparticles as the pre-germination process of rice. They implied that rice seed priming with phytosynthesized silver nanoparticle solutions at the concentration of 10 and 20 mg/L could stimulate starch metabolism (i.e., α-amylase activity) and water uptake by creating the seed coat nanopores and nanocatalyst activity. This resulted in improvement in the germination rate and seedling vigor, as well as increase in soluble sugar content in germinated rice seed (Mahakham et al. 2017). Nevertheless, the potential effects of nanopriming technology are based on various factors such as priming temperature, concentration and type of nanoparticles, and duration of priming (Song and He 2021).

11.3 Physicochemical Properties of Germinated Rice

11.3.1 Color Characteristic

Color is one of the major organoleptic characteristics of cereal-based products and influences consumer acceptance (Xia and Li 2018). The germination process showed the obviously decreased redness and yellowness of GBR, whereas lightness was slightly increased compared to the non-germinated sample. This could be related to biochemical changes during germination. The pigmented compounds such as phenolic acids and carotenoids, which mainly presented in the bran layer of brown rice, could be diffused into an aqueous environment during soaking and germination processes. This causes the colored change in the GBR (Chinma et al. 2015; Xia et al. 2020).

11.3.2 Texture Property

Most consumers prefer cooked rice that is soft and palatable; however, cooked brown rice presents high hardness, chewiness, and gumminess. Simultaneously, it has low adhesiveness, springiness, cohesiveness, and resilience (Ren et al. 2020). Accordingly, the germination process received much attention to address this problem. Germination has been indicated to improve the taste and texture of cooked rice, particularly soften the texture, enhance sweetness, increase adhesiveness and springiness, and improve resilience properties. This is probably due to the partial hydrolysis of starch, non-starch polysaccharides, and proteins during germination, which turn into sugars, oligosaccharides, and amino acids. This incident would enhance the bio-functional substances and improve the palatable texture of cooked GBR (Patil and Khan 2011; Ren et al. 2020). Other reasons are the formation of cracks on the kernels during germination and drying steps as well as the reduction of amylose content during germination, which is caused by the action of amylolytic enzymes. These phenomena resulted in the less cooking time of GBR, and the texture of cooked GBR provided much more cohesiveness and softness (Cho and Lim 2016).

Post-germination processes such as drying and storage conditions also influenced the texture and sensory qualities of the cooked GBR. When GBR was dried in a hot-air dryer at 130 °C, hardness of cooked GBR was decreased compared to the conventional drying method. Meanwhile, the hardness of GBR dried by superheated steam at 130 and 150 °C with superficial drying medium velocities of 3.5 and 5.7 m/s was observed to be similar to that of drying by conventional processes. The observation of different hardness of cooked GBR could be related to the different number of fissures on the GBR kernels when dried with different drying processes. The formation of a larger quantity of fissure could lead to more water absorption into the kernels during cooking and then allowed a larger amount of amylose leaching. In consequence, there was decrease in the hardness and stickiness of cooked GBR (Srisang et al. 2011). Additionally, increase in storage duration and temperature has been observed to induce the organization of the rice structure, resulting in increased hardness of cooked GBR. A previous study has compared two storage temperatures of GBR grains between temperatures of 4 and 37 °C for several months. They found that GBR grains stored at 37 °C, once cooked, showed higher hardness than the sample stored at 4 °C. The hardness of cooked GBR also increased with increasing storage period. The storage duration and temperature not only affected the texture, but also influenced the flavor and consumer acceptability of the cooked GBR. GBR stored at 4 °C for up to eight months presented better taste, flavor, and appearance than that stored at 37 °C. Hence, storage at low temperature would be recommended for the GBR to preserve its texture and palatability (Parnsahkorn and Langkapin 2013).

11.3.3 Functional Properties

The functional properties of GBR including foaming capacity, foam stability, bulk density, water absorption capacity, and swelling power are affected by several factors. The increase in foaming capacity and foam stability after the germination process is probably due to the increasing portion of solubilized proteins and reduction in the surface tension of molecules during germination (Singh et al. 2015). Meanwhile, alteration of dietary fiber, disintegration of polysaccharide molecules, and increase in the protein content of GBR are parameters effecting changes in its bulk density and increasing its water absorption capacity (Chinma et al. 2015). However, the swelling power of starch can reflect the interaction of water molecules and starch chains in the crystalline and amorphous regions. It is also associated with the degree of crystallinity of starch. Hence, changes in starch molecular structures and starch content during the germination process could affect the swelling power of GBR granules (Chinma et al. 2015; Musa et al. 2011).

11.3.4 Thermal Properties

The hydrogen bonds in starch granules would be disintegrated during germination by hydrolytic enzymes, leading to weakening of the intermolecular bonds in the crystalline domain of starch and shortening of the chain length of starch molecules. These phenomena cause reduction in the thermal properties of the GBR (Chungcharoen et al. 2014). The reduction in thermal properties can affect the decrement in the gelatinization temperature. Additionally, production of small dextrin and some reducing sugars during germination was observed as a parameter for reducing the enthalpy value of GBR (Chinma et al. 2015; Wu et al. 2013a). Furthermore, prolonging the duration of germination might cause more hydrolysis of hydrogen bonds and weaken the intermolecular bonding in the crystalline region, which subsequently reduces the enthalpy value of starch gelatinization of the GBR (Chungcharoen et al. 2014). In contrast, the study by Pinkaew et al. (2017) implied that increment in relative crystallinity and changes in amylopectin molecular properties during the germination process of brown rice contributed to an increase in starch gelatinization temperature and enthalpy value of GBR. Apart from gelatinization temperature and enthalpy, setback, referred to as the degree of reassociation between starch molecules, which is a determinant of starch retrogradation, was observed to be increased after germination. The setback viscosity may correspond to the breakdown viscosity of the samples since the GBR sample also presented higher breakdown viscosity than non-germinated sample. The amylose chain length and degree of amylose polymerization are factors reflecting setback viscosity as well (Musa et al. 2011). Vice versa, some previous studies found that higher starch degradation by α-amylase enzyme during germination decreased peak viscosity, breakdown, final viscosity, setback viscosity, and pasting temperature of GBR (Chinma et al. 2015).

11.3.5 Changes in the Crystalline Structure

Crystallinity depends on the degree of structural order of starch in the solid body, normally characterized as A-type, B-type, and C-type crystalline structures. The crystalline region contains the double helices of amylopectin, which is formed by two polyglycoside chains. The crystallinity of native starch granules typically presented approximately 14–45%, varying in the sources of starch (Dome et al. 2020). The types of crystalline structure and degree of crystallinity of starch are generally analyzed by X-ray diffraction (XRD) technique. Native rice starch is a type of thermally stable one, which is classified into an A-type crystalline structure. An A-type crystalline structure is known as highly susceptible to enzymatic hydrolysis than B-type and C-type structures. This might be attributed to B-type crystalline structure that is arranged in a double-helical hexagonally packaged structure, and C-type crystalline structure that is a mixture of A-type and B-type (Bertoft 2017; Sivakamasundari et al. 2021). Nevertheless, cooking or some processing could influence changes in the crystalline pattern of rice and consequently impact starch digestion of rice.

The crystalline pattern of rice is not affected by the germination process. Brown rice before and after germination showed a similar crystalline pattern of A-type structure, but the germination process has been found to slightly influence the changes in major reflections and percentage of relative crystallinity (RC). The RC values of Thai brown rice cultivars varied from 26.79% to 31.33% (Pinkaew et al. 2017), whereas the RC of the Malaysian brown rice showed 18.86% (Musa et al. 2011). RC values of these rice were slightly increased after germination as 27.75–35.91% and 19.13% in germinated Thai brown rice and germinated Malaysian brown rice, respectively. This could correspond to the changes in amylopectin molecular properties. Since the starch crystallinity related to its amylopectin content, the different amylopectin molecular properties might generate different RC and major reflections of the crystalline structure (Pinkaew et al. 2017). Although the germination process did not show an obvious effect on the crystalline pattern, the post-germination treatment by high temperature of drying generated V-type crystalline structure. The germination process followed by drying at extremely high temperature (130 and 150 °C) has been indicated to disrupt the A-type crystalline structure. Simultaneously, the formation of the amylose-lipid complexes, exhibited by V-type crystalline structure, could be generated. The degree of V-type crystallinity increased as the germination time, drying temperature, and degree of starch gelatinization increased (Chungcharoen et al. 2014). The formation of amylose-lipid complexes (V-type) is known to be enzymatic resistant, which retards the amylase penetration to starch granules. Thus, the occurrence of this formation complexes including its degree of crystallinity would reflect the behavior of starch digestion in the GBR.

11.4 Nutritional Properties of Germinated Rice

GBR is known as a cereal product that provides numerous health benefits such as antimutagenic activity, antioxidative properties, and reduction of the risk of chronic diseases. The germination process of rice seeds has been confirmed to enhance the production of functional nutritive compounds, including macronutrients, micronutrients (vitamins and minerals), GABA content, gamma-oryzanol content, dietary fiber, amino acids, and bioactive compounds (Lin et al. 2015; Sookwong et al. 2014). Changes in the nutritional compositions of GBR, including its digestibility, are summarized in Fig. 11.2.

11.4.1 Macronutrients

Carbohydrate

Carbohydrate is the major macronutrient in rice grain, consisting of starch and non-starch polysaccharides (i.e., dietary fiber). Germination could induce the complex biochemical process and reduce total starch, amylose, and amylopectin contents

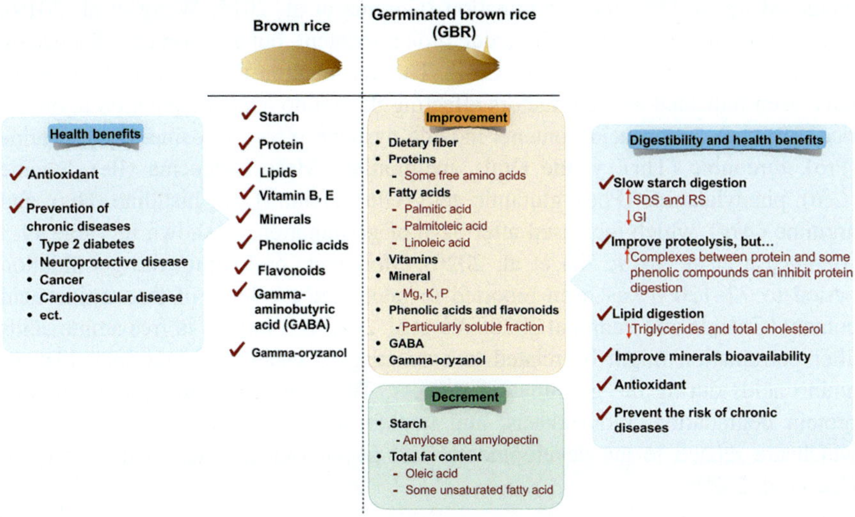

Fig. 11.2 Nutritional properties of brown rice and germinated brown rice (GBR). (Drawn by the author)

of the GBR. In particular, the amylose content is the main component of degradation during germination. Meanwhile, the reducing sugars in brown rice are outstandingly increased after germination (Chungcharoen et al. 2014). The accumulation of reducing sugars was found to be dependent upon the soaking conditions; brown rice soaked in low pH water for a long duration showed higher reducing sugar content. The optimum soaking condition to maximize the reducing sugar content has been observed to be at pH 3 of soaking water and soaking duration for 48 and 72 h, in germinated glutinous brown rice and non-glutinous brown rice, respectively (Wu et al. 2013b).

Apart from starch, dietary fibers are the edible carbohydrate polymers, also known as non-starch polysaccharides, that cannot be digested by digestive enzymes in the human small intestine. Dietary fibers can be classified by water solubility into soluble and insoluble dietary fibers. The dietary fibers in brown rice are found to be approximately 4–4.6 g/100 g, varying with rice cultivars. The amount of total dietary fibers could be increased to around 30 g/100 g after germination, and the soluble dietary fibers in GBR were increased by 61–79% (Jayadeep and Malleshi 2011; Weng et al. 2019). Nevertheless, the germination process under long-term atmospheric conditions decreases the insoluble dietary fibers. Thus, the germination process is recommended to terminate within a short period to prevent loss of insoluble dietary fiber (Cho and Lim 2016; Jayadeep and Malleshi 2011).

Protein

Protein is the second major macronutrient in rice. Brown rice contains 7.1–8.3 g/100 g of crude protein content (Saleh et al. 2019), and the value is significantly increased up to 25% after germination (Chinma et al. 2015; Weng et al. 2019). Germination not only affects the crude protein content, but also has an influence on the change in free amino acid of GBR. The germination time and rice genotypes have been indicated as main factors affecting the amino acid composition in brown rice. The free amino acid contents include cysteine (Cys), tyrosine (Tyr), proline (Pro), threonine (Thr), valine (Val), methionine (Met), isoleucine (Ile), leucine (Leu), phenylalanine (Phe), glutamic acid (Glu), lysine (Lys), histidine (His), and arginine (Arg), which increased after 32 h of germination, as shown in Table 11.3 (Benincasa et al. 2019; Xia et al. 2020). Moreover, prolonging the germination period to 72–120 h has been reported to maximize the level of free amino acid content (Chaijan and Panpipat 2020; Wu et al. 2013b). Changes in free amino acids after germination might be related to coexisting of catabolism and anabolism of amino acids during the germination process. They can act as the components of protein degradation, biosynthesis, and transformation of secondary metabolites, which are related to the developmental and physiological states of the seed cell (Xia et al. 2020).

Table 11.3 Individual free amino acid content of brown rice and germinated brown rice (GBR). Reprinted with permission from Xia et al. (2020)

	Concentration (mg/100 g)	
Free amino acids	Brown rice (no germination)	GBR for 32 h
Aspartate (Asp)	70.33 ± 2.54	38.55 ± 0.78
Serine (Ser)	38.70 ± 1.04	31.70 ± 0.75
Asparagine (Asn)	135.58 ± 3.83	26.16 ± 0.55
Glutamic acid (Glu)	99.89 ± 3.36	153.98 ± 3.24
Glutamine (Gln)	28.21 ± 0.74	26.20 ± 0.52
Glycine (Gly)	21.86 ± 0.62	16.37 ± 0.33
Alanine (Ala)	99.38 ± 2.93	80.04 ± 1.86
Cysteine (Cys)	1.35 ± 0.07	3.43 ± 0.13
Tyrosine (Tyr)	9.51 ± 0.23	21.63 ± 0.50
Proline (Pro)	16.59 ± 1.68	17.61 ± 0.38
Threonine (Thr)	13.00 ± 0.47	17.70 ± 0.36
Valine (Val)	12.95 ± 0.36	29.86 ± 0.74
Methionine (Met)	2.14 ± 0.08	7.85 ± 0.19
Isoleucine (Ile)	4.66 ± 0.14	12.92 ± 0.36
Leucine (Leu)	7.52 ± 0.18	23.39 ± 0.69
Phenylalanine (Phe)	4.64 ± 0.11	16.67 ± 0.45
Tryptophan (Trp)	18.54 ± 0.48	18.52 ± 0.38
Lysine (Lys)	16.38 ± 0.51	37.04 ± 0.86
Histidine (His)	12.65 ± 0.38	31.95 ± 0.71
Arginine (Arg)	50.10 ± 1.68	79.24 ± 2.17
Ornithine (Orn)	1.16 ± 0.03	1.16 ± 0.02

Data expressed as mean ± SD of three independent experiments

Lipids

Lipids or fat content, the third major macronutrient of brown rice, mainly exist in the outer bran layer of brown rice, accounting for 20% dry basis. They are responsible for rice qualities during processing and storage (Saleh et al. 2019). According to several studies, the germination process was found to reduce the fat content in brown rice, with approximately 55% reduction. This may be attributed to lipolysis and fat usage, which serve as an energy source for biochemical and physicochemical changes during the germination of brown rice (Chinma et al. 2015; Jayadeep and Malleshi 2011). Besides, the oleic acid content, an unsaturated fatty acid, that is abundantly found in brown rice has also been reported to decrease after 72 h of the germination process. Other fatty acids such as palmitic acid, palmitoleic acid, and linoleic acid were increased during the initial stage of germination, but their concentrations were also dramatically decreased after 72 h of germination (Wu et al. 2013b). This indicates that germination time plays a major role in the fatty acid content in GBR.

11.4.2 Micronutrients

Vitamins

Brown rice is well known to be a good source of vitamin B1 (thiamine), and vitamin B1 deficiency in people can cause beriberi disease and heart failure. Vitamin B1 can be improved by the germination process, especially by applying a long period of germination. The germination period of brown rice from 60 to 96 h showed notable enhancement in the concentration of vitamin B1. In particular, germination for 96 h was suggested as the optimum condition for yielding the highest vitamin B1 content in the GBR. Additionally, other germination conditions such as the intensity of light and the number of rinses can also be the important factors affecting the content of vitamin B1 (Chaijan and Panpipat 2020; Weng et al. 2019).

Vitamin E content is also one of the predominant vitamins in brown rice. Total vitamin E content in brown rice has been reported to be ranging from 1.15 to 3.02 mg/100 g, depending on rice genotypes. The concentration was slightly increased to 2.92–3.06 mg/100 g after germination for 1–3 days. The increase in vitamin E content of GBR might be due to the synthesis of tocopherols and tocotrienol content during germination as a seed growth accelerator, and thereby improvement of vitamin E bioactivity in GBR (H. Y. Kim et al. 2012; Yodpitak et al. 2019). The contents of tocotrienol and tocopherol in the GBR have been reported as 0.42–0.73 mg/100 g and 0.48–1.23 mg/100 g, respectively (Lin et al. 2015). However, the usual germination process could not much improve the vitamin E content in GB; thus, thus some advanced technology was used to alter the vitamin E content in GBR. For instance, application of cold plasma treatment on brown rice before germination has been revealed to significantly enhance the total vitamin E content in GBR; the value increased to 5.97–6.83 mg/100 g, varying with rice cultivars (Yodpitak et al. 2019).

Minerals

Several minerals in brown rice could be improved after germination such as magnesium, potassium, and phosphorus, whereas iron, zinc, sodium, and calcium remained unchanged. Changes in the mineral content of the GBR have been confirmed to be associated with phytic acid content. Phytic acid acts as a mineral inhibitor, and its structure has the ability to strongly chelate with cations of minerals, especially divalent cations, and consequently minerals could form insoluble salts (Chinma et al. 2015). Hence, the reduction in phytic acid content could be a factor to improve the mineral contents in the GBR. In addition to phytic acid, hydrothermal treatment reduced mineral contents in the GBR. This could be because hydrothermal treatment could accelerate the diffusion of minerals into the extracellular medium during soaking and/or germination processes (Xia and Li 2018).

11.4.3 Gamma-Aminobutyric Acid (GABA)

Gamma-aminobutyric acid (GABA) is a nonprotein amino acid that is naturally found in foods. It is widely distributed along with eukaryotes and prokaryotes, and it is also known as the predominant inhibitory neurotransmitters in the sympathetic nervous system. Consumption of foods enriched in GABA content could help reduce blood pressure and cholesterol; improve the kidney, liver, and cardiovascular functions; inhibit cancerous growth; and prevent Alzheimer's disease (Ren et al. 2020). GABA content is influenced by various factors such as harvesting period, soaking and germination conditions, and rice genotypes. Rice harvested in the early season has been suggested to have more GABA content than the rice harvested in semilate and late seasons, wherein indica rice contains slightly higher GABA content than japonica rice (Wu et al. 2013b). Considering treatment/process conditions, soaking can cause hypoxia because the available oxygen in water for the grains is insufficient, which could help to accumulate the GABA in plant tissues. However, excessive high temperature and time for soaking would also reduce the GABA content in grains (Lin et al. 2015; Weng et al. 2019). Thus, optimum soaking temperature and time are required; the soaking condition at 30 °C for 3 h, followed by germination at around 35 °C for 21–96 h, has been reported as a suitable condition to maximize the GABA content in the GBR (Benincasa et al. 2019; Chaijan and Panpipat 2020; Patil and Khan 2011; Zhang et al. 2014). The levels of GABA content in GBR could be improved up to 12-fold as the germination time increases, compared to the non-germinated one (Benincasa et al. 2019; H. Y. Kim et al. 2012).

The increment in the activity of glutamate decarboxylase (GAD) along with the reduction of glutamate content during germination can also be an important factor influencing the accumulation of GABA (Xia et al. 2020). The free glutamic acid, which is a substrate of GABA synthesis, could be released during soaking and germination processes, leading to activation of the GAD activity by converting the glutamate to GABA (Chaijan and Panpipat 2020; Lin et al. 2015). Additionally, glutamic acid and chitosan were indicated to possess a synergistic effect on GABA synthesis. Brown rice was suggested to germinate in glutamic acid and chitosan mixture solutions at a concentration of 50 ppm chitosan in 5 mM glutamic acid. According to this condition, GABA content in brown rice could be increased by 13-fold after germination compared with non-germinated brown rice (Oh 2003). Apart from the GAD activity, the activity of α-amylase to disintegrate high-molecular-weight polymers during germination results in the formation of bio-functional substances and thereby improvement of GABA content (Chungcharoen et al. 2014).

11.4.4 Gamma-Oryzanol

Gamma-oryzanol is a component existing in the primary plant cell walls, consisting of at least ten compounds with ester bonds between trans-ferulic acid and triterpenes. Among these compounds, cycloartenyl ferulate, 24-methylene cycloartanyl ferulate, and campestanyl ferulate are the predominant constituents of gamma-oryzanol in the GBR, accounting for 90%, which the accumulation during germination is varied in rice genotypes (Cho and Lim 2016; Wu et al. 2013b). Because of the numerous nutritional health effects of gamma-oryzanol (e.g., antioxidant and hypolipidemic properties, and help in the balance of the nervous system), several processes or treatments are developed to achieve the highest amount of gamma-oryzanol. Germination is a simple process to accumulate the gamma-oryzanol content. Total gamma-oryzanol content is generally found in brown rice, accounting for approximately 4–6 mg/100 g, which could be enhanced by 1.2- to 1.5-fold after 24 h of the germination process (H. Y. Kim et al. 2012; Yodpitak et al. 2019). However, the germination process at 36 °C for 72 h was indicated as the optimum condition for maximizing the accumulation of gamma-oryzanol (Lin et al. 2015).

11.4.5 Phenolic Compounds

Phenolic compounds are generally known as a plant secondary metabolite, containing one or more aromatic rings that bind with one or more hydroxyl groups. Phenolic compounds can act as antioxidants against free radical and ultraviolet radiation (Shao and Bao 2015). Total phenolic content (TPC) in brown rice was reported as 3.56–7.86 mg gallic acid equivalent (GAE)/100 g, varying among different rice cultivars. The germination process could stimulate the biosynthesis of phenolic compounds and consequently increase the concentration of TPC in GBR (Chinma et al. 2015; Yodpitak et al. 2019). Germination time is a major factor affecting the increment of TPC, which the concentration could be increased with the increasing of germination time. Specifically, the soluble phenolic compounds would be more produced because starch is hydrolyzed by saccharolytic enzymes during germination, hence greater production of a soluble fraction of phenolic compounds from hydroxycinnamate sucrose esters (Sookwong et al. 2014; Tian et al. 2004). Apart from the period of germination, the increase in the concentration of phenolic compounds in the GBR can also be related to the temperature of soaking and germination processes, storage conditions, and rice cultivars (Yodpitak et al. 2019).

Phenolic acids are the major phenolic compounds contained in brown rice; particularly the soluble or free fraction of phenolic acids were increased by 58–87% after germination. The free phenolic compounds were notably increased since 17 h of germination and then gradually increased by periods of germination. The germination condition in the darkness at 20 °C for 30 h has been reported as an optimum condition to maximize the free phenolic acids in germinated Chinese brown rice (Ti et al. 2014). Meanwhile, a previous study on germinated Ecuadorian brown rice found that the phenolic contents would reach the highest concentration when rice was germinated at 28 °C for 96 h (Cáceres et al. 2014a, b). The insoluble or bound fraction of phenolic acids was gradually increased during periods of germination in a range of 23–45%, which the increasing ratio is lower than that of the free fraction. This could be related to the large proportion of bound phenolics being covalently conjugated to cell wall components through ester bonds, resulting in difficulty in extraction (Ti et al. 2014). The optimum condition to maximize the level of bound phenolic contents was observed after 48 h of the germination process. The increment during prolonged germination time may be associated with polymerization and oxidation reactions of phenolic acids and production of enzymes to break down the cell wall and synthesize the phenolic acids (Donkor et al. 2012; Ti et al. 2014).

The phenolic acids that are mostly found in brown rice are composed of ferulic acid, *p*-coumaric acid, protocatechuic acid, caffeic acid, sinapic acid, and syringic acid (Shao and Bao 2015). Ferulic and *p*-coumaric acids are the predominant phenolic acids in brown rice, which mainly existed in the bound fraction. The increment of ferulic acid was found to be consistent with the level of GABA content. The levels of both ferulic and *p*-coumaric acids could be increased by 3.4- and 2.8-fold, respectively, after 48 h of germination compared to non-germinated brown rice (Ren et al. 2020; Ti et al. 2014). The concentrations of syringic and sinapic acids in the GBR, which were detected in both free and bound fractions, could also be enhanced by around six- and tenfold, respectively (Cho and Lim 2018; Ravichanthiran et al. 2018). Nevertheless, chlorogenic acid remained at a similar level as non-germinated rice (Cho and Lim 2018). Other phenolic acids that mainly existed in the free fractions such as protocatechuic and caffeic acids were found to be increased during the germination process (Ti et al. 2014).

In addition to phenolic acids, flavonoids can be perceived as important antioxidant phenolics in brown rice, which is mainly present in the seed coat, embryo, and aleurone layer. The antioxidant activity of flavonoids is generated by phenolic hydroxyls (Ravichanthiran et al. 2018). Flavones, a subclass of flavonoids, are the soluble fraction of flavonoids that are mostly found in brown rice, accounting for 54% of the flavonoid signal distribution. Another soluble fraction of flavonoids in rice is tricin and its conjugated derivatives, which present as 24% of the flavonoid signal distribution. However, the concentration of flavones is absolutely stable during the germination process (Galland et al. 2014; Shao and Bao 2019). As previous research has shown, the soluble fraction of flavonoids in brown rice was gradually increased from 61.1 to 75.4 mg catechin equivalent (CE)/100 g, and the insoluble fraction was slightly increased from 63.9 to 79 mg CE/100 g after 48 h of germination (Ti et al. 2014).

11.4.6 Antioxidant Activity

Numerous bioactive compounds present in brown rice grain show a potential effect as an antioxidant. The enhancement of antioxidant activity could prevent the oxidative stress related to chronic diseases. Thus, germination is one of the potential techniques for the improvement of antioxidants by promoting their biochemical transformation (Xia et al. 2020). Temperature and time of germination were indicated as key factors affecting the change in the antioxidant activity in GBR. The levels of antioxidant activity in brown rice could be elevated by two- and fourfold after rice was germinated for 48 and 96 h, respectively. Moreover, rice germinated at higher temperature (34 °C) showed higher antioxidant activity than that germinated at lower temperature (28 °C) (Cáceres et al. 2014a, b). As the germination temperature and time increased, the accumulation of phenolic compounds could also be induced more. Higher accumulation of these compounds would lead to peroxyl-scavenging activity, resulting in high antioxidant activity. Furthermore, the hydrolytic enzyme might generate a soluble fraction of phenolic compounds during germination, consequently with a high potential effect of antioxidant activity (Tian et al. 2004). Rice cultivars also enhance the capacity of antioxidant activity of brown rice after germination by approximately 13–75%; the concentration was varied between glutinous and non-glutinous rice cultivars (Yodpitak et al. 2019).

Antioxidant activity can be analyzed by several methods based on the interaction between antioxidant compounds and reagents to generate electron or hydrogen radical donation. Among these methods, the ferric reducing antioxidant power (FRAP) and DPPH assay (to measure hydroxyl radical scavenging activity) are generally used to investigate the antioxidant activity in various plant materials (M. Kaur et al. 2017). The antioxidant activity of GBR by the FRAP assay showed approximately 40–60%, whereas the DPPH assay showed approximately 42–85%, and the antioxidant ability varied according to different germination conditions and rice genotypes (Lin et al. 2015; Ti et al. 2014). Moreover, a recent study revealed that brown rice germinated at 37 °C in the darkness for 32 h possessed the antioxidant ability as scavenging activity against DPPH radicals, wherein the values of antioxidant activity by the DPPH assay were 41.01% higher than the values by the FRAP assay (Xia et al. 2020). Accordingly, the antioxidant activity of GBR extracts is associated with electron transfer, for stabilizing or terminating free radical reactions. The ability of antioxidants also depended on the chemical structure of substrates that reacted with reagents (M. Kaur et al. 2017).

11.4.7 Digestive Properties

Starch Digestibility

Starch is a major carbohydrate component in rice, contained in the endosperm fraction, which accounts for approximately 90%. The digestibility of starch can be classified into rapidly digestible starch (RDS), slowly digestible starch (SDS), and enzyme-resistant starch (RS). Rice starch digestibility is dependent on both intrinsic and extrinsic factors (Sivakamasundari et al. 2021; Toutounji et al. 2019). Starch digestibility can also be a parameter indicating the glycemic index (GI) of rice. GI of brown rice has been found in varying amounts, ranging from approximately 57 to 75, depending on cultivars and treatment processes (Rattanamechaiskul et al. 2014). Several studies have stated that germination has a potential effect on the reduction of postprandial blood glucose levels due to the inhibition of acylated sterol glucoside, α-amylase, and α-glucosidase activities. Thus, consumption of GBR might help control the level of postprandial blood glucose without increasing insulin secretion and reducing the level of GI (Donkor et al. 2012; B. Kaur et al. 2016; Patil and Khan 2011). The study by Xu et al. (2012) revealed that the proportion of SDS fraction in the GBR increased by 14%, which positively corresponded to the relative portion of the starch crystalline structure in the GBR. This formation structure could retard the accessibility of pancreatic amylase enzyme, resulting in a slower rate of starch hydrolysis in the gastrointestinal tract. Additionally, modification of the starch microstructure caused by germination and post-germination treatments (e.g., hydrothermal treatment and drying method) would slow the starch digestion rate of GBR, by increasing the SDS and RS levels (Li et al. 2017; Saleh et al. 2019). On the other hand, some previous research found that amylolytic enzyme was activated during germination and facilitated the liberating of soluble starch. This subsequently affected the increased portion of the readily digestible starch. For instance, brown rice germinated at 30 °C for 48 h and then cooked by boiling process showed 20–30% higher RDS than non-germinated rice (Chaijan and Panpipat 2020; Lemmens et al. 2019).

Protein Digestibility

The germination process could increase proteolysis, which the concentration would be increased as the germination temperature and time increased. Proteolysis is the hydrolysis of peptide bonds, which is referred to as the breakdown of proteins into smaller peptides and/or individual amino acid residues (Raju 2019). The digestibility of proteins in GBR has been observed as approximately 48–59%. Improvement in protein digestion was caused by increasing protease activity during germination, which facilitates various peptides and free amino acids released from germinated grains (Cáceres et al. 2014a, b; Lekjing and Venkatachalam 2020). Moreover, high protein digestibility may improve the amounts of bioactive peptides, which exert numerous positive effects on human health (Cáceres et al. 2014a, b). Nevertheless, protein digestibility is also related to the presence of soluble phenolic compounds, including tannins, and these phenolic compounds can form an insoluble complex with proteins including proteolytic enzymes and consequently decrease the digestibility of proteins (Lemmens et al. 2019). Thus, increase in soluble phenolic compounds during germination, as mentioned in the above section, might show an adverse effect on protein digestion in GBR.

Lipid Digestibility

The digestibility of lipids is generally based on lipase activity, and this enzymatic activity could be generated with the germination process. Specifically, increase in germination temperature accelerates the hydrolysis of lipid contents into glycerol and free fatty acids (Lemmens et al. 2019). Turning of lipids into free fatty acid composition after digestion can be an indicator of rice stability and rancidity, and oxidation susceptibility (Xia and Li 2018). However, in the clinical studies, the consumption of GBR for 1 month showed a prominent effect on lowering the postprandial plasma glucose, fasting plasma glucose, triglycerides, total cholesterol, high-density lipoprotein cholesterol, and low-density lipoprotein cholesterol in metabolic syndrome patients (Ren et al. 2020). Besides, intake of GBR for 4 months has proved that blood concentrations of fasting plasma glucose, triglycerides, total cholesterol, and fructosamine in type 2 diabetic patients were improved (Bui et al. 2014). According to these findings, consecutive consumption of GBR as a meal helps improve glycolipid metabolism and lipid level as well as control the body weight and blood pressure.

Mineral Bioavailability

The impact of germination on the mineral bioavailability of brown rice is indeterminate. Nevertheless, the level of phytic acid content has been affirmed as an antinutritional factor, which has showed a negative effect on the bioavailability of minerals (Benincasa et al. 2019). An increment in phytase activity during the germination process could reduce the level of phytic acid in the GBR, resulting in improvement in mineral bioavailability (Chinma et al. 2015). Besides, the complexation between minerals and other compositions in the food matrix can also be a parameter affecting the bioavailability of minerals. A previous study indicated that the formation of zinc complexations during the germination process could also be responsible for enhancing the level of in vitro soluble zinc (Wu et al. 2013b).

11.5 Summary and Future Direction

Germination is a simple process for producing GBR, and it can be both household and commercial production. The germination conditions, especially the duration of germination, could affect changes in the characteristics and nutritive values of GBR. Several advanced technologies have been applied to contribute to germination process and have showed the effectiveness in enhancing the qualities of the GBR. Other than germination time and temperature, the potential effects of the germination process can also be related to soaking solution and pH as well as the light conditions during germination (i.e., darkness or lightness). After germination, GBR has been proved to be a good source of numerous nutrients and bioactive substances, which provide positive immeasurable health impacts, particularly the prevention of chronic diseases. Additionally, GBR is inexpensive, and it can be cooked and consumed as staple food being tastier than ordinary brown rice.

Recently, a healthy trend has received much more interest. Based on the beneficial health effects, GBR can be used as an ingredient for the development of innovative food products. Simultaneously, the public should be more educated about the information-related health benefits of the innovative food products fortified with GBR through, e.g., social media and academy. Furthermore, collaboration between government organizations, research institutes, and food industries is essential, for examination and development of GBR products to achieve the maximum nutritive value. Various types of GBR products should be produced and developed to meet the requirement of consumers, which could encourage more consumption of GBR.

References

Adamu HA, Imam MU, Ooi D-J, Esa NM, Rosli R, Ismail M (2017) In utero exposure to germinated brown rice and its oryzanol-rich extract attenuated high fat diet-induced insulin resistance in F1 generation of rats. BMC Complement Altern Med 17(1):67. https://doi.org/10.1186/s12906-017-1571-0

Azmi NH, Ismail M, Ismail N, Imam MU, Alitheen NBM, Abdullah MA (2015) Germinated brown rice alters Aβ(1-42) aggregation and modulates Alzheimer's disease-related genes in differentiated human SH-SY5Y cells. Evid Based Complement Alternat Med 2015:153684. https://doi.org/10.1155/2015/153684

Babu PD, Subhasree R, Bhakyaraj R, Vidhyalakshmi R (2009) Brown rice-beyond the color reviving a lost health food-a review. Magnesium 187:13.10

Barba FJ, Terefe NS, Buckow R, Knorr D, Orlien V (2015) New opportunities and perspectives of high pressure treatment to improve health and safety attributes of foods. A review. Food Res Int 77:725–742. https://doi.org/10.1016/j.foodres.2015.05.015

Benincasa P, Falcinelli B, Lutts S, Stagnari F, Galieni A (2019) Sprouted grains: a comprehensive review. Nutrients 11(2):421. https://doi.org/10.3390/nu11020421

Bertoft E (2017) Understanding starch structure: recent progress. Agronomy 7(3):56

Bonto AP, Tiozon RN, Sreenivasulu N, Camacho DH (2021) Impact of ultrasonic treatment on rice starch and grain functional properties: a review. Ultrason Sonochem 71:105383. https://doi.org/10.1016/j.ultsonch.2020.105383

Bui TN, Le TH, Nguyen DH, Tran QB, Nguyen TL, Le DT et al (2014) Pre-germinated brown rice reduced both blood glucose concentration and body weight in Vietnamese women with impaired glucose tolerance. J Nutr Sci Vitaminol 60(3):183–187. https://doi.org/10.3177/jnsv.60.183

Cáceres PJ, Martínez-Villaluenga C, Amigo L, Frias J (2014a) Assessment on proximate composition, dietary fiber, phytic acid and protein hydrolysis of germinated Ecuadorian brown rice. Plant Foods Hum Nutr 69(3):261–267. https://doi.org/10.1007/s11130-014-0433-x

Cáceres PJ, Martínez-Villaluenga C, Amigo L, Frias J (2014b) Maximising the phytochemical content and antioxidant activity of Ecuadorian brown rice sprouts through optimal germination conditions. Food Chem 152:407–414. https://doi.org/10.1016/j.foodchem.2013.11.156

Chaijan M, Panpipat W (2020) Nutritional composition and bioactivity of germinated Thai indigenous rice extracts: a feasibility study. PLoS One 15(8):e0237844. https://doi.org/10.1371/journal.pone.0237844

Chaiyasut C, Sivamaruthi BS, Pengkumsri N, Saelee M, Kesika P, Sirilun S et al (2017) Optimization of conditions to achieve high content of gamma amino butyric acid in germinated black rice, and changes in bioactivities. Food Sci Technol 37:83–93

Chatchavanthatri N, Junyusen T, Moolkaew P, Arjharn W, Junyusen P (2020) Effect of soaking and sprouting treatment on germination rate of paddy. E3S Web Conf 187:04016

Chen HH, Chang HC, Chen YK, Hung CL, Lin SY, Chen YS (2016) An improved process for high nutrition of germinated brown rice production: low-pressure plasma. Food Chem 191:120–127. https://doi.org/10.1016/j.foodchem.2015.01.083

Chinma CE, Anuonye JC, Simon OC, Ohiare RO, Danbaba N (2015) Effect of germination on the physicochemical and antioxidant characteristics of rice flour from three rice varieties from Nigeria. Food Chem 185:454–458. https://doi.org/10.1016/j.foodchem.2015.04.010

Cho D-H, Lim S-T (2016) Germinated brown rice and its bio-functional compounds. Food Chem 196:259–271. https://doi.org/10.1016/j.foodchem.2015.09.025

Cho D-H, Lim S-T (2018) Changes in phenolic acid composition and associated enzyme activity in shoot and kernel fractions of brown rice during germination. Food Chem 256:163–170. https://doi.org/10.1016/j.foodchem.2018.02.040

Chung SI, Lee SC, Yi SJ, Kang MY (2018) Antioxidative and antiproliferative activities of ethanol extracts from pigmented giant embryo rice (*Oryza sativa* L. cv. Keunnunjami) before and after germination. Nutr Res Pract 12(5):365–370. https://doi.org/10.4162/nrp.2018.12.5.365

Chungcharoen T, Prachayawarakorn S, Tungtrakul P, Soponronnarit S (2014) Effects of germination process and drying temperature on gamma-aminobutyric acid (GABA) and starch digestibility of germinated brown rice. Dry Technol 32(6):742–753. https://doi.org/10.1080/07373937.2013.879387

Dang LTK, Therdthai N, Ratphitagsanti W (2019) Effects of ultrasonic and enzymatic treatment on physical and chemical properties of brown rice. J Food Process Eng 42(3):e13016. https://doi.org/10.1111/jfpe.13016

Ding J, Hou GG, Dong M, Xiong S, Zhao S, Feng H (2018a) Physicochemical properties of germinated dehulled rice flour and energy requirement in germination as affected by ultrasound treatment. Ultrason Sonochem 41:484–491. https://doi.org/10.1016/j.ultsonch.2017.10.010

Ding J, Ulanov AV, Dong M, Yang T, Nemzer BV, Xiong S et al (2018b) Enhancement of gamma-aminobutyric acid (GABA) and other health-related metabolites in germinated red rice (*Oryza sativa* L.) by ultrasonication. Ultrason Sonochem 40:791–797. https://doi.org/10.1016/j.ultsonch.2017.08.029

Dome K, Podgorbunskikh E, Bychkov A, Lomovsky O (2020) Changes in the crystallinity degree of starch having different types of crystal structure after mechanical pretreatment. Polymers 12(3):641. https://doi.org/10.3390/polym12030641

Donkor ON, Stojanovska L, Ginn P, Ashton J, Vasiljevic T (2012) Germinated grains—sources of bioactive compounds. Food Chem 135(3):950–959. https://doi.org/10.1016/j.foodchem.2012.05.058

do Espirito Santo Pereira A, Caixeta Oliveira H, Fernandes Fraceto L, Santaella C (2021) Nanotechnology potential in seed priming for sustainable agriculture. Nanomaterials (Basel, Switzerland) 11(2):267. https://doi.org/10.3390/nano11020267

Galland M, Boutet-Mercey S, Lounifi I, Godin B, Balzergue S, Grandjean O et al (2014) Compartmentation and dynamics of flavone metabolism in dry and germinated rice seeds. Plant Cell Physiol 55(9):1646–1659. https://doi.org/10.1093/pcp/pcu095

Huang H-W, Hsu C-P, Wang C-Y (2020) Healthy expectations of high hydrostatic pressure treatment in food processing industry. J Food Drug Anal 28(1):1–13. https://doi.org/10.1016/j.jfda.2019.10.002

Huang H-W, Wu S-J, Lu J-K, Shyu Y-T, Wang C-Y (2017) Current status and future trends of high-pressure processing in food industry. Food Control 72:1–8. https://doi.org/10.1016/j.foodcont.2016.07.019

Jayadeep A, Malleshi NG (2011) Nutrients, composition of tocotrienols, tocopherols, and γ-oryzanol, and antioxidant activity in brown rice before and after biotransformation. CyTA J Food 9(1):82–87. https://doi.org/10.1080/19476331003686866

Jirapa K, Jarae Y, Phanee R, Jirasak K (2016) Changes of bioactive components in germinated paddy rice (*Oryza sativa* L.). Int Food Res J 23(1):229–236

Kaur B, Ranawana V, Henry J (2016) The glycemic index of rice and rice products: a review, and table of GI values. Crit Rev Food Sci Nutr 56(2):215–236. https://doi.org/10.1080/10408398.2012.717976

Kaur M, Asthir B, Mahajan G (2017) Variation in antioxidants, bioactive compounds and antioxidant capacity in germinated and ungerminated grains of ten rice cultivars. Rice Sci 24(6):349–359. https://doi.org/10.1016/j.rsci.2017.08.002

Kim HY, Hwang IG, Kim TM, Woo KS, Park DS, Kim JH et al (2012) Chemical and functional components in different parts of rough rice (*Oryza sativa* L.) before and after germination. Food Chem 134(1):288–293. https://doi.org/10.1016/j.foodchem.2012.02.138

Kim MY, Lee SH, Jang GY, Park HJ, Li M, Kim S et al (2015) Effects of high hydrostatic pressure treatment on the enhancement of functional components of germinated rough rice (*Oryza sativa* L.). Food Chem 166:86–92. https://doi.org/10.1016/j.foodchem.2014.05.150

Komatsuzaki N, Tsukahara K, Toyoshima H, Suzuki T, Shimizu N, Kimura T (2007) Effect of soaking and gaseous treatment on GABA content in germinated brown rice. J Food Eng 78(2):556–560. https://doi.org/10.1016/j.jfoodeng.2005.10.036

Lekjing S, Venkatachalam K (2020) Effects of germination time and kilning temperature on the malting characteristics, biochemical and structural properties of HomChaiya rice. RSC Adv 10(28):16254–16265. https://doi.org/10.1039/D0RA01165G

Lemmens E, Moroni AV, Pagand J, Heirbaut P, Ritala A, Karlen Y et al (2019) Impact of cereal seed sprouting on its nutritional and technological properties: a critical review. Compr Rev Food Sci Food Saf 18(1):305–328. https://doi.org/10.1111/1541-4337.12414

Li Y, Su X, Shi F, Wang L, Chen Z (2017) High-temperature air-fluidization-induced changes in the starch texture, rheological properties, and digestibility of germinated brown rice. Starch - Stärke 69(9–10):1600328. https://doi.org/10.1002/star.201600328

Lin Y-T, Pao C-C, Wu S-T, Chang C-Y (2015) Effect of different germination conditions on antioxidative properties and bioactive compounds of germinated brown rice. Biomed Res Int 2015:608761. https://doi.org/10.1155/2015/608761

Mahakham W, Sarmah AK, Maensiri S, Theerakulpisut P (2017) Nanopriming technology for enhancing germination and starch metabolism of aged rice seeds using phytosynthesized silver nanoparticles. Sci Rep 7(1):8263. https://doi.org/10.1038/s41598-017-08669-5

Musa AS, Umar M, Ismail M (2011) Physicochemical properties of germinated brown rice (*Oryza sativa* L.) starch. Afr J Biotechnol 10(33):6281–6291

Oh SH (2003) Stimulation of gamma-aminobutyric acid synthesis activity in brown rice by a chitosan/glutamic acid germination solution and calcium/calmodulin. J Biochem Mol Biol 36(3):319–325. https://doi.org/10.5483/bmbrep.2003.36.3.319

Park D-J, Han J-A (2016) Quality controlling of brown rice by ultrasound treatment and its effect on isolated starch. Carbohydr Polym 137:30–38. https://doi.org/10.1016/j.carbpol.2015.10.045

Parnsahkorn S, Langkapin J (2013) Changes in physicochemical characteristics of germinated brown rice and brown rice during storage at various temperatures. Agric Eng Int CIGR J 15(2):293–303

Patil SB, Khan MK (2011) Germinated brown rice as a value added rice product: a review. J Food Sci Technol 48(6):661–667. https://doi.org/10.1007/s13197-011-0232-4

Patist A, Bates D (2008) Ultrasonic innovations in the food industry: from the laboratory to commercial production. Innovative Food Sci Emerg Technol 9(2):147–154. https://doi.org/10.1016/j.ifset.2007.07.004

Pinkaew H, Wang Y-J, Naivikul O (2017) Impact of pre-germination on amylopectin molecular structures, crystallinity, and thermal properties of pre-germinated brown rice starches. J Cereal Sci 73:151–157. https://doi.org/10.1016/j.jcs.2016.12.013

Rai-Kalal P, Tomar RS, Jajoo A (2021) Seed nanopriming by silicon oxide improves drought stress alleviation potential in wheat plants. Funct Plant Biol 48(9):905–915. https://doi.org/10.1071/FP21079

Raja K, Sowmya R, Sudhagar R, Moorthy PS, Govindaraju K, Subramanian KS (2019) Biogenic ZnO and Cu nanoparticles to improve seed germination quality in blackgram (*Vigna mungo*). Mater Lett 235:164–167. https://doi.org/10.1016/j.matlet.2018.10.038

Raju TS (2019) Proteolysis of proteins. In T. S. Raju (Ed.), *Co- and post-translational modifications of therapeutic antibodies and proteins* (pp. 183–202)

Rattanamechaiskul C, Soponronnarit S, Prachayawarakorn S (2014) Glycemic response to brown rice treated by different drying media. J Food Eng 122:48–55

Ravichanthiran K, Ma ZF, Zhang H, Cao Y, Wang CW, Muhammad S et al (2018) Phytochemical profile of brown rice and its nutrigenomic implications. Antioxidants 7(6):71

Ren C, Hong B, Zheng X, Wang L, Zhang Y, Guan L et al (2020) Improvement of germinated brown rice quality with autoclaving treatment. Food Sci Nutr 8(3):1709–1717. https://doi.org/10.1002/fsn3.1459

Saleh ASM, Wang P, Wang N, Yang L, Xiao Z (2019) Brown rice versus white rice: nutritional quality, potential health benefits, development of food products, and preservation technologies. Compr Rev Food Sci Food Saf 18(4):1070–1096. https://doi.org/10.1111/1541-4337.12449

Shao Y, Bao J (2015) Polyphenols in whole rice grain: genetic diversity and health benefits. Food Chem 180:86–97. https://doi.org/10.1016/j.foodchem.2015.02.027

Shao Y, Bao J (2019) 8 - Rice phenolics and other natural products. In: Bao J (ed) Rice, 4th edn. AACC International Press, pp 221–271

Singh AK, Rehal J, Kaur A, Jyot G (2015) Enhancement of attributes of cereals by germination and fermentation: a review. Crit Rev Food Sci Nutr 55(11):1575–1589. https://doi.org/10.1080/10408398.2012.706661

Sivakamasundari SK, Priyanga S, Moses JA, Anandharamakrishnan C (2021) Impact of processing techniques on the glycemic index of rice. Crit Rev Food Sci Nutr 1-22. https://doi.org/10.1080/10408398.2020.1865259

Song K, He X (2021) How to improve seed germination with green nanopriming. Seed Sci Technol

Sookwong P, Yodpitak S, Doungkaew J, Jurithayo J, Boonyawan D, Mahatheeranont S (2014) Application of oxygen-argon plasma as a potential approach of improving the nutrition value of pre-germinated brown rice. J Food Nutr Res 2(12):946–951

Srisang N, Varanyanond W, Soponronnarit S, Prachayawarakorn S (2011) Effects of heating media and operating conditions on drying kinetics and quality of germinated brown rice. J Food Eng 107(3):385–392. https://doi.org/10.1016/j.jfoodeng.2011.06.030

Ti H, Zhang R, Zhang M, Li Q, Wei Z, Zhang Y et al (2014) Dynamic changes in the free and bound phenolic compounds and antioxidant activity of brown rice at different germination stages. Food Chem 161:337–344. https://doi.org/10.1016/j.foodchem.2014.04.024

Tian S, Nakamura K, Kayahara H (2004) Analysis of phenolic compounds in white rice, brown rice, and germinated brown rice. J Agric Food Chem 52(15):4808–4813. https://doi.org/10.1021/jf049446f

Toutounji MR, Farahnaky A, Santhakumar AB, Oli P, Butardo VM, Blanchard CL (2019) Intrinsic and extrinsic factors affecting rice starch digestibility. Trends Food Sci Technol 88:10–22. https://doi.org/10.1016/j.tifs.2019.02.012

Waskow A, Howling A, Furno I (2021) Mechanisms of plasma-seed treatments as a potential seed processing technology. Front Phys 9:174. https://doi.org/10.3389/fphy.2021.617345

Weng X, Sun M, Gao H, Liu Z, Huang J, Liao X, Shen GX (2019) Germinated brown rice, a whole grain with health benefits for common chronic diseases. Nutr Food Sci J 2(1):119

Wu F, Chen H, Yang N, Wang J, Duan X, Jin Z, Xu X (2013a) Effect of germination time on physicochemical properties of brown rice flour and starch from different rice cultivars. J Cereal Sci 58(2):263–271. https://doi.org/10.1016/j.jcs.2013.06.008

Wu F, Yang N, Touré A, Jin Z, Xu X (2013b) Germinated brown rice and its role in human health. Crit Rev Food Sci Nutr 53(5):451–463. https://doi.org/10.1080/10408398.2010.542259

Xia Q, Li Y (2018) Mild high hydrostatic pressure pretreatments applied before soaking process to modulate wholegrain brown rice germination: an examination on embryo growth and physicochemical properties. Food Res Int 106:817–824. https://doi.org/10.1016/j.foodres.2018.01.052

Xia Q, Tao H, Huang P, Wang L, Mei J, Li Y (2017) Minerals *in vitro* bioaccessibility and changes in textural and structural characteristics of uncooked pre-germinated brown rice influenced by ultra-high pressure. Food Control 71:336–345. https://doi.org/10.1016/j.foodcont.2016.07.018

Xia Q, Tao H, Li Y, Pan D, Cao J, Liu L, Barba FJ (2020) Characterizing physicochemical, nutritional and quality attributes of wholegrain *Oryza sativa* L. subjected to high intensity ultrasound-stimulated pre-germination. Food Control 108:106827. https://doi.org/10.1016/j.foodcont.2019.106827

Xu J, Zhang H, Guo X, Qian H (2012) The impact of germination on the characteristics of brown rice flour and starch. J Sci Food Agric 92(2):380–387. https://doi.org/10.1002/jsfa.4588

Yodpitak S, Mahatheeranont S, Boonyawan D, Sookwong P, Roytrakul S, Norkaew O (2019) Cold plasma treatment to improve germination and enhance the bioactive phytochemical content of germinated brown rice. Food Chem 289:328–339. https://doi.org/10.1016/j.foodchem.2019.03.061

Zargarchi S, Saremnezhad S (2019) Gamma-aminobutyric acid, phenolics and antioxidant capacity of germinated indica paddy rice as affected by low-pressure plasma treatment. LWT 102:291–294. https://doi.org/10.1016/j.lwt.2018.12.014

Zhang Q, Xiang J, Zhang L, Zhu X, Evers J, van der Werf W, Duan L (2014) Optimizing soaking and germination conditions to improve gamma-aminobutyric acid content in japonica and indica germinated brown rice. J Funct Foods 10:283–291. https://doi.org/10.1016/j.jff.2014.06.009

Chapter 12
Impact of Heat-Moisture Treatment on the Hierarchical Structure and Functional Properties of Rice Starch

Hongwei Wang and Hua Zhang

Abstract The heat-moisture treatment (HMT) is an extensively welcomed hydrothermal technique that could affect the intrinsic multi-scale structures of rice starch, leading to the modification of the physicochemical and functional properties of starch. Thus, in this part, after briefly introducing the theory of HMT, the impact of HMT on rice starch hierarchical structures (including molecular, crystalline, lamellar, and granular structure) is discussed, and the functionality of HMT-treated starch (HMTS) affected by exogenous lipids, the influence of HMTS on the final product texture and volume, as well as the potential applications of HMT-treated starches are discussed in detail.

Keywords Heat-moisture · Hierarchical structures · Physicochemical · Digestibility

Abbreviations

BV	Breakdown
DSC	Differential scanning calorimetry
FV	Final viscosity
HMT	Heat-moisture treatment
HMTS	HMT-treated rice starch
LOA	Linoleic acid
OA	Oleic acid
PV	Peak viscosity
RS	Resistant starch
SB	Setbacks
SP	Swelling power
TV	Trough viscosity

H. Wang · H. Zhang (✉)
College of Food and Bioengineering, Zhengzhou University of Light Industry, Zhengzhou, Henan, China
e-mail: hwwang@zzuli.edu.cn; 2009046@zzuli.edu.cn

J. Tian et al. (eds.), *Science of Rice Chemistry and Nutrition*,
https://doi.org/10.1007/978-981-99-3224-5_12

12.1 Introduction

Rice is one of the most important cereals, which serves as the staple food for about 3.5 billion people worldwide (Muthayya et al. 2014). Rice contains an important source of carbohydrates, mainly starch, that is predominantly present in the endosperm cells of mature brown rice, accounting for approximately 90% (dry basis) of milled rice (Amagliani et al. 2016; Bao and Bergman 2018). Normally, the quality attributes of final products are closely associated with the intrinsic constituents of raw materials, and rice starch is the constituent that exerts the greatest effect on the processability of rice. Therefore, the structure of this fraction and its physicochemical properties are the primary characteristics used to select rice varieties and rice starch for specific industrial applications (Lee et al. 2017). For example, the eating and cooking quality of rice is influenced by the pasting and thermal properties of starch (Li and Gilbert 2018), while the textural quality of cooked rice in storage is influenced by the retrogradation properties of starch (Li et al. 2021). Currently, rice starch is widely used in foods (e.g., baby foods, bakery products, confectionery, ice cream, meat products, sauces, snack foods, soft drinks, and soups) with the effect of thickening, gelling, stabilizing, or processing aids (Chen et al. 2017a, b; Wang et al. 2020a, b, c). Also, it can be exploited in nonfood applications, including adhesives, cosmetics, medical, oil drilling, and pharmaceuticals (Amagliani et al. 2016; Bemiller and Huber 2015; Yi et al. 2021). However, native rice starch has certain limitations such as insolubility in water, propensity toward retrogradation, high viscosity even at low concentrations, and inability to withstand at high temperatures, which extremely restrict its applications in the food industry (Wang et al. 2020a, b, c; Zhang et al. 2015). To overcome these disadvantages and broaden their industrial applications, a significant number of studies have been carried out to explore techniques for starch modification over the last few decades (Bemiller and Huber 2015). Physical, enzymatic, and chemical approaches are the three main modification methods, which can significantly change the functional properties of starch (Han et al. 2020; Kumari et al. 2020; Masina et al. 2017). In contrast to other modifications, physical modification techniques such as irradiation, annealing, osmotic pressure pulsed electric field, and heat-moisture treatment are increasingly popular because they have less toxic chemical residues and seem more rapid and simpler to create desired clean-label foods while maintaining food safety (Schafranski et al. 2021; Zhu 2016, 2018).

Among the physical modification methods, heat-moisture treatment (HMT) is an extensively welcomed hydrothermal technique that could affect the intrinsic multi-scale structures of rice starch, leading to the modification of the physicochemical and functional properties of starch (He et al. 2020; Wang et al. 2018; Xie et al. 2020; Yang et al. 2019; Zheng et al. 2018; Zheng et al. 2020a). This chapter updates recent advances in HMT on rice starch hierarchical structure, functional properties, and its application in final products.

12.2 Heat-Moisture Treatment (HMT)

Numerous studies have investigated the influence of HMT on starch with different botanical sources since 1944 (Hoover and Manuel 1996; Kaur and Singh 2019; Sair and Fetzer 1944; Shi et al. 2018). Depending on the origin of starch, various changes in the physical and chemical properties may occur, e.g., structural changes in the granules and improvement in the rheological characteristics of the paste, texture of the gels, thermal stability, and enzymatic resistance (Hoover 2010; Schafranski et al. 2021; Wang et al. 2021b; Zavareze and Guerra Dias 2011). These transformations make starches more suitable for industrial applications. Therefore, HMT represents a good way to modify starch structure and functionalities and is preferred by the food industry.

Heat-moisture treatment (HMT), as a typical starch modification method, has various advantages including its effectiveness, flexibility in relation to heat sources, low cost, and non-generation of chemical residues. Generally, HMT refers to the exposure of starch granules to higher temperature (90–135 °C) for a certain time period ranging from 15 min to 16 h at restricted moisture content (10–35%) (Ambigaipalan et al. 2014; Wang et al. 2017a; Wang et al. 2016a). The nature of HMT is a heating process that occurs under high temperature and conditions of restricted moisture content; nonetheless, gelatinization hardly occurs during HMT due to the restricted moisture level. In other words, water acts as a plasticizer in the functioning of starch polymers, and its incorporation may enhance starch chain distance and destroy the internal hydrogen bonding. And thermal energy is converted to kinctic energy continuously by the activated water molecules, accelerating the entanglement slippage of the starch amorphous domain and letting it shift from a glassy to a more flexible fluidlike state. Furthermore, the generated pressure helps increase the thermal energy, which is continuously converted into kinetic energy by water molecules. This also promotes the slippage, entanglement, and interactions of starch molecular chains, thus causing the rupture and/or reordering of starch hierarchical structure (e.g., crystalline and helical structures), eventually affecting the physicochemical and functional properties of HMT-treated starches (HMTS).

Recently, the impacts of HMT on hierarchical structures and physicochemical and functional properties of rice starch have been carried out in abundance. Also, its interactions with other constituents (e.g., lipids) under HMT have been evaluated. Additional information is described below regarding the effect of HMT on the structural, physicochemical, and functional properties of rice starch, also including the interactions between rice starch and other constituents.

12.3 Impact of HMT on Rice Starch Hierarchical Structures

12.3.1 Impact of HMT on the Molecular Structure

When rice starch is subjected to HMT, the degradation of starch molecular chains (mainly amylopectin) is usually observed due to the thermal energy and high pressure emerging during HMT. For example, the range of M_w of native rice starch was higher, and the fractions in the ranges of 1×10^7–3×10^7 g/mol and $> 3 \times 10^7$ g/mol accounted for 41.44% and 58.56%, respectively. After HMT, the molecular distribution of rice starch moved toward lower M_w. At 30% *MC*, M_w was in the range of 2×10^6–5×10^6 g/mol (Wang et al. 2018). Also, after HMT, both the amylopectin chain distribution and unit chain distributions of waxy and normal rice starch decreased, especially under high temperature (>120 °C). According to the report of Jiranuntakul et al. (2012), smaller molecules were observed after HMT, which signified the degradation of amylopectin molecules, and this may be due to the fact that most of the branch points are located in the amorphous lamella, i.e., α-1,6-glycosidic linkages could be easier degraded by HMT. Similarly, Chen et al. (2017a, b) and Alimi and Workneh (2018) reported that CH- bond stretching in starch was influenced by HMT, which also reflected that the α-(1,6)-glycosidic bond could be easily altered. Also, these changes in amylopectin are prone to increase the levels of amylose; however, some studies declared that HMT decreased the apparent amylose level in rice starch (Bian and Chung 2016). Presumably, some amylose interacts with newly generated chains during HMT through strengthened molecular bonds, or HMT favors the entanglement of amylose with lipids, thus diminishing its affinity with iodine and interfering with the amylose content measurement.

The molecular degradation of HMT-treated starches tends to decline due to the harsh condition of HMT, especially under enhanced temperature and moisture content. More interestingly, repeated HMT could lead to higher molecular weight than continuous treatments when the same total treating time is used (Zhang et al. 2020). As reported, chain characteristics (length, type, and amount) of starch are a factor influencing molecular degradation during HMT. A thermostable polymorph is readily formed for starch with a higher amount of short external chains; nevertheless, the long internal chains across crystalline lamellae may conduce to the disintegration of compact regions and make starch less resistant (Lin et al. 2019a). It is hypothesized that repeated HMT facilitates the generation of many shorter chains at the earlier cycling period, and cooling/moisture reconditioning process at intervals makes it accessible for these chains to redistribute and reassociate into structurally perfect fractions, which are difficult to degrade during subsequent cycling. It would be exciting to detect the dynamic changes in chain length versus cycling times. Considering continuous HMT, HMT-treated starches have lowered the proportion of long chains (DP ≥37) and decreased average chain length (Bian and Chung 2016), implying the possibility of the attack to α-(1,4)-glycosidic bond. However, its exact relationship with the structural stability of treated starch has not been validated.

12.3.2 *Impact of HMT on Ordered Molecular Structure*

Starch is composed of two main components: slightly branched amylose and highly branched amylopectin. Also, these two biopolymers are organized on different length scales to form a hierarchical structure, comprising the semicrystalline lamellae, long-range crystallites, ordered molecular structure, and helical conformations (Li et al. 2020; Zhang et al. 2013). Thereinto, the molecular chains of starch (amylose and amylopectin) can self-assemble to form double- and single-helical structures. The double-helical structure can be formed with suitable chain length and minimum amylopectin DP 10 (Gidley and Bulpin 1987), while the chains with DP 14–24 (Shi and Seib 1992) or DP 12–22 (Vandeputte et al. 2003) form double helices much more readily than short or long amylopectin. The solid-state ^{13}C CP/MAS NMR can be used to identify the helical structural changes of native and HMT rice starches. Our previous study (Wang et al. 2018) indicated that HMT disrupted the intermolecular hydrogen bonding of rice starch, leading to a decrease in double-helix fractions. Also, the percentage of amorphous starch was reduced. However, the content of single helices (i.e., V-type) increased, suggesting that HMT facilitated the formation of V-type single-helical conformation. These may be due to the mobility of molecular chains, which allowed the formation of new ordered molecular structures, as shown by the increased amounts of single helices and densely packed amorphous starch regions. Figure 12.1 shows the schematic model

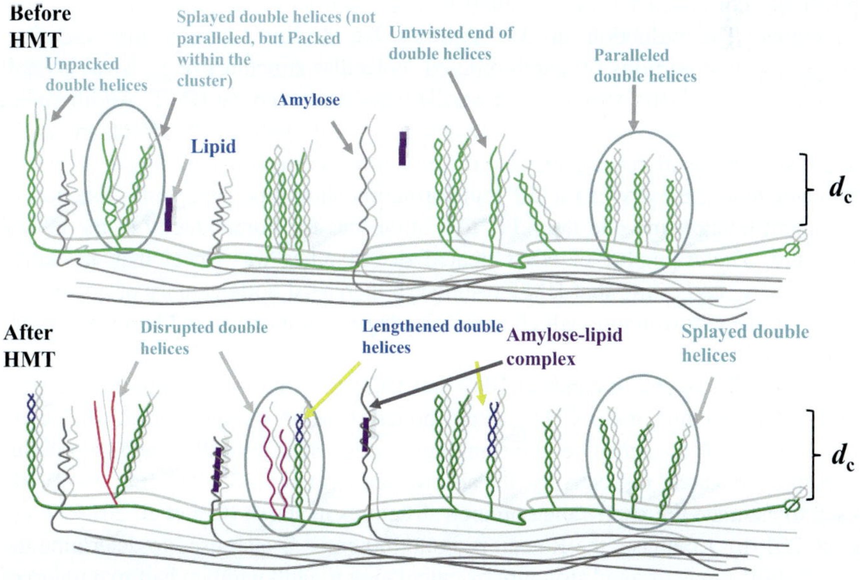

Fig. 12.1 Schematic model of the backbone structure of ordered molecular structure of rice starch before and after HMT. (This schematic graph was drawn by the authors)

of the backbone structure of the ordered molecular structure of rice starch before and after HMT.

Furthermore, differential scanning calorimetry (DSC) can also be used for measuring the ordered molecular structure of starch through supervising its thermal stability. The typically recorded parameters from DSC contain gelatinization temperatures (T_o, T_p, and T_c, i.e., onset, peak, and conclusion temperature, respectively), temperature range (ΔT, $T_c - T_o$), and enthalpy (ΔH). Normally in DSC thermogram of starch, there is only a single endothermic peak near T_p. Upon HMT, a shoulder peak can be seen coexisting with the main peak, such as rice starch with 13% moisture treated at 120 °C for 60 min (Arns et al. 2015). The biphasic transition suggests that HMT evokes the formation of two-type crystals possessing distinct thermal stability, which may be related to the separation and/or reorganization of molecular chains within starch granules. As for rice starch containing more lipids, another endothermic peak is located near 100 °C (Lin et al. 2019a, b). The phenomenon is likely to result from V-type polymorph melting. Overall, HMT increased the T_o, T_p, and T_c but decreased ΔH in comparison with the native counterpart (Liu et al. 2015; Yan et al. 2020). These effects triggered by HMT become more obvious under more intense conditions, e.g., treatment with higher moisture content or for a longer holding time (Sui et al. 2016; Van Hung et al. 2020). The elevation in gelatinization temperatures means HMTS has enhanced thermal stability. HMT possibly promotes starch molecule diffusion and facilitates the intertwining of amylose and amylopectin via intra/intermolecular linkages, which could not only confine the flexibility of the amorphous domain and retard its expansion, but also render the crystalline structure more robust (Wang et al. 2020a, b, c). The transition enthalpy (ΔH) was positively correlated with the amount of long- and short-range ordered molecular structures. The reduction in ΔH signifies the decline of molecular ordering, suggesting that portions of starch ordered molecular structures (e.g., weak crystallites and double-helix structures or small) were destroyed by HMT. Additionally, variations in rice amylose content can also affect the modification effect, and high-amylose starch is more susceptive to the heat-moisture treatment (Yang et al. 2019). Fundamentally, the driver for ΔT transformation should be synergistic function of treatment temperature and humidity that influences the homogeneity of the helical structure. The ΔT value of HMT-treated rice starches was lower than its native counterpart, suggesting that HMT resulted in increased homogeneity of crystallites and/or molecular structure, which meant the degradation of less stable or smaller-in-size crystallites.

The FTIR absorption bands at 1047 and 1022 cm^{-1} are ascribed to the presence of C–O–H bending and –CH2-related modes (Lozano-Vazquez et al. 2021). As those two bands are sensitive to the amount of ordered or crystalline and amorphous structures of starch, respectively, the absorbance ratio of 1047/1022 cm^{-1} is usually used to estimate the short-range ordered structures in starch (Chi et al. 2017; Wang et al. 2017b). Likewise, the Raman peak at 480 cm^{-1} is also used to determine the starch short-range ordered structure by calculating the full width at half maximum of the band, in which the peak is related to the –CH2- mode (Mutungi et al. 2012; Wang et al. 2016b). A higher $R_{1047/1022}$ or a lower full width at half maximum (*FWHM*) indicates that the starch sample contains more short-range ordered structures. After HMT, the value of $R_{1047/1022}$ decreased as the enhancement of HMT intensity, but the value of *FWHM* increased, indicating that ordered molecular structures at short-range scale could be destroyed during HMT. This transformation could be due to the synergistic action of thermal energy and water molecules, which disrupted intra- and intermolecular hydrogen bonding in starch granules, leading to the irregular packing or unwinding of helical structures. As revealed by previous results related to FTIR and Raman, these phenomena became more severe with enhancing HMT intensity.

12.3.3 Impact of HMT on the Crystalline Structure

As the helical structures are oriented in a specific mode, starch can be densely packed with A-, B-, C-, and V-type polymorphs. According to Miao et al. (2009), crystallinity differences between starches may be attributed to the following: crystal size, number of crystalline regions that are influenced by amylopectin content and chain length, orientation of the double helices within the crystalline, and extent of interaction between the double helices. Generally, rice starch presents a typical A-type polymorph and exhibits obvious diffraction intensity at 2θ of ca. 15.3°, 17.1°, 18.2°, and 23.5°, which is characteristic of cereal starches (Wang et al. 2018). However, starch from mutant rice grains displays a B-type X-ray pattern with obvious diffraction peaks at 5.6°, 14.4°, 17.2°, 22.2°, and 24° in 2θ (Lin et al. 2019a, b). The treatment conditions will influence the result of HMT on the starch crystalline structure. For example, the birefringence cross of HMT-treated rice starch is fuzzier and weaker than that of native starch under polarized light, and it even disappears completely under high-intensity conditions (Xie et al. 2020). Like most untreated rice starches, HMT-treated starches also have a typical A-type polymorph, the crystal unit of which is monoclinic and contains closely packed left-handed double helices. However, HMT often results in the transformation of the less thermodynamically stable B-polymorphic structure (with hexagonal packing of double helices and about 36 water molecules inside each cell) into a more stable monoclinic structure of A-type polymorphs (with about six water molecules inside the helices) (Zhang et al. 2014). The HMT-induced changes in the diffraction pattern of starches can be attributed to dehydration and to movement of a pair of double helices into the central channel. This movement during HMT could disrupt starch crystallites and/or change the crystalline orientation. Apart from A-type polymorph, V-type diffraction pattern has been found in plentiful HMT-treated starches, particularly those obtained by HMT with high moisture levels (Wang et al. 2018). V-type polymorph was reported to be derived from single helical complexes of amylose with endogenous lipids. This diffraction pattern of HMT-treated starches is more conspicuous as high-amylose-containing starches (Yang et al. 2019). Moreover, the occurrence of V-type polymorph is associated with the increased percentage of single helices, which is at the expense of double helices decoupling through internal hydrogen bonding breakage. Another factor influencing V-type crystallite formation is the availability of residual lipids in modified starch, but their compositional changes have been less studied. Figure 12.2 shows the schematic of the starch crystalline structure before and after HMT.

Additionally, noticeable alterations in crystallinity degree have been observed by many researchers, and the intensity of peaks can vary (Arns et al. 2015; Hung et al. 2016; Zavareze et al. 2010). Based on the above studies, HMT could cause the disassociation of hydrogen bonds, which caused the transformations of lattice and double-helix orientation, and the reduced tightness or unwinding of the double helix, eventually leading to the reduction in the crystallinity degree, and the decreasing extent is remarkably dependent on the amylose content (Yang et al. 2019). Besides, hydrogen bonding, another factor in keeping crystals stable, might partially break down during the HMT process. Therefore, it is not surprising for the decreased crystallinity of HMT-treated rice starches. However, others have identified an increase in the relative crystallinity of rice starch (Hormdok and Noomhorm 2007). With regard to the opposite change, it is speculated that sufficient mobilization of starch branches during HMT improves the orientation of less stable helical chains, induces recombination of smaller crystals, or promotes the conversion of the amorphous region into a new crystalline state, thus leading to the crystallinity increase. As concluded by Liu et al. (2019), the improved crystallinity of HMT-modified barley starch was accompanied by the reduction in the electron density difference between crystalline and amorphous lamellae. Hence, semicrystalline lamellae damage may also be a contributor, and specific correlations remain to be elucidated.

12.3.4 Impact of HMT on the Lamellar Structure

Starch semicrystalline structures are packed by alternating crystalline and amorphous lamellae. Along with the crystallinity loss, the thickness of both crystalline and amorphous lamellae gets enhanced after modification (Wang et al. 2018; Xie et al. 2020; Yang et al. 2019), implying that HMT destroys the regularity and compactness of the starch semicrystalline system. After HMT, increases in the thicknesses of semicrystalline (d), crystalline (d_c), and amorphous (d_a) lamellae were observed as the moisture content rose (especially at 30%). These results suggested that HMT could simultaneously increase both the crystalline (d_c) and amorphous (d_a) lamellae of rice starch. The increase in d_c might mainly result from the swelling in amorphous lamellae or the disassociation of some tailing ends of double helices, and the amorphous parts were less tightly packed after HMT. Regarding the increase in d_c, it was speculated that water and thermal energy induced the packing of some tailing ends of double helices or the partial breakage of the hydrogen bonds of starch molecular helices in the starch granule interior, which led to an irregular rearrangement of double helices in crystalline lamellae. In a word, the synergy of heat and moisture could evoke unwinding of helical tailing ends in swollen starch granules, expediting molecular chain movement and

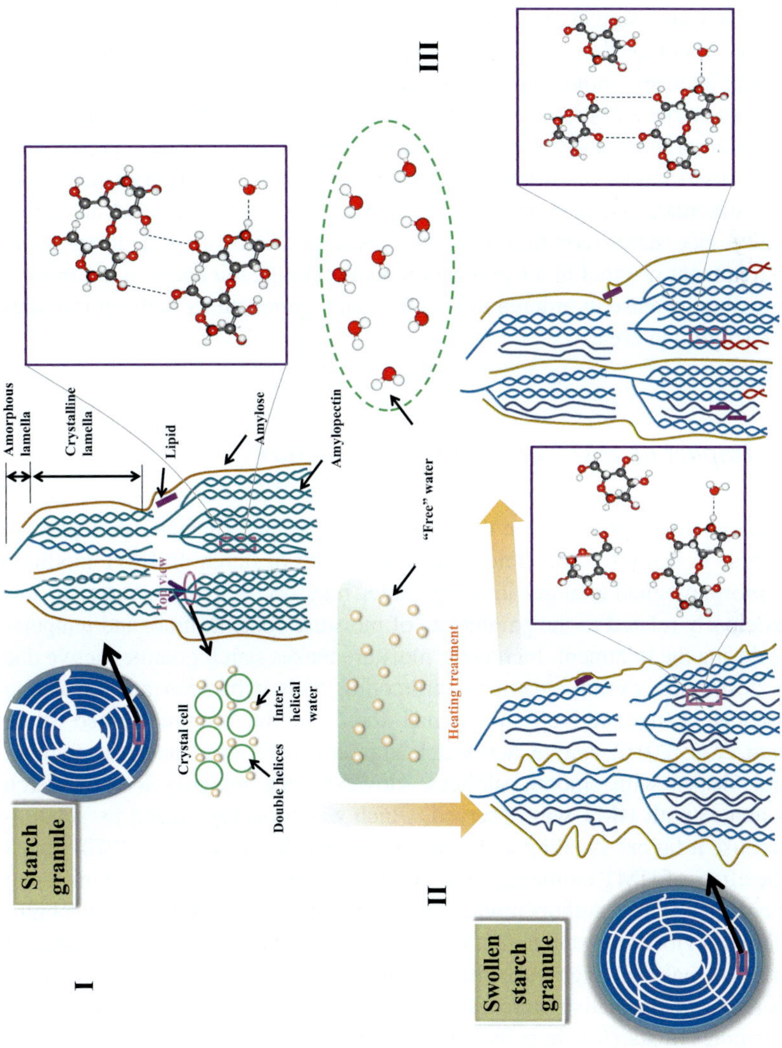

Fig. 12.2 The schematic of starch crystalline structure before and after HMT. (Drawn by the authors)

rearrangement into a more disordered structure. Also, the area (A_{peak}) of the lamellar peak and the difference in electron density of semicrystalline lamellae system reflect the ordering of the lamellar structure. The peak area (A_{peak}) of HMT-treated rice starches declined with the increase of *MC*, suggesting a lower ordering degree of the lamellar regions. Moreover, for those HMT-treated starches, the intensity in the small q range (0.1–0.6 nm^{-1}) increased after HMT, while the peak definition reduced, which also suggested a declining ordering degree of lamellar structure. Furthermore, in comparison with native starch, HMT starches had slightly higher lamellar peak intensity, indicating the disruption of the amorphous background. The result confirmed that an order of HMT-treated starches existed for the compactness difference in amorphous and crystalline lamellae. These alterations were presumably due to the dissociation of ordered molecular structures, the disorganization of crystalline structure, and their realignment (Qiao et al. 2016; Zhang et al. 2016). Additionally, rice starch complexation with oleic acid/linoleic acid during heat-moisture treatment resulted in a significant reduction in the thickness of amorphous lamellae and an increase in ordered molecular chain aggregations in the amorphous amylose background area.

12.3.5 Impact of HMT on Granular Structure

Granule morphology, granule size distribution, and surface characteristics play an important role in many foods and nonfood applications of starch. In HMT-modified starches, morphological changes in the granules may occur, and the extent of this change is closely related to the parameters of moisture content, time, and temperature used during the treatment. Increasing moisture makes starch granules active due to water absorption during treatment, promoting their expansion and providing morphological changes driven by thermal force. In the aforementioned study (Amagliani et al. 2016), native rice starch granules were small and had angular, polyhedral shapes with smooth surfaces. However, the integrity loss of starch granules treated with HMT was observed, which was probably caused by gelatinization because it favors swelling and granular fusion. Zavareze et al. (2010) investigated the effect of HMT on the granular morphology of rice starches with different amylose contents. These authors reported that the HMT (25% moisture) of the high- and medium-amylose-rice starch granules slightly affected the format and degree of agglomeration, making the granules more aggregated and the surface more irregular, as compared with the native starches (Fig. 12.3). Yang et al. (2019) studied the granular features of starch in heat-moisture-treated polished rice grains with varying amylose contents; they observed that in its native form, it was angular, polygonal, and with few cracks in a smooth surface. Because of HMT, all rice starch granules underwent agglomeration and melting of the surface, to a lesser extent in starches with a low concentration of amylose. These results were attributed to the amylose–amylose interactions that help to prevent disturbances in the granular structure of the starch. However, recent studies have shown that modification by HMT did not

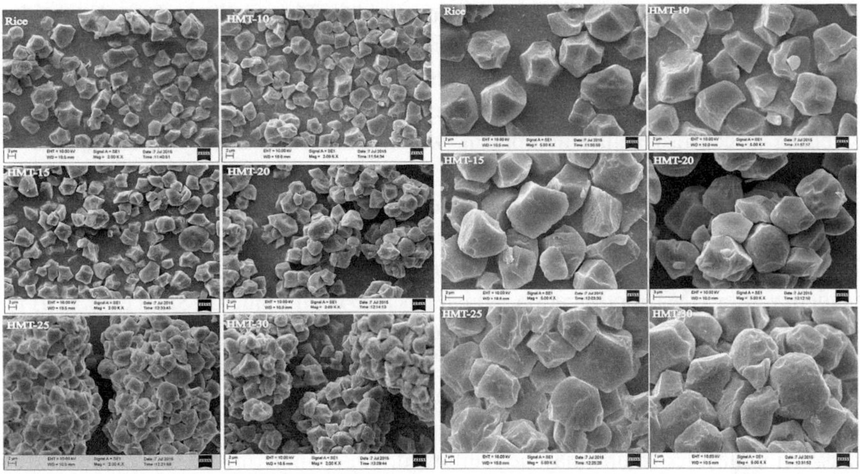

Fig. 12.3 SEM graphs of native and HMT-treated rice starch granules. Insights into the multi-scale structure and digestibility of heat-moisture-treated rice starch. Reprinted with permission from Wang et al. (2018)

significantly alter the morphology of rice starch granules; this may be caused by different species and HMT conditions (Wang et al. 2021a, b).

The particle size of HMT-treated granules was not altered compared with native ones, as reviewed by Zavareze and Guerra Dias (2011). However, recently, many papers have reported the shift in the HMT-treated starch size distribution curve toward a larger size range (Wang et al. 2018). The granular size of HMT-treated starch is proved to increase with the enhancement of the HMT moisture level (Wang et al. 2020a, b, c). Increasing moisture content makes starch granules active by absorbing more water during treatment, which promotes their expansion and favors morphological changes driven by thermal force. Other than micromorphology, the surface characteristics of HMT-treated starch at the nanometer scale are also distinct from native starch. HMT greatly reduced the number of protrusions on the rice starch surface. More research should be carried out on nano-level granular morphology of HMT-treated starch from various rice grains.

12.3.6 Impact of HMT on the Physicochemical and Functional Properties of Rice Starch

Impact of HMT on Swelling Power (SP) and Solubility

In most food systems, the pasting behaviors and rheological properties of the partially or fully swollen starch granules depend on starch swelling (Wang et al. 2020a, b, c). Water molecules can be captured by hydroxyl groups within starch through hydrogen bonding under certain conditions. The swelling of starch granules is quantitatively related to the processing temperature, being accompanied by leaching of amylose molecules and lipids, normally bound in the granule, into the continuous phase. The presence of amylose and lipids in the granule retards starch swelling in contrast with amylopectin, which contributes mainly to the uptake of water. Here, swelling power measures the water-retaining capacity of starch at specific temperatures; solubility indicates the amount of leached amylose from starch granules because of diffusion and dissociation during swelling. Because of HMT, both the swelling power and solubility of rice starch exhibit a downward tendency after HMT modification, which has also been found by Silva et al. (2017). The higher the moisture content or temperature during HMT processing, the lower the swelling power and solubility of HMT-treated starch. A possible reason for the decrease in these two indices is that the amylose/amylopectin array within granules is disrupted. When subjected to heat-moisture conditions, the internal interplay between molecular chains may be reinforced, thus hampering the hydration of free hydroxyl groups. Side-chain clusters could be packed more compactly during hydrothermal modification, which contributes to the establishment of rigid crystalline architecture that impedes water percolation into starch granules (Nakamura et al. 2017; Piecyk et al. 2018). Also, HMT-induced amylose–lipid complex can obstruct starch swelling, thereby decreasing amylose leaching and solubility. Another cause of *SP* decrease can be the transformation in the amorphous areas where the amylose gets a helical form. HMT applied by Sun et al. (2013) at different treatment temperatures and times on early indica rice with different amylose contents induced a decrease in *SP* and solubility, with the severity depending on amylose content, due to the structural reorganization, water restriction, and amylose–lipid and amylose–amylopectin interactions.

Impact of HMT on Pasting Properties

Pasting properties reveal the clear viscosity changes when the starch suspension is heated on the basis of a controlled heating-cooling procedure under a certain agitating force. The pasting profile of HMT-treated starch has been primarily determined using a Rapid Visco Analyzer and Brabender Viscograph. From the RVA and BV curves, related parameters can be observed, such as the pasting temperature, which corresponds to the temperature at which the viscosity begins to increase. The peak time, or maximum viscosity time, is the heating time in which maximum paste viscosity occurs. The maximum viscosity reached during the heating period is regarded as the peak viscosity. The breakdown refers to the difference between peak viscosity and minimum viscosity. The difference between final viscosity and minimum viscosity is called setback, a process known as retrogradation. The final viscosity is the viscosity value obtained at the end of the cooling cycle (Balet et al. 2019; Cozzolino 2016).

The pasting parameters of starches can be affected by HMT modification. Previous researchers (Arns et al. 2015; Sun et al. 2013; Wang et al. 2018; Yang et al.

2019) studied the effects of HMT on the physical-chemical and structural properties of rice starch; they found that regardless of the moisture and time used, the values of peak viscosity (PV), trough viscosity (TV), final viscosity (FV), breakdown (BD), and setbacks (SB) for treated normal starch were lower than those for native starch. Specifically, PV is positively correlated with the swelling ability of starch (Zheng et al. 2020a, b), and the lower swelling power of HMT-treated starch may hinder the leaching of amylose for viscosity attenuation. Changes in FV, signifying the potential of HMTS to yield cold viscous paste, are largely concerned with the reordering of leached hydrocarbon chains in an aqueous system upon cooling (Suriya et al. 2019). BD refers to the fragility of hot starch paste, while SB stands for the syneresis tendency of cold starch paste. HMT-treated starch has lower BD and SB, indicating that HMT improved mechanical stability under thermal conditions and is more retrogradation resistant. Both indices might be modulated by amylose–amylopectin restructuration, concentration of noncleaved swollen granules, and amylose leaching (Sun et al. 2014). Contrary to the preceding results, some studies reported an increase in pasting viscosity values following HMT, e.g., the PV and TV of red rice starch (Van Hung et al. 2020). Apart from the special structural changes, explanations for the disagreement should also consider interactions between starch and other minor residual components throughout the isolation process.

Enhancing the severity of HMT leads to a greater reduction in these viscosity values, such as increasing the treating temperature of rice starch from 85 °C to 120 °C. The viscosity reduction seems more pronounced for starch granules with higher amylose content, as interpreted in Zavareze et al. 2010; HMT could induce higher amylose level in rice starch and disrupt its chain length distribution, which are possible reasons for the declined pasting viscosities of HMT-treated starch. In the case of rice starch with low amylose content, HMT resulted in little or no significant reduction in pulp properties, mainly due to the lack of amylose in these granules. However, pasting temperature (PT) of starches is increased more or less after HMT modification, and higher PT of HMT-treated starch may arise from granule agglomeration, crystallinity enhancement, generation of more intermolecular cross-links, reduced space between molecules, or fortified associative forces inside chains (Sudheesh et al. 2020), which causes additional heat uptake during the pasting process. Likewise, the influence of HMT on moisture was observed at the peak of viscosity, which gradually decreased as the moisture level increased, indicating a low restricted expansion capacity.

Impact of HMT on Gelling Properties

The textural properties of gels depend on the constituents of starch and amylose, the volume and deformation of the granules, and the interaction between the continuous and dispersed phases (Gunaratne et al. 2018). Many factors influence dough and starch gel rheology and texture, such as amylose content, size and deformation of starch granules, granule swelling, shape and volume, and interactions between the continuous and dispersed phases. Hormdok and Noomhorm (2007) studied the effect of HMT on rice starch (27% amylose) and concluded that this treatment increased the gel hardness of starch treated with 15% and 20% moisture. And the increase in gel hardness has been attributed to the increased cross-linking between starch chains in the particular amylose portion. For treatments under 25% moisture, they found no difference with native starch and attributed this fact to a possible partial gelatinization, resulting in a less rigid gel due to the partial collapse of the structure of the granules.

Impact of HMT on Starch Nutritional Functionalities

Digestibility of HMT-Treated Rice Starch (HMTS)

Currently, the consumption of high-energy starchy food leads to health problems related to metabolic disorders. Thus, the digestion behaviors of rice starch play an important role in its functionality. It is generally accepted that starch digestion starts in the mouth and mainly occurs in the small intestine where pancreatic α-amylase, dextrinase, amyloglucosidase, α-glucosidase, and maltase are embedded (Chi et al. 2021a, b). Based on the extent and rate of starch digestion, starch can be classified into rapidly digestible starch (RDS), slowly digestible starch (SDS), and resistant starch (RS) (Wang et al. 2021a, b). RDS increases human blood glucose levels immediately after starch ingestion, providing necessary energy for the human body and maintaining normal physiological functions of the human brain and central nervous system (Chi et al. 2021a, b). Glucose freed from the digestion of RDS-enriched foods contributes to a rapid increase in the human postprandial blood glucose concentration and a high glycemic response, which can induce diet-related metabolic chronic diseases such as type 2 diabetes mellitus. SDS is digested slowly but completely in the small intestine to maintain a stable glucose concentration in the human bloodstream (Zhong et al. 2021). RS cannot be digested in the human upper gastrointestinal tract but is degraded by microflora in the colon through anaerobic fermentation (Lockyer and Nugent 2017), which is a benefit to the regulation of gut microbiota and the prevention of metabolic diseases. Therefore, controlling starch digestion is essential for fortifying its nutritional value, particularly in the staple food industry. In addition, HMT has been widely applied to increase the level of SDS and/or RS.

Most noteworthy, gelatinized HMT-treated starch generally has lower RDS but higher SDS and RS content than the gelatinized native counterpart. Wang et al. (2018) showed that HMT at different moisture levels induced structural reordering of rice starch, which leads to its slow digestion in vitro than the unmodified samples. HMT determined an increase of SDS from 3.5 to 10.5% and resistant starch (RS) from 2.6% to 14.5% at 30% moisture of starch samples. Also, the increased degree of RS and SDS fractions depends on the moisture content of the samples during HMT. This may be due to the fact that the motion of molecular chains allowed the formation of new aggregation structures, as shown by increased amounts of single helices (a highly ordered starch matrix) and densely packed amorphous starch molecules. Besides, the disassociation and rearrangement of starch molecular helices led to increased thicknesses of crystalline and thermal stability of the ordered molecular structure. With the formation of a new ordered molecular aggregation architecture, HMT-treated rice starch was more resistant to the enzyme attack. Moreover, Ferreira-Silva et al. (2017) studied the digestion behavior of rice starch after HMT at different moisture levels. The results revealed that both SDS and RS contents were higher than in the untreated samples due to the interactions between starch molecular chains. The digestibility of starch is related to the granule size, *SP*, and thermal and pasting characteristics (Ferreira-Silva et al. 2017). A lower *SP* and *SL* indicate the increase in interactions among starch chains and consequently the increase in digestion resistance (Ferreira-Silva et al. 2017). The glycemic index depends on the amylose–amylopectin ratio and was found to be greater for the HMT rice flour than for the native, the high amylose rice variety presenting lower values compared to the low amylose one due to the amylose lower enzymatic susceptibility (Ashwar et al. 2016; Ruiz et al. 2018). Kim et al. (2017) reported an increase in RS and a decrease in RDS of HMT rice flour cake compared with the untreated one, and the SDS did not significantly change. Figure 12.4 shows the schematic representation of the changes in the multi-scale structure and digestibility of native and HMT-treated rice starches.

However, some recent studies have shown that HMT could improve the hydrolysis rate and extent of native granular starch. Native rice starch had a dense packing structure that could hinder the diffusion and binding activity between enzyme and starch molecules, consequently resulting in limited digestion rate and degree (i.e., lowest RDS) (Xie et al. 2020; Zheng et al. 2018). After HMT, the crystalline and molecular ordered structures of native rice starch were destroyed, which may facilitate the attack of amylolytic enzymes to starch chains, therefore increasing RDS and decrease of SDS. Similar results have been found by Anderson et al. (2002), who found a small but significant increase in digestibility for waxy rice starch (10.5% moisture) treated at the melting temperature (T_m) for 30 and 60 min. However, at the same moisture content (10.5%), the digestibility of non-waxy rice starch remained unchanged on HMT. The above authors also reported an increase in digestibility in non-waxy (25%) and waxy rice (10%) starches (adjusted to 20% moisture w/b) heated in a differential scanning calorimeter and held at the temperature of melting (T_m) for 60 min. The results from all of the above studies showed that the impact of HMT on digestibility is influenced by starch moisture content and by structural (crystallite disruption, change in polymorphic form, crystallite reorientation, amylose-lipid interaction, starch chain interaction) and morphological (formation of fissures and cracks on the granule surface) changes on HMT.

Nutritional Function of HMT-Treated Rice Starch (HMTS)

Obesity is one of the most prevalent chronic metabolic diseases in the world, which is mainly induced by an energy imbalance that results in excessive body fat accumulation. It has been increasingly known that there are many factors contributing to

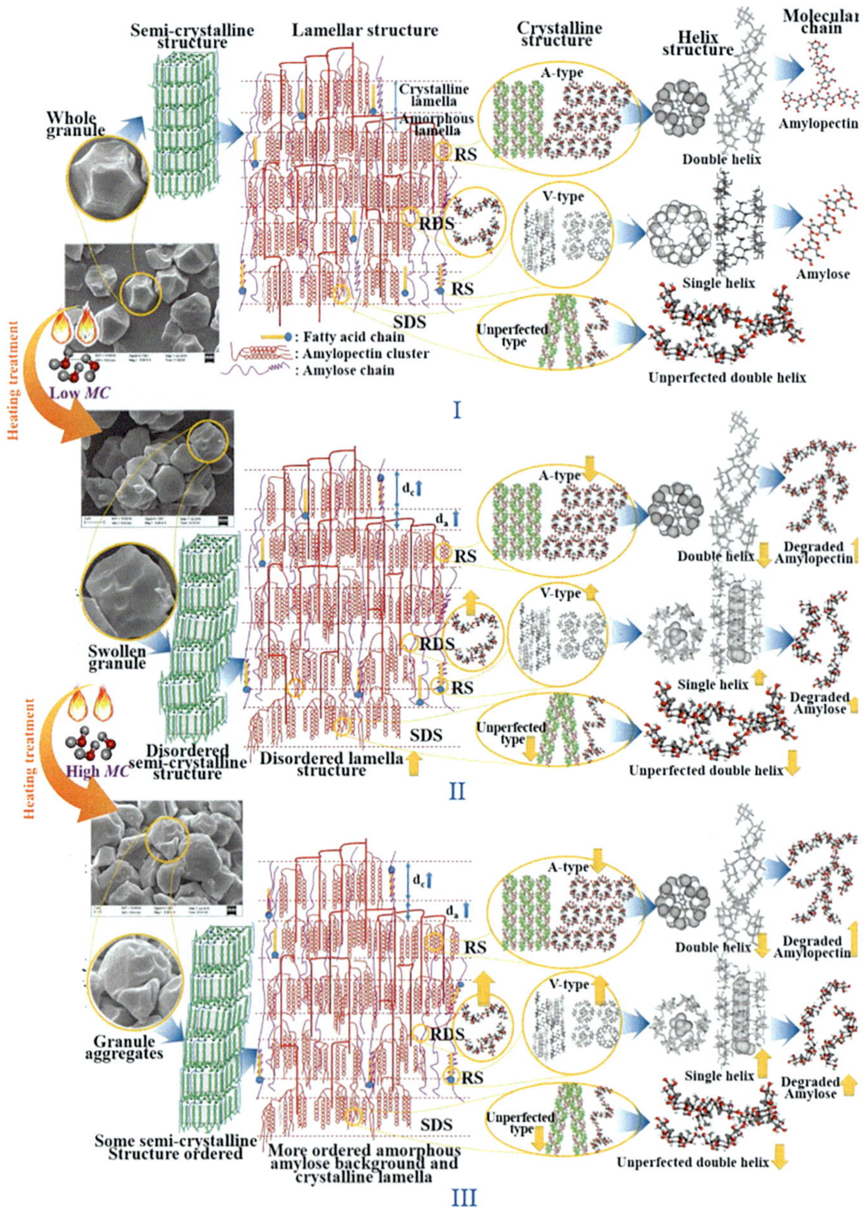

Fig. 12.4 Schematic representation for the changes of multi-scale structure and digestibility of native and HMT-treated rice starches. (Drawn by the author)

obesity development and there is no doubt that more and more fine dietary structures involving high carbohydrate and fat content play an important role (Raikos and Ranawana 2019). Reasonably optimized food components in the diet have been one of the widely accepted hotspots in the field of nutrition medicine and food science. From a prospective observational study on nutritional food, there is a growing body of evidence that foods with a high content of SDS and RS are protective against the development of diet-induced obesity.

Starch is the main component in foods, which plays an essential role in human nutrition. And the digestion rate and extent of starch are primary factors affecting the postprandial glucose and insulin responses. Many studies have highlighted that SDS and RS could lower the glycemic index (GI) and improve insulin sensitivity. In particular, RS was fermented by the gut microbiota into short-chain fatty acids (SCFAs) that exerted a prebiotic effect on colon microbes, the adjustment of lipid metabolism, and the reduction in the risk of ulcerative colitis. Zheng et al. (2018) have investigated the nutritional functions of HMT-modified rice starch through the physiological and biochemical indices, proliferation and distribution of intestinal microflora, and functional diversity by putative metagenome analysis. They found that high SDS content samples can decrease the blood glucose, serum lipid, oxidative stress, and liver function metabolic levels/indices of obesity mice. Meanwhile, the species diversity index was observed to be higher in the high-SDS group, and *Bifidobacteria* was identified as a kind of bacteria related to the relatively higher content of RS. Also, they further evaluated the impact of HMT-treated rice starch-oleic acid complex supplementation on body weight, serum lipid profiles, hepatic metabolism, and composition of gut microbiota of rats fed a high-fat diet, and the beneficial effect of HMT-treated complex might be related to the higher resistant starch (RS) level (Zheng et al. 2020a, b). Interestingly, rice starch-oleic acid complex supplementation contributed to the proliferation and growth of butyrate-producing bacteria. The related results revealed that the genera *Turicibacter* and *Romboutsia* were positively correlated with HDL-c and SOD levels. Meanwhile, based on the metagenomic data, the *Bifidobacteria* genus might be the main primary degrader after rice starch-oleic acid complex intake, which was associated with the change in key starch degradation enzymes. However, the specific mechanism of how SDS or RS formed during HMT degradation in the gut and how it affects and regulates metabolic pathways in gut microbiota should be further studied. And some pathways influenced by these two fractions were annotated to reveal the functional metabolism in the body.

12.4 Functionality of HMT-Treated Starch (HMTS) Was Affected by Exogenous Lipids

Only a few studies have evaluated the effect of exogenous lipids on the functionalities of HMT-treated rice starch. Incorporating exogenous lipids into the HMT system usually leads to two endothermic peaks in the DSC curve, which corresponds to the melting of crystalline double helices and amylose–lipid complex, respectively. Several fatty acids, which may create oil layer coating starch granules during HMT, could delay the emergence of the first peak. Regarding the second peak, the phase transition varies with the type and amount of lipids as well as HMT parameters. For example, the rice starch-oleic acid complex treated by HMT had a T_p of 105.1–110.0 °C, while the value of treated complex containing linoleic acid was 91.70–97.66 °C (He et al. 2020). Chen et al. (2017a, b) demonstrated that rice starch complexed with maize oil required higher temperatures for dissociation, i.e., a higher gelatinization temperature was obtained in HMT-modified rice starch-maize oil complex, especially at higher moisture content. On the one hand, these features were attributed to amylose–amylose and amylose–amylopectin interactions during HMT. These interactions reduced the mobility of the amylopectin chains and increased dissociating temperatures consequently. On the other hand, the formation of amylose–lipid complex combined with a new superficial oil layer during HMT restricted water penetration into the granules and retarded the granule swelling. Furthermore, the enthalpy changes (ΔH) of starch–lipid complex showed an increasing trend with the harsh conditions during HMT (He et al. 2020), suggesting that the presence of exogenous lipids could increase more ordered structure during HMT (Fig. 12.5).

The synergistic effect of HMT and fatty acid complexation could increase the PT of rice starch but decrease its pasting viscosities. More remarkably, the SB could dwindle to a negative value. Like the gelatinization profile changes, it indicates the

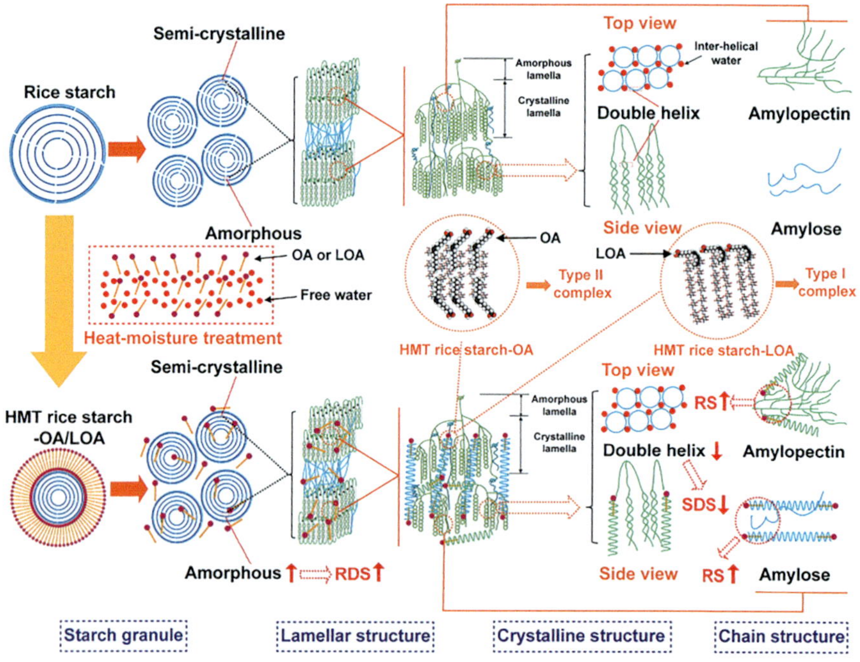

Fig. 12.5 Mechanism of structure changes and enzymatic resistivity with the effect of oleic acid (OA) or linoleic acid (LOA) on rice starch during HMT. (Drawn by the authors)

formation of a stabilized structure during HMT through the mutual entanglement or linking in starch-water-lipid ternary system. Junction zones in cold paste may be blocked by the amylose–lipid complex, which also favors the decrease in SB. Enhancing HMT intensity makes their interactions stimulated, which could be reflected by the pasting profile changes of treated rice starch-corn oil complex as the HMT moisture content rose from 20% to 30%. As well, the principle has been demonstrated for the incorporation of saturated fatty acids. Further investigations could analyze the role of fatty acid characteristics (e.g., the amount of unsaturated bonds and chain length) in their interactions with starch during HMT. Blending HMTS and food ingredients would spur its application in frozen and bakery foods, owing to the further enhanced thermal/mechanical stability and anti-retrogradation capacity (reduced BD and SB).

Elevating HMT moisture content could promote the conversion of RDS to SDS and RS when rice starch coexists with corn oil (Chang et al. 2014). This is because greater moisture content enhances the diffusion of lipids to bind with amylose during treatment. Compared with the original amorphous amylose, the amylose–lipid complex possesses a greatly reinforced architecture (in orderly packed single spiral state) that makes enzyme attack more difficult. Meanwhile, similar changes in the digestion profile of HMTS have been found with the addition level of fatty acids. The type of fatty acids also has a profound impact. It can be supported by the changing degree induced by oleic acid and linoleic acid addition, respectively, when the same level (2%) is applied. As confirmed by He et al. (2020), fatty acids of higher saturation degree formed a stable amylose–lipid complex more smoothly due to its easier access to the helical cavity. Whether HMTS from other cereals obeys this principle needs further exploration. Compared with HMT-modified isolated starch, HMTS generated from flour has prominently lower RDS but higher SDS or RS content, such as that from rice flour (Xiao et al. 2017).

Though the functionalities of HMT-treated rice starch can be affected by coexisting lipids through non-covalent interactions, how other rice ingredients (e.g., non-starch carbohydrates, protein) affect the techno-functional and processing properties of HMTS is still urgently needed to be solved.

12.5 Influence of HMTS on the Final Product Texture and Volume

The textural properties of the final products are one of the most important factors that affect the consumer's behavior. The texture parameters are usually determined through a texture analyzer. Firmness, also known as hardness, represents the peak force to snap the bakery product and is correlated with the sensory characteristics (Chompoorat et al. 2018). Cohesiveness is a measure of the sample internal linkage strength. Springiness, also called elasticity, represents the rate at which the deformed sample is regaining its form and size. Resilience is close to springiness, but it is calculated as a ratio of energies and it refers to how well the sample fights to return to its initial position. The energy needed to chew a food sample until it becomes ready to swallow represents the chewiness.

Concerning baked foods, the presence of HMTS has also caused the darkening effect on products, which becomes more obvious when HMT intensity gets enhanced (Fathi et al. 2016). Selecting proper HMT conditions is vital for developing products with sound appearance and sensory acceptability. For example, the volume of bread was enhanced when prepared from HMTS-containing wheat flour generated at 55 or 65 °C, whereas it revealed a smaller volume and unwelcome shape as the HMT temperature reached 95 °C (Cetiner et al. 2017). A larger volume and softer texture could be achieved for pound cake after the supplementation of HMT-millet flour modified at 100 °C, while excessively severe HMT led to an opposite change and even caused deterioration in odor and taste (Fathi et al. 2016). The attractive color and flavor of the cake (from sole rice) got remarkably enhanced after HMT on rice flour, and further enhancement was obtained as the HMT time rose from 1 to 2 h (Kim et al. 2017). More efforts must reveal the detailed correlations between the sensory attributes of products and HMT parameters. Also, Chung et al. (2014) showed that wheat flour replacement with germinated and HMT-treated brown rice in cookies determined a decrease in hardness compared to the untreated samples, due to the interactions between cookie components. As the substitution level was higher, the hardness was lower. The properties of cookies are influenced by the composite matrix structure, which comprises protein aggregates, lipids, and sugars, and scattered among the starch granules (Chung et al. 2014). Another study made by Bourekoua et al. (2016) revealed that the volume of gluten-free rice and corn bread was improved by the hydrothermal treatments, probably due to the higher viscosity, which allows better gas retention. The hardness and chewiness of breads with rice and cord additions were lower than those with untreated flours. Springiness is an indicator of aeration and product elasticity, and the HMT induced lower values of this parameter. A negative effect was also observed for the cohesiveness, which represents the capacity of the material to be deformed before breaking down (Bourekoua et al. 2016).

Fundamentally, the hardness of baked foods made of HMT flours decreased compared to the control, except that one made of flour subjected to higher temperature (120 °C) and moisture content (30%) with the volume diminishing can be caused by the excessive viscosity of the batter that prevented the oven expansion (Gularte et al. 2012; Marston et al. 2016). Specific gravity decrease and consistency increase may be due to the reinforcements of batter structure and due to the air bubbles incorporated (Marston et al. 2016). Lower hardness values were reported for cake made of HMT rice flour at high temperature and long times due to the structural rearrangement during HMT; the degraded starch chains and other small molecules that appear during heating may be responsible for inhibiting starch chain association (Kim et al. 2017). HMT rice flour cake presented lower springiness, cohesiveness, chewiness, and gumminess compared to the control, indicating a weaker structure because of the retarded retrogradation, which increased cake elasticity (Kim et al. 2017). HMTS-based flour can extend the shelf life of baked products.

12.6 Potential Applications of HMT-Treated Starches

Hydrothermal treatments can be used to obtain food products with higher nutritional values and low gluten content (Chung et al. 2014; Wang et al. 2016a, b). The enhancement of the specific volume, texture, and crumb structure recommends hydrothermal treatments to obtain bakery products for persons having celiac disease (Zavareze et al. 2010; Bourekoua et al. 2016). Value-added cakes with high quality and nutritional value can be obtained using these techniques. HMT-modified starches have greater thermal stability and shear resistance; they can be used in sauces, confectionery, canned products, and pasta. The application of amaranth and cornstarch in HMT-treated pasta was studied; it was observed that the pasta had good expansion, minimum cooking time, and firm texture, all of which are desirable characteristics in pasta manufacturing. Additionally, HMT enhances the water absorption of starch, leading to a better gel formation and strength, which is beneficial in sauce production. HMT reduces the retrogradation tendency, which is an advantage for the extension of bakery products' shelf life. The increase in starch-based system viscosity after hydrothermal treatments is another advantage that leads to higher loaf volume and gas retention in bakery products (Bourekoua et al. 2016; Fathi et al. 2016). Moreover, HMT-modified starches are important as expanders in the production of biscuit, as a binder in breaded foods; in bakery products, as emulsion stabilizers in beverages, creams, and canned foods; and as thickeners in various food products. Thus, the modification of starches by HMT makes it possible to obtain starches with appropriate characteristics, such as gelatinization, heat stability, solubility, and textures, for the most diverse industrial applications, mainly in the food sector. As driven by the increased consumption of processed and convenience foods, the modified starch market is highly concerned.

Hydrothermal treatments affect food product nutritional value and digestibility. During moisture heating, the RDS content is reduced and the RS and SDS increase. The RS increase depends on the moisture level, and thus, it is important to optimize this parameter of the hydrothermal treatment to obtain better results. The optimization is not a time-consuming method and should depict the complete effects of the factors in the process, considering the combined interactions among these factors. By applying these treatments, the fiber content of the starch-based food products can be enhanced. The non-wheat flours can be added in bakery products after hydrothermal treatments, without strongly negatively influencing the quality of the final product (Wang et al. 2021a, b).

Finally, HMT has been applied to improve the characteristics of biodegradable films obtained from starch and rice flour. The authors showed that this treatment can improve the properties of the films, resulting in better tensile strength, increased stretch resistance, and greater rigidity and extensibility compared to non-hydrothermally treated films (Majzoobi et al. 2015). This physical modification was also applied to edible films prepared with sweet potato starch. The films modified by HMT showed an increase in thickness, tensile strength, and elongation; however, solubility decreased (Ismailoglu and Basman 2016).

12.7 Conclusions and Prospects

HMT enables rice starch to undergo glass transition, inducing considerable changes in its hierarchical structure. Generally, compared with native starch, HMTS usually has lowered molecular weight, and the molecular degradation is closely associated with chain characteristic change or chemical bond breakage. HMT disrupts the molecular interactions located in the exterior or interior of starch granules, especially for hydrogen bonds. The disassociation of hydrogen bonds disrupts molecular structure, such as decreased degree of short-range ordered structure and ordered molecular structure. HMTS retains the original A-type polymorph and generates the newly formed V-type crystals. The relative crystallinity greatly varies with amylose content and HMT conditions, which is essentially evoked by the adjustments in lamellae and helical structure as well as molecular reordering. HMTS has changed surface characteristics at micro- and nanoscale along with larger granular size. Following these structural fluctuations, the functionalities of cereal starch get changed to varying extent based on the intensity and type of HMT. In particular, most papers have verified that HMTS has enhanced gelatinization/pasting temperatures and declined BD/SB values. This means that HMT endows it with a more stable and resistant architecture, which could broaden the application in easy-to-aging products (e.g., frozen or bakery foods). Edible (gelatinized) HMTS has a higher level of RS or SDS, providing an innovative approach for the exploitation of functional ingredients. The functionalities of HMTS are affected by coexisting lipids through non-covalent interactions or simply attachment/encapsulation. Supplementation of HMTS-containing flour would ease the sensory quality and shelf life of foods if proper HMT parameters are adopted. Although there is substantial data about the characteristic structure and physicochemical properties of HMTS in the literature, some aspects still require improvements for the better and effective usage in starchy food systems. For example, the investigation of HMT parameters at present mainly involves treating moisture content, duration, temperature, and repeating times. Other factors may also be taken into account, such as treatment pressure, heating matrix (e.g., oil, steaming, or oven), and cooling procedure. Given the structural identification, more efforts can be on the nano-level change of HMTS granules and the specific location of degraded molecular chains. Quantitative correlation analysis would be useful for grasping the exact associations among the granular, crystalline, helical, and molecular structure of HMTS. Until now, only the interactions with exogenous lipids have been studied, and more research should be carried out on the interactions with other food ingredients including non-starch polysaccharides and protein. More comparative studies should be conducted on the characteristics of HMTS generated from isolated starch versus cereal flour/whole grains, to systematically understand the interplay with endogenous components. It would be significant to prepare dough or food products with sole HMTS, and make contrasts with those from HMTS + gluten, thus elucidating the role of HMTS in food quality. HMTS further modified by other methods may facilitate its applications in the production of foods with improved consumer acceptability and nutritional attributes.

References

Alimi BA, Workneh TS (2018) Structural and physicochemical properties of heat moisture treated and citric acid modified acha and iburu starches. Food Hydrocoll 81:449–455

Amagliani L, O'Regan J, Kelly AL, O'Mahony JA (2016) Chemistry, structure, functionality and applications of rice starch. J Cereal Sci 70:291–300

Ambigaipalan P, Hoover R, Donner E, Liu Q (2014) Starch chain interactions within the amorphous and crystalline domains of pulse starches during heat-moisture treatment at different temperatures and their impact on physicochemical properties. Food Chem 143:175–184

Anderson AK, Guraya HS, James C, Salvaggio L (2002) Digestibility and pasting properties of rice starch heat-moisture treated at the melting temperature (T_m). Starch - Stärke 54(9):401–409

Arns B, Bartz J, Radunz M, Evangelho JAD, Pinto VZ, Zavareze EDR, Dias ARG (2015) Impact of heat-moisture treatment on rice starch, applied directly in grain paddy rice or in isolated starch. LWT Food Sci Technol 60(2):708–713

Ashwar BA, Gani A, Wani IA, Shah A, Masoodi FA, Saxena DC (2016) Production of resistant starch from rice by dual autoclaving-retrogradation treatment: invitro digestibility, thermal and structural characterization. Food Hydrocoll 56:108–117

Balet S, Guelpa A, Fox G, Manley M (2019) Rapid visco analyser (RVA) as a tool for measuring starch-related physiochemical properties in cereals: a review. Food Anal Methods 12(10): 2344–2360

Bao J, Bergman CJ (2018) Chapter 10 - Rice flour and starch functionality. In: Sjöö M, Nilsson L (eds) Starch in food, 2nd edn. Woodhead Publishing, pp 373–419

Bemiller JN, Huber KC (2015) Physical modification of food starch functionalities. Annu Rev Food Sci Technol 6(1):19

Bian L, Chung HJ (2016) Molecular structure and physicochemical properties of starch isolated from hydrothermally treated brown rice flour. Food Hydrocoll 60:345–352

Bourekoua H, Benatallah L, Zidoune MN, Rosell CM (2016) Developing gluten free bakery improvers by hydrothermal treatment of rice and corn flours. LWT Food Sci Technol 73:342–350

Cetiner B, Acar O, Kahraman K, Sanal T, Koksel H (2017) An investigation on the effect of heat-moisture treatment on baking quality of wheat by using response surface methodology. J Cereal Sci 74:103–111

Chang F, He X, Fu X, Huang Q, Jane JL (2014) Effects of heat treatment and moisture contents on interactions between lauric acid and starch granules. J Agric Food Chem 62(31):7862–7868

Chen X, Du X, Chen P, Guo L, Xu Y, Zhou X (2017a) Morphologies and gelatinization behaviours of high-amylose maize starches during heat treatment. Carbohydr Polym 157:637–642

Chen X, He X, Fu X, Zhang B, Huang Q (2017b) Complexation of rice starch/flour and maize oil through heat moisture treatment: structural, in vitro digestion and physicochemical properties. Int J Biol Macromol 98:557–564

Chi C, Li X, Huang S, Chen L, Zhang Y, Li L, Miao S (2021a) Basic principles in starch multi-scale structuration to mitigate digestibility: a review. Trends Food Sci Technol 109:154–168

Chi C, Li X, Zhang Y, Chen L, Li L, Miao S (2021b) Progress in tailoring starch intrinsic structures to improve its nutritional value. Food Hydrocoll 113:106447

Chi C, Li X, Zhang Y, Chen L, Li L, Wang Z (2017) Digestibility and supramolecular structural changes of maize starch by non-covalent interactions with gallic acid. Food Funct 8(2):720–730

Chompoorat P, Rayas-Duarte P, Hernandez-Estrada ZJ, Phetcharat C, Khamsee Y (2018) Effect of heat treatment on rheological properties of red kidney bean gluten free cake batter and its relationship with cupcake quality. J Food Sci Technol Mysore 55(12):4937–4944

Chung HJ, Cho A, Lim ST (2014) Utilization of germinated and heat-moisture treated brown Rices in sugar-snap cookies. LWT-Food Sci Technol 57(1):260–266

Cozzolino D (2016) The use of the rapid visco analyser (RVA) in breeding and selection of cereals. J Cereal Sci 70:282–290

Fathi B, Aalami M, Kashaninejad M, Mahoonak AS (2016) Utilization of heat-moisture treated proso millet flour in production of gluten-free pound cake. J Food Qual 39(6):611–619

Gidley MJ, Bulpin PV (1987) Crystallisation of malto-oligosaccharides as models of the crystalline forms of starch: minimum chain-length requirement for the formation of double helices. Carbohydr Res 161(2):291–300

Gularte MA, de la Hera E, Gómez M, Rosell CM (2012) Effect of different fibers on batter and gluten-free layer cake properties. LWT Food Sci Technol 48(2):209–214

Gunaratne A, Gan R, Wu K, Kong X, Collado L, Arachchi LV, Kumara K, Pathirana SM, Corke H (2018) Physicochemical properties of mung bean starches isolated from four varieties grown in Sri Lanka. Starch - Stärke 70(3–4):1700129

Han Z, Shi R, Sun DW (2020) Effects of novel physical processing techniques on the multi-structures of starch. Trends Food Sci Technol 97:126–135

He H, Zheng B, Wang H, Li X, Chen L (2020) Insights into the multi-scale structure and in vitro digestibility changes of rice starch-oleic acid/linoleic acid complex induced by heat-moisture treatment. Food Res Int 137:109612

Hoover R (2010) The impact of heat-moisture treatment on molecular structures and properties of starches isolated from different botanical sources. Crit Rev Food Sci Nutr 50(9):835–847

Hoover R, Manuel H (1996) The effect of heat–moisture treatment on the structure and physico-chemical properties of normal maize, waxy maize, dull waxy maize and amylomaize V starches. J Cereal Sci 23(2):153–162

Hormdok R, Noomhorm A (2007) Hydrothermal treatments of rice starch for improvement of rice noodle quality. LWT Food Sci Technol 40(10):1723–1731

Hung PV, Vien NL, Lan Phi NT (2016) Resistant starch improvement of rice starches under a combination of acid and heat-moisture treatments. Food Chem 191:67–73

Ismailoglu SO, Basman A (2016) Physicochemical properties of infrared heat-moisture treated wheat starch. Starch - Stärke 68(1):67–75

Jiranuntakul W, Puttanlek C, Rungsardthong V, Puncha-arnon S, Uttapap D (2012) Amylopectin structure of heat-moisture treated starches. Starch - Stärke 64(6):470–480

Kaur M, Singh S (2019) Influence of heat-moisture treatment (HMT) on physicochemical and functional properties of starches from different Indian oat (*Avena sativa* L.) cultivars. Int J Biol Macromol 122:312–319

Kim MJ, Oh SG, Chung HJ (2017) Impact of heat-moisture treatment applied to brown rice flour on the quality and digestibility characteristics of Korean rice cake. Food Sci Biotechnol 26(6): 1579–1586

Kumari S, Yadav BS, Yadav RB (2020) Synthesis and modification approaches for starch nanoparticles for their emerging food industrial applications: a review. Food Res Int 128:108765

Lee S, Lee JH, Chung HJ (2017) Impact of diverse cultivars on molecular and crystalline structures of rice starch for food processing. Carbohydr Polym 169:33–40

Li C, Hu Y, Li E (2021) Effects of amylose and amylopectin chain-length distribution on the kinetics of long-term rice starch retrogradation. Food Hydrocoll 111:106239

Li H, Gilbert RG (2018) Starch molecular structure: the basis for an improved understanding of cooked rice texture. Carbohydr Polym 195:9–17

Li N, Wang L, Zhao S, Qiao D, Jia C, Niu M, Lin Q, Zhang B (2020) An insight into starch slowly digestible features enhanced by microwave treatment. Food Hydrocoll 103:105690

Lin CL, Lin JH, Lin JJ, Chang YH (2019a) Progressive alterations in crystalline structure of starches during heat-moisture treatment with varying iterations and holding times. Int J Biol Macromol 135:472–480

Lin L, Huang J, Zhang L, Zhang C, Liu Q, Wei C (2019b) Effects of inhibiting starch branching enzymes on molecular and crystalline structures of starches from endosperm different regions in rice. Food Chem 301:125271

Liu H, Guo X, Li W, Wang X, Lv M, Peng Q, Wang M (2015) Changes in physicochemical properties and in vitro digestibility of common buckwheat starch by heat-moisture treatment and annealing. Carbohydr Polym 132:237–244

Liu K, Zhang B, Chen L, Li X, Zheng B (2019) Hierarchical structure and physicochemical properties of highland barley starch following heat moisture treatment. Food Chem 271:102–108

Lockyer S, Nugent AP (2017) Health effects of resistant starch. Nutr Bull 42(1):10–41

Lozano-Vazquez G, Alvarez-Ramirez J, Lobato-Calleros C, Vernon-Carter EJ, Hernández-Marín NY (2021) Characterization of corn starch-calcium alginate xerogels by microscopy, thermal, XRD, and FTIR analyses. Starch - Stärke 73(7–8):2000282

Majzoobi M, Pesaran Y, Mesbahi G, Golmakani MT, Farahnaky A (2015) Physical properties of biodegradable films from heat-moisture-treated rice flour and rice starch. Starch - Stärke 67: 1053–1060

Marston K, Khouryieh H, Aramouni F (2016) Effect of heat treatment of sorghum flour on the functional properties of gluten-free bread and cake. LWT-Food Sci Technol 65(1):637–644

Masina N, Choonara YE, Kumar P, Du Toit LC, Govender M, Indermun S, Pillay V (2017) A review of the chemical modification techniques of starch. Carbohydr Polym 157:1226–1236

Miao M, Zhang T, Jiang B (2009) Characterisations of kabuli and desi chickpea starches cultivated in China. Food Chem 113(4):1025–1032

Muthayya S, Sugimoto JD, Montgomery S, Maberly GF (2014) An overview of global rice production, supply, trade, and consumption. In: DeRegil LM, Laillou A, MoenchPfanner R, PenaRosas JP (eds) Technical considerations for rice fortification in public health, vol 1324, pp 7–14

Mutungi C, Passauer L, Onyango C, Jaros D, Rohm H (2012) Debranched cassava starch crystallinity determination by Raman spectroscopy: correlation of features in Raman spectra with X-ray diffraction and ^{13}C CP/MAS NMR spectroscopy. Carbohydr Polym 87(1):598–606

Nakamura S, Okumura H, Sugawara M, Noro W, Homma N, Ohtsubo K (2017) Effects of different heat–moisture treatments on the physicochemical properties of brown rice flour. Biosci Biotechnol Biochem 81(12):2370–2385

Piecyk M, Drużyńska B, Ołtarzewska A, Wołosiak R, Worobiej E, Ostrowska-Ligęza E (2018) Effect of hydrothermal modifications on properties and digestibility of grass pea starch. Int J Biol Macromol 118:2113–2120

Qiao DL, Yu L, Liu HS, Zou W, Xie FW, Simon G, Petinakis E, Shen ZQ, Chen L (2016) Insights into the hierarchical structure and digestion rate of alkali-modulated starches with different amylose contents. Carbohydr Polym 144:271–281

Raikos V, Ranawana V (2019) Reformulation as a Strategy for Developing Healthier Food Products Challenges, Recent Developments and Future Prospects: Challenges, Recent Developments and Future Prospects[M]. 23–45. https://doi.org/10.1007/978-3-030-23621-2

Ruiz E, Srikaeo K, Revilla L (2018) Effects of heat moisture treatment on physicochemical properties and starch digestibility of rice flours differing in amylose content. Food Appl Biosci J 6(3):140–153

Sair L, Fetzer WR (1944) Water sorption by cornstarch and commercial. Modifications of starches. Ind Eng Chem 36(4):316–319

Schafranski K, Ito VC, Lacerda LG (2021) Impacts and potential applications: a review of the modification of starches by heat-moisture treatment (HMT). Food Hydrocoll 117:106690

Shi M, Gao Q, Liu Y (2018) Corn, potato, and wrinkled pea starches with heat–moisture treatment: structure and digestibility. Cereal Chem 95(5):603–614

Shi YC, Seib PA (1992) The structure of four waxy starches related to gelatinization and retrogradation. Carbohydr Res 227:131–145

Silva WMF, Biduski B, Lima KO, Pinto VZ, Hoffmann JF, Vanier NL, Dias ARG (2017) Starch digestibility and molecular weight distribution of proteins in rice grains subjected to heat-moisture treatment. Food Chem 219:260–267

Sudheesh C, Sunooj KV, Alom M, Kumar S, Sajeevkumar VA, George J (2020) Effect of dual modification with annealing, heat moisture treatment and cross-linking on the physico-chemical, rheological and in vitro digestibility of underutilised kithul (*Caryota urens*) starch. J Food Meas Charact 14(3):1557–1567

Sui Z, Yao T, Ye X, Bao J, Kong X, Wu Y (2016) Physicochemical properties and starch digestibility of in-kernel heat-moisture-treated waxy, low-, and high-amylose rice starch. Starch - Stärke 68:1–9

Sun Q, Han Z, Wang L, Xiong L (2014) Physicochemical differences between sorghum starch and sorghum flour modified by heat-moisture treatment. Food Chem 145:756–764

Sun Q, Wang T, Xiong L, Zhao Y (2013) The effect of heat moisture treatment on physicochemical properties of early indica rice. Food Chem 141(2):853–857

Suriya M, Reddy CK, Haripriya S (2019) Functional and thermal behaviors of heat-moisture treated elephant foot yam starch. Int J Biol Macromol 137:783–789

Van Hung P, Binh VT, Nhi PHY, Phi NTL (2020) Effect of heat-moisture treatment of unpolished red rice on its starch properties and in vitro and in vivo digestibility. Int J Biol Macromol 154:1–8

Vandeputte GE, Vermeylen R, Geeroms J, Delcour JA (2003) Rice starches. I. Structural aspects provide insight into crystallinity characteristics and gelatinisation behaviour of granular starch. J Cereal Sci 38(1):43–52

Wang H, Ding J, Xiao N, Liu X, Zhang Y, Zhang H (2020a) Insights into the hierarchical structure and digestibility of starch in heat-moisture treated adlay seeds. Food Chem 318:126489

Wang H, Liu Y, Chen L, Li X, Wang J, Xie F (2018) Insights into the multi-scale structure and digestibility of heat-moisture treated rice starch. Food Chem 242:323–329

Wang H, Wang Z, Li X, Chen L, Zhang B (2017a) Multi-scale structure, pasting and digestibility of heat moisture treated red adzuki bean starch. Int J Biol Macromol 102:162–169

Wang H, Xiao N, Ding J, Zhang Y, Liu X, Zhang H (2020b) Effect of germination temperature on hierarchical structures of starch from brown rice and their relation to pasting properties. Int J Biol Macromol 147:965–972

Wang H, Xu K, Liu X, Zhang Y, Xie X, Zhang H (2021a) Understanding the structural, pasting and digestion properties of starch isolated from frozen wheat dough. Food Hydrocoll 111:106168

Wang H, Xu K, Ma Y, Liang Y, Zhang H, Chen L (2020c) Impact of ultrasonication on the aggregation structure and physicochemical characteristics of sweet potato starch. Ultrason Sonochem 63:104868

Wang H, Zhang B, Chen L, Li X (2016a) Understanding the structure and digestibility of heat-moisture treated starch. Int J Biol Macromol 88:1–8

Wang Q, Li L, Zheng X (2021b) Recent advances in heat-moisture modified cereal starch: structure, functionality and its applications in starchy food systems. Food Chem 344

Wang S, Sun Y, Wang J, Wang S, Copeland L (2016b) Molecular disassembly of rice and lotus starches during thermal processing and its effect on starch digestibility. Food Funct 7(2): 1188–1195

Wang S, Wang S, Liu L, Wang S, Copeland L (2017b) Structural orders of wheat starch do not determine the in vitro enzymatic digestibility. J Agric Food Chem 65(8):1697–1706

Xiao Y, Liu H, Wei T, Shen J, Wang M (2017) Differences in physicochemical properties and in vitro digestibility between tartary buckwheat flour and starch modified by heat-moisture treatment. LWT-Food Sci Technol 86:285–292

Xie X, Qi L, Xu C, Shen Y, Wang H, Zhang H (2020) Understanding how the cooking methods affected structures and digestibility of native and heat-moisture treated rice starches. J Cereal Sci 95:103085

Yan Y, Feng L, Shi M, Cui C, Liu Y (2020) Effect of plasma-activated water on the structure and in vitro digestibility of waxy and normal maize starches during heat-moisture treatment. Food Chem 306:125589

Yang X, Chi C, Liu X, Zhang Y, Zhang H, Wang H (2019) Understanding the structural and digestion changes of starch in heat-moisture treated polished rice grains with varying amylose content. Int J Biol Macromol 139:785–792

Yi D, Maike W, Yi S, Xiaoli S, Dianxing W, Wenjian S (2021) Physiochemical properties of resistant starch and its enhancement approaches in rice. Rice Sci 28(1):31–42

Zavareze EDR, Guerra Dias AR (2011) Impact of heat-moisture treatment and annealing in starches a review. Carbohydr Polym 83(2):317–328
Zavareze EDR, Storck CR, de Castro LAS, Schirmer MA, Dias ARG (2010) Effect of heat-moisture treatment on rice starch of varying amylose content. Food Chem 121(2):358–365
Zhang B, Chen L, Li X, Li L, Zhang H (2015) Understanding the multi-scale structure and functional properties of starch modulated by glow-plasma: a structure-functionality relationship. Food Hydrocoll 50:228–236
Zhang B, Li X, Liu J, Xie F, Chen L (2013) Supramolecular structure of A- and B-type granules of wheat starch. Food Hydrocoll 31(1):68–73
Zhang B, Zhao K, Su C, Gong B, Ge X, Zhang Q, Li W (2020) Comparing the multi-scale structure, physicochemical properties and digestibility of wheat A- and B-starch with repeated versus continuous heat-moisture treatment. Int J Biol Macromol 163:519–528
Zhang B, Zhao Y, Li X, Zhang P, Li L, Xie F, Chen L (2014) Effects of amylose and phosphate monoester on aggregation structures of heat-moisture treated potato starches. Carbohydr Polym 103:228–233
Zhang T, Li X, Chen L, Situ W (2016) Digestibility and structural changes of waxy rice starch during the fermentation process for waxy rice vinasse. Food Hydrocoll 57:38–45
Zheng B, Wang H, Shang W, Xie F, Li X, Chen L, Zhou Z (2018) Understanding the digestibility and nutritional functions of rice starch subjected to heat-moisture treatment. J Funct Foods 45: 165–172
Zheng B, Wang T, Wang H, Chen L, Zhou Z (2020a) Studies on nutritional intervention of rice starch- oleic acid complex (resistant starch type V) in rats fed by high-fat diet. Carbohydr Polym 246:116637
Zheng M, Xiao Y, Yang S, Liu H, Liu M, Yaqoob S, Xu X, Liu J (2020b) Effects of heat-moisture, autoclaving, and microwave treatments on physicochemical properties of proso millet starch. Food Sci Nutr 8(2):735–743
Zhong Q, Li M, Atiba EM, Yin Y, Yang Q, Zhang L, Sun Z (2021) Effects of physicochemical and structural properties of single and double feedstuffs derived from different botanic sources on in vitro starch digestion and glucose release kinetics. J Anim Physiol Anim Nutr 105(3): 493–506
Zhu F (2016) Impact of γ-irradiation on structure, physicochemical properties, and applications of starch. Food Hydrocoll 52:201–212
Zhu F (2018) Modifications of starch by electric field based techniques. Trends Food Sci Technol 75:158–169

Chapter 13
Perspective of Future Rice Processing

Yukiharu Ogawa

Abstract As a summary of this book, the nutritional properties of rice, the structure changes at different levels, and the processing technologies and the usage of rice by-products are mentioned in brief again. Particular, future rice foods with more healthy and functional attributes are also looked into the distance. And confirmed that advanced processing methods to enhance the healthy and functional properties of rice should be developed and applied to rice products in the future.

Keywords Rice processing technology · Healthy benefits · Future food

The COVID-19 pandemic has changed the international situation, and the destabilization by the armed conflicts among nations in the grain-growing regions would influence the global food supply-demand balance. However, as of 2021, world rice production is forecast to be 518.2 million tons of milled rice 2021 annually, which will be up 0.9% from the 2020 production (FAO 2021). Also, global rice production is expected to reach 567 million tons by 2030, while Asian countries, except Japan and Korea, will contribute to most of the additional global output (OECD/FAO 2021). Therefore, rice has still been one of the most important staple foods, mainly in Asian countries.

As a foodstuff, the nutritional properties are the most crucial attribute to understanding its value. Therefore, in general, that is the primary concern of food science, as described in many previous publications as the essential information of foods as physical matter. However, food is a complex material with structural attributes at tissue and molecular levels, usually called "food matrix." Nutritional ingredients are enclosed in or form such structural complexes themselves. Thus, the structural changes, particularly at the postprandial digestion time series, can influence and vary its nutritional properties. These property changes are connected to the bioaccessibility and bioavailability of food ingredients, which could reveal the actual nutritional values of food. In other words, the nutritional properties of food need to

Y. Ogawa (✉)
Graduate School of Horticulture, Chiba University, Chiba, Japan
e-mail: ogwy@faculty.chiba-u.jp

J. Tian et al. (eds.), *Science of Rice Chemistry and Nutrition*,
https://doi.org/10.1007/978-981-99-3224-5_13

be considered under the dynamic phenomena. For example, the nutritional properties of starch foods can be evaluated by not only static total starch amount but also changes in the postprandial blood glucose level, which is related to the structural food attributes. Such dynamic phenomena are essential factors in assessing the health-related properties of starch foods.

Rice is usually consumed as a cooked grain form that remains a plant seed tissue structure. Therefore, it can be regarded as an appropriate food to evaluate the influence of structural attributes on nutritional value. Such structural characteristics can also be varied by the processing and cooking methods. This book consists of "Rice Chemistry" part, which mainly describes the chemical components of rice, and "Rice Processing" part, which sectionally focuses on the dynamic phenomena, including the effect of processing on the digestive properties.

In the "Rice Chemistry" part, the effective usage of rice bran, which is a by-product of rice postharvest processing, can be recognized as a rice-associated future perspective. Because brown rice grain is a typical Caryopsis, the morphological structure consists of the pericarp, seed coat, germ, and endosperm. The pericarp, seed coat, germ, and part of the aleurone layer of the endosperm are the main causes of the low sensory traits of cooked rice; thus, they are usually removed. Rice bran is also recognized as a source of high-quality plant-based protein. The growing stage of the rice kernel can also be considered as a protein source. Hydrolysates from rice proteins can be applied to nutraceutical products. However, fully effective processing methods, particularly in the extraction of rice bran proteins with high recovery and efficiency on economical production cost, have remained to be developed. The bioactive components of rice bran, such as polyphenols, deserve further and deeper investigation. Some processing methods can increase the content of polyphenols in rice, such as germination, microwave processing, and ultrasound processing. However, the mechanisms of the health benefits of rice polyphenols and their synergistic effects with other bioactive components need to be investigated. The functional components in rice bran tend to be lost during the preparation of dietary fiber collection. Because rice bran has unstable factors such as lipids and enzymes, a stabilization method for the rice bran needs to be developed. To promote the application of dietary fiber in rice bran in the food industry, in vivo experiments need to be applied.

In the "Rice Processing" part, advanced processing and cooking methods can promote a rice-related future perspective. Rice postharvest processing still significantly influences various changes in rice, especially the physicochemical and nutritional properties. The impact of postharvest processing on human health benefit-related traits has presented an enormous opportunity to improve rice production and reduce postharvest rice losses. Rice cooking also has a significant effect on the overall preference, mainly texture. Although the cooking method needs to be optimized based on the rice variety and required preferences, typical cooking procedures such as washing, watering, soaking, heating, and steaming are also essential factors in the cooked rice quality. The relationship between cooking methods and cooked rice quality, including digestive properties, must be investigated. Newly applicable processing methods tend to fulfill the convenience and preference of users. Frozen and semi-aseptic-packed types of cooked rice can be preserved longer time. These rice meals can be processed by microwave heating, one of the applicable processing methods for freshly cooked rice. The demand for low-GI staple foods has increased with the rise of lifestyles and health consciousness in the modern world. Low GI in processed rice, especially in aseptically packaged rice, can be an essential quality factor. Germination is a simple processing method. The conditions of germination, especially duration, affect changes in characteristics and nutritional values. Germinated rice has proved a good source of numerous nutrients and bioactive substances, which provide positive immeasurable health impacts. It can be used as an ingredient in innovative food products. Various types of germinated rice are produced and developed to meet the requirements of consumers. Heat moisture treatment can vary the rice starch at a molecular level. The heat moisture-treated starch retained the original A-type polymorph and generates the newly formed V-type crystals. The relative crystallinity greatly varies with amylose content and heat moisture conditions, which is essentially evoked by the adjustments in lamellae and helical structure and molecular reordering. Coexisting lipids affect the function of treated starch through non-covalent interactions or simply attachment/encapsulation. The treated starch would ease the sensory quality and shelf life of foods if proper treatment parameters are adopted. The treatment parameters mainly involve moisture content, duration, temperature, and repeat times. Although only interactions with exogenous lipids have been elucidated, the interactions with other food ingredients, including non-starch polysaccharides and protein, also need to be investigated. The interplay with endogenous components would be significant in preparing dough or food products with sole heat moisture-treated starch.

Today, more healthy and functional attributes are required. Processed rice products should fit the requirements of social and medical backgrounds, such as modifying or adjusting starch digestibility, which is related to decreasing type II diabetes. Thus, rice processing must focus on increasing healthy and functional properties. Therefore, advanced processing methods to enhance the healthy and functional properties of rice must be developed and applied to rice products.

References

FAO (2021). Food outlook—biannual report on global food markets. Food Outlook, November 2021. Rome. https://doi.org/10.4060/cb7491en

OECD/FAO (2021) OECD-FAO Agricultural Outlook 2021–2029, FAO, Rome/OECD Publishing, Paris. https://doi.org/10.1787/19428846-e

GPSR Compliance

The European Union's (EU) General Product Safety Regulation (GPSR) is a set of rules that requires consumer products to be safe and our obligations to ensure this.

If you have any concerns about our products, you can contact us on ProductSafety@springernature.com

In case Publisher is established outside the EU, the EU authorized representative is:

Springer Nature Customer Service Center GmbH
Europaplatz 3
69115 Heidelberg, Germany

Batch number: 10731141

Printed by Printforce, the Netherlands